ARITHMETIC

PARTS I, II AND III COMPLETE

With Answers

ARITHMETIC

BY THE LATE

C. GODFREY, M.V.O., M.A.

AND

E. A. PRICE, B.A.

PARTS I, II AND III COMPLETE

With Answers

CAMBRIDGE

AT THE UNIVERSITY PRESS

1957

CAMBRIDGE
UNIVERSITY PRESS

University Printing House, Cambridge CB2 8BS, United Kingdom

Cambridge University Press is part of the University of Cambridge.

It furthers the University's mission by disseminating knowledge in the pursuit of education, learning and research at the highest international levels of excellence.

www.cambridge.org
Information on this title: www.cambridge.org/9781107594777

© Cambridge University Press 1957

This publication is in copyright. Subject to statutory exception and to the provisions of relevant collective licensing agreements, no reproduction of any part may take place without the written permission of Cambridge University Press.

First edition 1915
Reprinted 1919, 1921, 1923, 1925
Second edition 1927
Reprinted 1931, 1936, 1942, 1943, 1944, 1946, 1949, 1952, 1957
First paperback edition 2015

A catalogue record for this publication is available from the British Library

ISBN 978-1-107-59477-7 Paperback

Cambridge University Press has no responsibility for the persistence or accuracy of URLs for external or third-party internet websites referred to in this publication, and does not guarantee that any content on such websites is, or will remain, accurate or appropriate.

PREFACE

WE began with the intention of writing a text for *The Winchester Arithmetic*—a collection of examples published in 1905. But eventually it proved desirable entirely to rearrange and revise the examples after nine years' experience in using them; and also to add very largely to their number. A great proportion of the added examples are drawn from Osborne examination papers, and for these we have to thank our colleagues. If any examples from other text-books have slipped in by this route, we regret it and offer our apologies.

We draw attention to the **large type** used, which conforms to the standards laid down for various ages by the British Association Committee on "The Influence of School-books upon Eyesight." The boldest type is used in the examples in Part I, which presumably will be used by pupils of 9—12. The type used in the rest of the book is smaller, but marks an advance on that hitherto used in the majority of school-books.

The book has perhaps a rather practical flavour, but we have not departed so far from the traditional course as to make it unsuited to the needs of schools that take the ordinary public examinations. We have however lightened it by omitting matter that may now fairly be considered obsolete.

The scope of the work is sufficiently shown in the table of contents.

It is supposed that before using this book the pupil will have mastered the **Simple and Compound Rules**, as presented in a beginners' book such as *The Groundwork of Arithmetic* by Miss Punnett (Longmans). Our chapters on these topics are intended for purposes of revision and practice in computation.

Parts I and II correspond broadly to the first two stages of the syllabus of Mathematical Teaching issued by the Headmasters' Conference.

We have followed the Mathematical Association Committee in recommending the 'shop' method of subtraction, and multiplication 'from the left.' But when a boy has once learnt to subtract, multiply and divide, it is unwise to make him adopt new methods. There is a time for thinking, and a time for automatic work: a boy should learn to perform the simple operations automatically, with as little thought as possible. He should not be disturbed by doubts as to which way his work should slope.

Part I contains chapters on **unitary method**, and on **areas and volumes**, in which these topics are treated **without introducing any fractions.**

With regard to the question whether **vulgar** or **decimal fractions** should be taught first, we have so arranged the work that teachers can adopt whichever order they prefer: those who begin with decimals are recommended to prelude with the two paragraphs on 'the meaning of a fraction,' which stand at the beginning of the chapter on vulgar fractions.

We may perhaps say one word in explanation of our attitude towards **problems.** There have been many complaints of the behaviour of text-book writers in this matter.

It is said that whenever examiners succeed in putting a little novelty into their papers by inventing a new type of problem, the next text-book writer by means of a special section effectually sterilizes the new type and purges it of all educational value. From the examiners' point of view, this tendency is certainly regrettable. From an educational point of view, it is doubtful whether much is gained by solving arithmetical problems by means of 'pattern' solutions; presumably a problem is intended to make the pupil devise an appropriate solution, rather than to imitate a pattern: and many teachers hold that much of the time spent on these *chinoiseries* might go to something of more definitely mathematical interest, such as Trigonometry or Mechanics. Nevertheless, some of the stock types of problem do introduce a 'principle' (perhaps this is too dignified a word) of some interest; in these cases, some degree of classification seems permissible; and these types only have received special treatment in this book.

The **Revision Papers** at the end of each Part are supposed to provide straightforward out-of-school work for about 45 minutes. Those at the end of Part III are intentionally made rather long, to leave a certain choice: for instance, some teachers may instruct their pupils to omit questions on square root tables or cylinders, others may prefer to omit compound interest.

The **Miscellaneous Exercises** are less straightforward than the revision papers.

Mr A. W. Siddons, of Harrow, has very kindly read the proof-sheets and suggested many improvements; and some of our colleagues at Osborne have helped us in various ways. Our thanks are also due to the Controller of H.M. Stationery

Office for permission to use questions from certain Civil Service Examination Papers: and to the Cambridge University Press and Mr Siddons for permission to copy the tables of Squares, Square-roots, Reciprocals and Logarithms from *Four Figure Tables* by Godfrey and Siddons.

C. G.
E. A. P.

April, 1915.

NOTE TO THE SECOND EDITION

A few minor alterations and corrections have been introduced in this new edition, and an additional Exercise in Stocks and Shares inserted at the end of the last Chapter. The type has been completely re-set.

E. A. P.

October, 1926.

¶ *Exercises distinguished thus*, ¶*Ex.* 1, *are intended for discussion in class.*

TABLE OF CONTENTS

PART II. FRACTIONS

PART III

WEIGHTS AND MEASURES

Measures of Length

British	*Metric*
12 inches = 1 foot	10 millimetres (mm.) = 1 centimetre (cm.)
3 feet = 1 yard	10 cm. = 1 decimetre (dm.)
1760 yards = 1 mile	10 dm. = 1 metre (m.)
	10 m. = 1 dekametre (Dm.)
	10 Dm. = 1 hektometre (Hm.)
	10 Hm. = 1 kilometre (Km.)

Notes. 1 inch = 2·540 cm.
1 metre = 39·37 inches.
1 kilometre = about $\frac{5}{8}$ mile.
A surveyor uses 100 links = 22 yards = 1 chain.

Measures of Area

British	*Metric*
144 square inches = 1 sq. foot	100 sq. mm. or mm.^{2} = 1 cm.^{2}
9 sq. ft. = 1 sq. yd.	100 cm.^{2} = 1 dm.^{2}
4840 sq. yds. = 1 acre	100 dm.^{2} = 1 m.^{2}
640 acres = 1 sq. mile	&c.

Note. 10 square chains = 1 acre.

Measures of Volume

British	*Metric*
1728 cubic inches = 1 cu. ft.	1000 mm.^{3} = 1 cm.^{3} (1 c.c.)
27 cu. ft. = 1 cu. yd.	&c.

Measures of Weight

British (Avoirdupois)		*Metric*	
16 ounces (oz.)	= 1 pound (lb.)	10 milligrams	= 1 centigram
14 lbs.	= 1 stone	10 cgm.	= 1 dgm.
28 lbs.	= 1 quarter (qr.)	10 dgm.	= 1 gm.
4 qrs. / 8 stone / 112 lbs.	= 1 hundredweight (cwt.)	1000 gm.	= 1 Kgm.
20 cwts.	= 1 ton		

Notes. 1 cubic foot of water weighs 1000 oz. = $62\frac{1}{2}$ lbs. (more accurately 62·4 lbs.).

1 gram is the weight of 1 c.c. of pure water under certain conditions of temperature and pressure.

1 kilogram = about $2\frac{1}{5}$ lbs.

Measures of Capacity

British		*Metric*	
2 pints	= 1 quart	1000 c.c.	= 1 litre
4 quarts	= 1 gallon	10 decilitres	= 1 litre
8 gallons	= 1 bushel	&c.	

Notes. A litre is about $1\frac{3}{4}$ pints.
1 gallon of water weighs about 10 lbs.

Money

British		*French*	
4 farthings	= 1 penny (*d.*)	100 centimes	= 1 franc
12*d.*	= 1 shilling (1*s.* or 1/-)	*German*	
20/-	= 1 pound (£)	100 pfennig	= 1 mark
		American	
		100 cents	= 1 dollar

Notes. A franc is worth about 10*d.*
A mark ,, ,, 1*s.*
A dollar ,, ,, 4*s.* $2\frac{3}{4}$*d.*

Time

60 seconds (60″)	= 1 minute (1′)
60 minutes	= 1 hour
24 hours	= 1 day
365 days	= 1 year
366 days	= 1 leap year

PART I

MAINLY INTEGERS

CHAPTER I

THE FOUR SIMPLE RULES

Number and Quantity

§ 1. 4 tons is a **quantity;** so also 4 feet, 10 minutes, 7 pints are quantities. Sometimes we speak of "concrete quantities" instead of saying simply "quantities."

On the other hand 4, 10, 7, etc., are **numbers,** sometimes called "abstract numbers."

Notice that to describe a quantity we need a number together with a **unit.** 4 is the number; "a ton" is the unit: "4 tons" is the quantity.

ADDITION

§ 2. In adding several columns of figures, the subsequent checking is facilitated if we write down the sum of each column in full and subsequently add these sums. Each column should be added *downwards* and the sum put down in the appropriate place, nothing being "carried" to the next column. The sum of each column should be checked by adding *upwards* before the final result is found.

```
 75643
 37875
 46943
 79529
 83214
 63925
 ------
     29
    20
   39
  33
 35
 ------
 387129
```

EXERCISE I a

Find the sum of each of the following:

1.	123 456 789	**2.**	987 654 321	**3.**	246 802 468	**4.**	135 791 357
5.	321 778 416	**6.**	5789 879 7658	**7.**	476 1395 2764	**8.**	357 3478 5987
9.	2984 739 1564	**10.**	6474 4693 737	**11.**	1493 5437 7631	**12.**	2175 5432 6957
13.	8751 3179 4473	**14.**	9231 7489 5484	**15.**	6418 7464 7874	**16.**	79196 4956 78843

17.	**18.**	**19.**	**20.**
38495	49845	5883	3713
59555	5479	64357	95256
7792	35285	66974	89384

21.	**22.**	**23.**	**24.**
1573	4214	9828	7453
2946	5972	2482	6787
9763	6465	2926	7384
4329	9369	3856	2439

25.	**26.**	**27.**	**28.**
73257	36673	99142	43462
4372	64465	5364	7597
89499	2978	27943	6391
5876	23089	12397	72785

29.	**30.**	**31.**	**32.**
67964	35864	2227	6957
6439	3972	73757	4397
25793	12586	6946	12954
2979	14375	45387	34638
14698	3947	22569	26279

33.	**34.**	**35.**	**36.**
62469	34865	99863	2257
79573	6395	3537	86397
3268	44935	98656	57379
4875	2874	89832	8762
47358	76967	74372	84652

37.	**38.**	**39.**
987321	235427	921324
476294	984905	874321
575348	876584	478593
648778	329109	123958
599812	876584	642187
477667	321497	601798
532179	642198	320558

40.	41.	42.
325497	321742	218754
654832	598621	325421
238456	964215	998754
697584	642176	958632
738496	898732	469835
510307	465426	210864
246890	721987	727374
753798	998698	858687

43.	44.	45.
413052	295453	789762
298742	628761	295438
107534	921864	769476
298821	585497	987635
745566	987964	879659
668991	729853	642870
742198	471198	938756
877089	729876	748362
964276	641595	843363
849493	789421	798675

46.	47.	48.
482346	842364	284563
732631	132235	334253
546549	654742	264754
849683	483456	485364
382074	301983	839013
746459	459746	549467
654047	465024	464502
385892	315995	159593
935657	534642	653424
451723	809813	368722

EXERCISE I b

Find the sum of the following without writing in columns:

1.	$1 + 3 + 5 + 7$	**2.**	$2 + 4 + 6 + 8$
3.	$7 + 9 + 6 + 5$	**4.**	$6 + 5 + 11 + 3$
5.	$4 + 3 + 7 + 13$	**6.**	$9 + 9 + 8 + 11$
7.	$12 + 14 + 13$	**8.**	$15 + 17 + 19$
9.	$23 + 41 + 62$	**10.**	$31 + 65 + 79$
11.	$29 + 95 + 87$	**12.**	$14 + 73 + 69$
13.	$17 + 23 + 52 + 28$	**14.**	$59 + 47 + 69 + 73$
15.	$65 + 59 + 36 + 92$	**16.**	$34 + 27 + 73 + 42$
17.	$295 + 692 + 112$	**18.**	$253 + 475 + 465$
19.	$357 + 574 + 268$	**20.**	$435 + 543 + 372$
21.	$731 + 371 + 173$	**22.**	$297 + 378 + 635$
23.	$345 + 695 + 753$	**24.**	$882 + 772 + 993$
25.	$2 + 45 + 712 + 63$	**26.**	$17 + 243 + 9 + 25$

27. $69 + 13 + 3042 + 7$

28. $461 + 305 + 17 + 301$

29. $1156 + 24 + 329 + 17$

30. $217 + 3010 + 27 + 8$

EXERCISE I c

Problems on Addition

1. *A*, on a cycling tour, rides 64 miles on the first day, 58 on the second, 89 on the third, none on the fourth and fifth and 67 on each of the sixth and seventh days. *B* rides 35 miles each day. How far does each go?

2. A battalion consists of six companies, of which the numbers are 125, 103, 98, 101, 100, 108. What is the strength of the battalion?

3. In 1911 the population of England and Wales was 36,070,492; of Scotland 4,760,904; and of Ireland 4,390,219. What was the population of the United Kingdom?

4. A boy is 15 years old now: what will he be 30 years hence? 45 years hence? 5 years hence?

5. The figures on the map give the number of miles between the towns. Find the distance:

(1) From London to Berwick via Cambridge.
(2) „ „ „ via Leicester.
(3) From London to Birmingham via Oxford.
(4) „ „ „ via Warwick.
(5) „ „ „ via Rugby.
(6) From London to Carlisle via Leicester.
(7) „ „ „ via Rugby and Manchester.
(8) From London to Carlisle via Rugby and Liverpool.
(9) From Dover to Chester.
(10) From Brighton to Cambridge.
(11) From London to Portsmouth via Guildford.
(12) „ „ Bristol.

(13) From London to Winchester via Basingstoke.

(14) „ „ „ via Guildford.

(15) „ „ Dartmouth.

(16) From Land's End to Berwick via London and Cambridge.

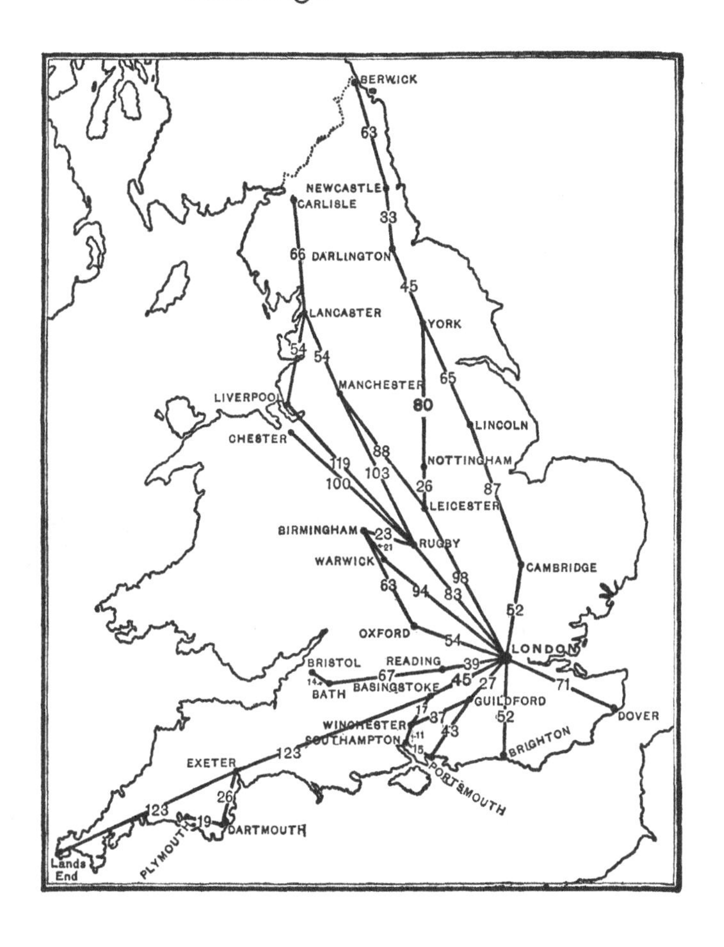

SUBTRACTION

§ 3. The method known as "complementary addition" is recommended*. The example shown should, if done orally, read:

"2 and **1**, 3; 5 and **9**, 14;
10 and **6**, 16; 8 and **7**, 15;
2 and **1**, 3."

35643
17952
17691

The result should in every case be verified by adding the last two lines.

EXERCISE I d

Subtract the second from the first number:

1.	173 141	**2.**	296 254	**3.**	769 635	**4.**	697 243
5.	795 243	**6.**	715 209	**7.**	624 317	**8.**	594 489
9.	237 128	**10.**	357 275	**11.**	485 296	**12.**	573 397
13.	3561 2342	**14.**	2973 1641	**15.**	3762 1759	**16.**	4697 3785
17.	7958 5449	**18.**	6305 2974	**19.**	3807 2918	**20.**	7643 2079
21.	7006 5384	**22.**	8017 3759	**23.**	6347 5078	**24.**	7073 3987
25.	129436 47687	**26.**	93862 79494	**27.**	63986 41889	**28.**	70073 58792

* See Preface, page vi, line 12.

29. 99176
89887

30. 579431
287549

31. 357924
218076

32. 6986954
2897076

33. 1384691
975879

34. 6372571
4652885

35. 2321097
886745

36. 6007583
878694

EXERCISE I e

Problems on Subtraction

1. An army of 10,000 men loses 2,165 in a certain battle. How many are left?

2. A battleship carrying 804 men is torpedoed; 83 are saved. How many were lost?

3. The sum of two numbers is 117; one of them is 52. What is the other?

4. The sum of two numbers is 140; the greater is 82. What is the lesser number?

5. The sum of two numbers is 1368; the lesser is 257. What is the greater?

6. If it is the morning of the 4th of July, how many complete days are left in July?

7. If it is the evening of the 18th of June, how many complete days are left in June?

8. How many hours are there between 3 p.m. and midnight? Between 10 a.m. and midnight? Between 6 a.m. and 4 p.m.?

9. A man has £28 and pays away £6, £5, £8, and £3. How much has he left?

10. In the House of Commons on a certain evening 452 members were present. At the first division 261 voted for the Government, 191 against. At the next division the votes were 230 for and 222 against. What was the Government majority in each case, and how many must have changed sides?

11. Of 12 apples, 8 are rotten, 6 are worm-eaten, and 3 are sound. How many must be both rotten and worm-eaten?

12. I am 24 years old now; in how many years shall I be 37? 48? 70?

13. Put down a number of three figures; under this write the same digits in the reverse order; subtract the smaller number from the greater. Under the answer write down the number found by reversing its digits, and add the two together. Try this with different numbers, e.g.:

$$\begin{array}{r} 541 \\ 145 \\ \hline 396 \end{array} \qquad \begin{array}{r} 396 \\ 693 \\ \hline 1089 \end{array}$$

ORDER OF OPERATIONS IN ADDITION AND SUBTRACTION. BRACKETS

§ 4. In interpreting such an expression as $7 + 5 - 3 - 4$ the rule is that we begin at the left, and perform the operations in succession; thus, to 7 add 5, then subtract 3, then subtract 4. The result is 5.

EXERCISE I f

Write down the value of:

1. $7 + 3 - 4$.
2. $7 - 4 + 3$.
3. $5 + 8 - 4 - 3$.
4. $6 + 4 - 5 - 2$.
5. $8 + 7 - 6 + 3$.
6. $8 + 6 - 4 - 7$.
7. $6 + 9 - 7 - 5$.
8. $3 + 7 - 4 + 1 - 5$.
9. $6 + 7 - 5 + 4 - 3$.
10. $7 - 6 + 8 - 5 - 1$.
11. $23 - 17 + 5$.
12. $31 + 21 - 40$.
13. $51 - 27 + 32$.
14. $103 + 29 - 17$.
15. $29 + 74 - 82 - 5$.
16. $61 - 43 + 21 + 13$.
17. $19 + 29 - 35 + 21$.
18. $47 + 43 - 29 - 31$.
19. $71 - 48 + 22 - 27$.
20. $34 - 25 + 16 - 17$.

§ 5. If there are many numbers to be dealt with, the work can be shortened. Thus in

$$7 + 10 - 4 + 12 - 16 + 3 - 5$$

the result is *increased* by the addition of 10, 12, 3; and *diminished* by the subtraction of 4, 16, 5. Suppose that the numbers represent points for and against, at some game; the points "for" are, altogether,

$$7 + 10 + 12 + 3 = 32.$$

The points "against" are, altogether,

$$4 + 16 + 5 = 25.$$

The result is 32 – 25, or 7 points "for." We express this as follows:

$$7 + 10 - 4 + 12 - 16 + 3 - 5$$
$$= 7 + 10 + 12 + 3 - (4 + 16 + 5)$$
$$= 32 - 25 = 7.$$

The **brackets** round $4 + 16 + 5$ mean that the sum inside the brackets is to be worked before the subtraction is performed.

EXERCISE I g

Find the value of:

1. $5 + 8 - (4 + 3)$.

2. $6 + 4 - (5 + 2)$.

3. $8 + 7 - 6 + 3$.

4. $8 + 7 - (6 + 3)$.

5. $8 + 7 - 6 - 3$.

6. $8 + 7 - (6 - 3)$.

7. $8 + 7 + 6 + 3$.

8. $8 + 7 + (6 + 3)$.

9. $5 + 7 - 4 - 3$.

10. $5 + 7 - (4 - 3)$.

11. $5 + (7 - 4) - 3$.

12. $(5 + 7) - 4 - 3$.

13. $(5 + 7) - (4 - 3)$.

14. $8 + 6 - 5 + 4$.

15. $8 + 6 - (5 + 4)$.

16. $8 + (6 - 5) + 4$.

17. $(8 + 6) - 5 + 4$.

18. $(8 + 6) - (5 - 4)$.

19. $8 + 6 - 5 - 4$.

20. $8 + (6 - 5) - 4$.

21. $5 + 9 - 3 + 6 - 4 + 7$.

22. $5 + 7 - 9 + 3 + 4 - 6$.

23. $11 - 7 - 2 + 14 - 6 + 1$.

24. $23 + 17 - 20 - 9 + 6 - 5$.

25. $41 - 3 + 17 - 11 - 21$.

26. $29 + 34 - 51 - 3 + 12$.

27. $102 - 22 + 31 - 17$.

28. $43 + 21 - 50 - 4 + 31$.

29. $19 - 11 + 46 + 12 - 37 - 11$.

30. $70 - 37 - 21 + 28 - 39$.

Multiplication

§ 6. To multiply anything (a number or a quantity—see Chap. I, § 1) by 7 means to find the result of adding together 7 of these things. Accordingly, the multiplier can only be a number; it cannot be a quantity.

The sign × means "multiplied by"; thus 3 ft. × 7 = 21 ft. and (7 × 3) ft. = 21 ft., but 7 × 3 ft. is unmeaning: and 3 ft. × 7 ft. is unmeaning. It is nonsense to speak of taking anything "7 feet times."

Examples of the arrangement recommended in a multiplication sum are given below. Unit figure of multiplier is below unit figure of multiplicand, and right-hand figure in each product falls below the figure used as a multiplier. The left-hand figure of the multiplier is used first*.

Example. Multiply 322123 by 312.

```
  322123
     312
--------
966369
 322123
  644246
--------
100502376
```

Example. Find the value of 9573 × 3058.

```
    9573
    3058
--------
28719
  47865
   76584
--------
29274234
```

Example. Find the value of 329 × 406 × 73. This expression is called the **continued product** of 329, 406 and 73.

```
   329
   406
------
 1316
   1974
------
 133574
     73
-------
935018
 400722
-------
9750902
```

* See Preface, page vi, line 12.

EXERCISE I h

Find the product (or continued product) of:

1.	516 × 211.	2.	468 × 217.
3.	713 × 123.	4.	887 × 91.
5.	898 × 789.	6.	7723 × 107.
7.	3015 × 215.	8.	7698 × 315.
9.	7879 × 698.	10.	4321 × 784.
11.	6987 × 356.	12.	4578 × 486.
13.	6905 × 785.	14.	1795 × 326.
15.	4989 × 786.	16.	7777 × 489.
17.	6435 × 298.	18.	7786 × 479.
19.	456 × 4674.	20.	908 × 7879.
21.	6154 × 5789.	22.	9865 × 2574.
23.	3987 × 9863.	24.	4091 × 9147.
25.	6382 × 5943.	26.	6432 × 9593.
27.	7893 × 2974.	28.	5837 × 6975.
29.	4659 × 9436.	30.	6569 × 9985.
31.	23 × 49 × 73.	32.	69 × 412 × 28.
33.	514 × 213 × 17.	34.	317 × 395 × 734.
35.	626 × 727 × 91.	36.	731 × 495 × 563.
37.	779 × 875 × 497.	38.	937 × 798 × 578.
39.	635 × 839 × 766.	40.	954 × 478 × 678.

EXERCISE I i

Problems on Multiplication

1. In a house of 7 rooms, each room has 3 windows and each window 12 panes. How many panes are there in the house?

2. A wall is built of 12 courses, or layers, of bricks. Each course is 3 bricks wide and 1480 bricks long. Find the number of bricks in the wall.

3. A book contains 357 pages; on every page there are 35 lines of print, and every line contains on an average 9 words. How many words are there in the book?

4. Calculate from the tables on page xiii:

(1) The number of pounds in a ton.

(2) The number of seconds in a year.

5. Light travels at the rate of 186,000 miles in a second. How far does it travel in 492 seconds?

6. Find the value of 513 motor cars of which 212 are worth £450 each, and the rest are worth £250 each.

7. There were on the British railways in 1902, 21,975 locomotives, 49,521 passenger coaches and 705,437 goods wagons. An engine costs £2500, a coach £450, and a wagon £100. Find the total value of the rolling stock.

8. Ten out of a hundred workmen receive 30/- a week, fifteen receive 25/- a week, and the remainder 15/- a week. How many shillings are paid to them (1) in one week, (2) in a year?

9. What is the weight in pounds of 2564 bales of cotton if each bale weighs 487 pounds?

10. Find the cost of 14295 mines if each cost £279.

11. If a man consumes 18 ounces of bread a day, how many ounces will 57260 men consume in 23 weeks?

12. How many feet will a locomotive travel in 195 minutes if the driving wheel is 19 feet in circumference and revolves 173 times in a minute?

INDICES

§ 7. 5×5 (= 25) is called the **square** of 5, and written 5^2.

$5 \times 5 \times 5$ (= 125) is called the **cube** of 5, and written 5^3.

$5 \times 5 \times 5 \times 5$ (= 625) is called the **fourth power** of 5, or 5 **to the power of** 4; it is written 5^4.

The number 2 in 5^2 is called the **index;** similarly 3, 4 are the **indices** in 5^3, 5^4 respectively.

It is suggested that the values of 2^2, 2^3, 2^4 up to 2^{10}, 3^2, 3^3, 3^4 up to 3^{10}, etc., etc. should be found orally in class.

EXERCISE I j

Find the value of:

1. The square of 23. **2.** The cube of 13.

3. The 4th power of 11. **4.** 12^4.

5. 9^5. **6.** 15^3. **7.** 23^4.

8. 104^3. **9.** 217^3. **10.** 22^4.

11. $2^3 \times 3^2 \times 5$. **12.** $3^3 \times 5^2 \times 7$. **13.** $5^3 \times 7^2$.

14. $2^4 \times 3^3 \times 7^2$. **15.** $3^4 \times 11^3$. **16.** $5^3 \times 7^3$.

17. $2^3 \times 3^2 \times 17^2$. **18.** $19 \times 4^3 \times 3^4$.

19. $2^5 \times 3^4 \times 5^3 \times 7^2$. **20.** $2^4 \times 5^4 \times 7 \times 11^3$.

Division

§ 8. When the divisor can be expressed as the product of several factors less than 13 it is sometimes considered convenient to use repeated short division, but the determination of the remainder in this case requires so much thought that the pupil is recommended to use long division (unless the divisor is an integer less than 13 followed by zeros).

Example. Divide 764945 by 600.

```
600)7649,45
    1274,  545
```

Quotient 1274, remainder 545.

Example. Divide 49769 by 253.

```
       196
253)49769
    253
    2446
    2277
     1699
     1518
      181
```

Quotient 196, remainder 181.

EXERCISE I k

Find the quotient and remainder when:

1. 56821 is divided by 3, 5, 7, 11, 12.
2. 48537 ,, 70, 700, 7000.
3. 92700 ,, 110, 1100, 11000.
4. 128463 ,, 120, 1200, 12000.
5. 3612 ,, 17, 37, 311.
6. 4216 ,, 14, 18, 116.
7. 7714 ,, 23, 47, 99.
8. 2136 ,, 117, 213, 412.
9. 32154 ,, 47, 513, 4216.
10. 52134 ,, 49, 93, 1127.
11. 41186 ,, 51, 157, 4001.
12. 5216321 ,, 56, 117, 62153.

EXERCISE 11

Find the quotient and remainder in each of the following:

1. $2469 \div 47$. **2.** $93561 \div 329$.

3. $69432 \div 437$. **4.** $11127 \div 293$.

5. $46759 \div 965$. **6.** $73496 \div 879$.

7. $47365 \div 294$. **8.** $63571 \div 795$.

9. $54789 \div 297$. **10.** $47765 \div 498$.

11. $395649 \div 4986$. **12.** $273571 \div 9007$.

13. $654327 \div 1906$. **14.** $724357 \div 6984$.

15. $273546 \div 2473$. **16.** $498764 \div 7981$.

17. $555778 \div 4694$. **18.** $632846 \div 5734$.

19. $798623 \div 2986$. **20.** $119478 \div 3872$.

PARTITION AND QUOTITION

§ 9. A distinction should be made between the following processes.

Example. How many shares of £25 can be given out of a total of £1000? The answer is 40.

Shortly, £1000 ÷ £25 = 40, or £1000 contains £25 exactly 40 *times*.

Dividend and divisor are both sums of money (quantities) and the answer is a number. This is an example of **Quotition** (Latin *quot* = how many times).

Example. If £1000 is divided into 40 equal shares what is the value of each share? The answer is £25.

Shortly, £1000 ÷ 40 = £25.

The dividend is a sum of money (quantity) and the divisor is a number; the answer is a quantity. This is an example of Sharing or **Partition**.

EXERCISE I m

Problems on Partition and Quotition

1. A battalion 624 strong contains 6 companies of equal strength. Each company is divided into 4 equal sections and each section into 2 squads. How many men are there in a company, section, squad?

2. £7392 prize money is to be distributed into equal portions among a crew of 672. How much will each man get?

3. 500 presents are to be divided equally amongst 123 children. How many presents will each child get and how many will be left over?

4. A man deals 100 cards to 13 people. How many will get 7 cards? and how many 8?

5. There are 10,000 rounds of ammunition in the magazine. How many rounds per man can be issued to a regiment of 615 men, and how many rounds will remain?

6. How many pieces of wood 10 inches long can be cut from a plank 67 inches long, and what will be the length of the piece left over?

7. How many pieces of paper 13 inches long can be cut from a length of 142 in. and what will remain?

8. A steamer of coal capacity 700 tons is being coaled by a barge which can carry 180 tons. How many journeys must the barge make, and how many tons must it carry on the last journey?

ORDER OF OPERATIONS. BRACKETS

§ 10. In simplifying expressions in which the signs $+$, $-$, $\times$, $\div$ occur, the operations $\times$ and $\div$ should be completed before $+$ and $-$ are dealt with; e.g. $4 + 9 \times 6$ means "multiply 9 by 6 and add the result to 4"; the answer is 58.

Also $7 \times 5 + 3 \times 4 - 8 \div 2 = 35 + 12 - 4 = 43$.

Brackets are used to indicate operations which should be completed first; e.g. $8 \times (5 - 3)$ or more commonly $8\,(5 - 3)$ means "multiply 8 by the difference between 5 and 3," and the answer is 16. Also $(5 - 3) \times (7 + 3)$ or more commonly $(5 - 3)\,(7 + 3) = 2 \times 10 = 20$.

EXERCISE I n

Find the value of:

1. $50 + 3 \times 6$.
2. $50 \times 3 + 6$.
3. $50 - 6 \times 3$.
4. $(50 - 6) \times 3$.
5. $37 - 8 + 4$.
6. $37 - 8 \times 4$.
7. $(37 - 8) \times 4$.
8. $37 \times 8 - 4$.
9. $37 \times (8 - 4)$.
10. $(37 - 4) \div 3$.
11. $5 \times (6 - 4)$.
12. $9 \times 8 - 7 \times 8$.
13. $(9 - 7) \times 8$.
14. $6 \times 5 - 4 \times 5$.
15. $8 \times 8 - 7 \times 8$.
16. $(8 - 7) \times 8$.
17. $8 \times 9 - 8 \times 7$.
18. $8 \times (9 - 7)$.
19. $8 \times (8 - 7)$.
20. $8 \div (9 - 7)$.
21. $7\,(3+2) - 5\,(4-3)$.
22. $7\,(3+2) - 5 \times 4 - 3$.
23. $5\,(4-2) - 4\,(7-5)$.
24. $(5 + 4)\,(5 + 4)$.
25. $5 \times 5 + 4 \times 4$.
26. $8 \times 8 - 3 \times 3$.
27. $(8 - 3)\,(8 - 3)$.
28. $(8 - 3)\,(8 + 3)$.
29. $(7 - 5)\,(7 + 5)$.
30. $7 \times 7 - 5 \times 5$.
31. $25 \times 13 + 121$.
32. $25 + 13 \times 121$.
33. $(84 + 26) \div 11$.
34. $20\,(17 - 14) \times 13$.

35. $5(25 - 13) - 6(31 - 27)$.

36. $18 \times (29 - 4) + 45(17 - 11)$.

37. $23 + 17(31 - 23) - 16 \times 7$.

38. $31 \times 43 + 19(17 - 5 \times 3)$.

39. $(31 \times 43 + 19)(17 - 5 \times 3)$.

40. $(31 + 19)43 - (17 - 3) \times 5$.

41. $21 \times 40 \div (15 \times 28)$.

42. $34 \times 39 \div (13 \times 17)$.

43. $259 \times 76 \div (14 \times 38)$.

44. $174 \times 279 \div (87 \times 18)$.

45. $235 \times 1173 \div (115 \times 141)$.

§ 11. The squares of numbers given below will be found useful for class practice in mental arithmetic. The following are some of the exercises that may be set:

(i) Add up the numbers in each row or column.

(ii) Add 4 to each number. Subtract....from each.

(iii) Go along a row giving the product (or sum) of each pair.

(iv) Multiply the sum of two consecutive numbers by the first of these.

(v) Give the sum (or difference) of the squares of each pair.

A.

5	6	9	4	1
0	1	4	9	6
3	2	8	4	5
1	0	8	5	7
9	1	4	3	5

B.

11	8	3	5	12
5	10	9	1	9
2	7	6	12	3
4	8	11	0	6
7	2	9	5	12

C.

13	2	12	5	15
7	14	6	11	8
16	2	13	8	6
1	15	6	5	13
11	3	8	12	7

EXERCISE 10

Problems on the Four Rules

1. A sheet of stamps contained 18 rows with 12 stamps in each row. How many stamps were there in the sheet?

2. At the beginning of a year a battalion was 847 strong. During the year the strength was increased by 127 men. Find the strength of the battalion at the end of the year.

3. How many compartments are there in a train that carries 256 passengers, if there are eight passengers in each compartment?

4. How many pages are there in a book consisting of six volumes, if each volume contains 432 pages?

5. A liner sets out on a voyage of 3572 miles. After she has sailed 1079 miles, what is the length of the remaining portion of the voyage?

6. On the day of publication of a new book, one library purchased 350 copies, another 425 copies, a third 575 copies, and the publisher found he had 650 remaining. How many copies had he at first printed?

7. A bushel of English wheat weighs 63 lbs. What is the weight in lbs. of 56 bushels?

8. How many railway coaches each seating 56 passengers will be required to seat a party of 952 excursionists?

9. A certain chapter of a book begins at the top of page 579, and ends at the bottom of page 607. How many pages are there in the chapter?

10. A "hundred" of herrings number 132. How many herrings are there in 47 "hundreds"?

11. There are four rooms in a picture gallery. In the first there are 137 pictures, in the second 253, in the third 178, and in the fourth 266. How many pictures are there in all four rooms together?

12. If 2553 lbs. of potatoes are put into sacks and each sack holds 69 lbs., how many sacks are required?

13. A motor car travelling 15 miles an hour took 8 hours to go from one town to another. How far were the towns apart?

14. How many days are there from March 26th to September 29th inclusive?

CHAPTER II

REDUCTION. COMPOUND RULES. PRACTICE

Reduction

§ 12. The following are suggestions for oral work and should be supplemented by the teacher. See Tables, p. xii.

EXERCISE II a (oral)

1. How many pence are there in: 1/3, 2/6, 7/-, 6/-, 4/6, 8/-, 8/2, 8/4, 4/2, 1/8, 10/-, 10/6, 12/6, 15/-, 4/1, 5/6, 5/8, 4/11, £1?

2. How many shillings are there in £1, £2, £5, £4. 10*s*. 0*d*., £5. 12*s*. 0*d*., £3. 15*s*. 0*d*., £1. 19*s*. 0*d*., £1. 18*s*. 0*d*., £3. 17*s*. 0*d*.?

3. Express in pounds, shillings and pence: 15*d*., 28*d*., 30*d*., 50*d*., 100*d*., 200*d*., 25/-, 28/-, 52/-, 75/-, 48/6, 39/5, 170/-, 180/-.

4. How many halfpennies are there in 6*d*., 1/3, 2/6?

5. How many farthings are there in 2*d*., $3\frac{3}{4}$*d*., 6*d*., $5\frac{1}{2}$*d*., 1/0, 2/6?

6. How many half-crowns are there in £1, £5. 5*s*. 0*d*., £10?

7. Express in pounds, shillings and pence:

(*a*) 100 farthings, (*b*) 600 halfpennies, (*c*) 100 half-crowns, (*d*) 27 half-crowns.

8. How many inches are there in 2 ft., 1 yd., 3 ft. 6 in.; 5 ft. 10 in.; 2 yds., 1 ft. 8 in.?

9. How many feet are there in 3 yds., 15 yds., 10 yds. 2 ft.; 5 yds. 1 ft.; 22 yds., 100 yards, 1 mile, quarter-mile, half-mile?

10. Express in yards, feet, and inches: 30 in., 50 in., 100 in., 200 in., 18 ft., 27 ft., 28 ft., 92 ft.

11. How many lbs. are there in 2 stone, 4 stone 8 lbs., 1 cwt., 2 cwts., 1 ton?

12. Express in tons, cwts., stones, and lbs.: 56 lbs., 60 lbs., 2240 lbs., 8 stone, 20 stone, 50 cwts., 100 cwts.

13. How many pints are there in 4 quarts, 1 gallon, 9 gallons?

14. Express in gallons, quarts, and pints: 50 pints, 100 pints, 40 quarts.

15. How many minutes are there in half-an-hour, 12 hours, 24 hours, between 2 p.m. and 5 p.m., between 10 a.m. and 1 p.m.?

§ 13. No general rules for reduction are given. In each case the most appropriate process should be adopted.

Example. Reduce £23. 7*s*. 7*d*. to pence.

```
 £23. 7s. 7d.
  20
 467 shillings
  12
5611 pence.
```

Example. Express 9765 pence in £. *s*. *d*.

```
12|9765 pence
20| 813s. 9d.
   £40. 13s. 9d.
```

Example. Reduce 13 tons 11 cwts. 24 lbs. to lbs.

To find the number of cwts. multiply 13 by 20 and add 11. To find the number of lbs. multiply 271 by 112 and add 24.

```
   13 tons 11 cwts. 24 lbs.
   20
  271 cwts.
  112
271
 271
  542
   24
30376 lbs.
```

Example. Reduce 17 cwts. 3 qrs. 13 lbs. to lbs.

Multiply 17 by 4 and add 3.
Multiply 71 by 28 and add 13.

```
  17 cwts. 3 qrs. 13 lbs.
   4
  71 qrs.
  28
 142
 568
  13
2001 lbs.
```

Example. Express 13579 oz. in cwts., lbs., oz.

Divide 13579 by 16; remainder is 11.

Divide 848 by 112; remainder is 64.

```
     848 lbs.            7 cwts.
16)13579 oz.      112)848 lbs.
   128                784
   ---                ---
    77                 64 lbs.
    64
    ---
    139
    128
    ---
     11 oz.

7 cwts. 64 lbs. 11 oz.
```

Example. Reduce 1 day 15 hrs. 29 secs. to secs.

```
24 + 15 = 39 hrs.
          60
        ----
        2340 mins.
          60
      ------
      140429 secs.
```

Example. Express 23779 secs. in hrs., mins., secs.

```
60|2377,9 secs.
 60|39,6 mins.  19 secs.
    6 hrs.  36 mins.  19 secs.
```

EXERCISE II b

Reduce:

1. £2. 1*s*. 7*d*. to pence.
2. £12. 11*s*. 4*d*. to pence.
3. £5. 7*s*. 7½*d*. to farthings.
4. £7. 18*s*. 3*d*. to halfpence.
5. £43. 19*s*. 1*d*. to pence.
6. £101. 5*s*. 9*d*. to pence.
7. £17. 13*s*. 4*d*. to farthings.
8. £17. 9*s*. 6½*d*. to halfpence.

9. £211. 12*s*. 11*d*. to pence.

10. £59. 19*s*. 7$\frac{3}{4}$*d*. to farthings.

Express in £. *s*. *d*.:

11. 2357 pence. **12.** 5367 farthings.

13. 35000 pence. **14.** 17680 halfpence.

15. 13121 farthings. **16.** 73845 farthings.

17. Make a table showing the wages for 1 day of 8 hrs., 6 days, 30 days, 78 days and 310 days at 2*d*., 2$\frac{1}{2}$*d*., 3*d*., 3$\frac{1}{2}$*d*., 4*d*., 4$\frac{1}{2}$*d*., 5*d*., 5$\frac{1}{2}$*d*., 6*d*. and 1*s*. per hr. (Each day contains 8 working hours.)

EXERCISE II c

Reduce:

1. 1 cwt. 1 qr. 1 lb. to oz.

2. 3 cwts. 2 qrs. 11 lbs. to lbs.

3. 1 ton 3 cwts. 17 lbs. to oz.

4. 13 tons 18 cwts. 15 lbs. to lbs.

5. 7 cwts. 101 lbs. to oz.

6. 3 qrs. 21 lbs. 5 oz. to oz.

7. 17 stone 5 lbs. to oz.

8. 123 stone 11 lbs. to lbs.

9. 3 cwts. 5 st. 7 lbs. to lbs.

10. 7 cwts. 19 lbs. 12 oz. to oz.

Express in tons, cwts., lbs., oz.

11. 29543 oz. **12.** 19200 oz.

13. 22400 oz. **14.** 2956 lbs.

15. 14735 lbs. **16.** 100000 oz.

17. 29875 lbs. **18.** 78439 oz.

19. 14900 lbs. **20.** 10000 lbs.

EXERCISE II d

Reduce each of the following to seconds:

1.	1 day.	**2.**	2 days 13 hrs.
3.	17 hrs. 14 mins. 5 secs.	**4.**	3 days 2 mins. 47 secs.
5.	3 hrs. 35 mins.	**6.**	10 days 23 hrs.
7.	1 day 43 mins. 15 secs.	**8.**	3 days 11 hrs. 14 mins.
9.	21 days 7 hrs. 31mins.	**10.**	4 days 23 hrs. 29 secs.

Express in weeks, days, hours, minutes, seconds:

11.	734965 secs.	**12.**	7469 mins.
13.	246357 secs.	**14.**	96437 hrs.
15.	255587 mins.	**16.**	3249732 secs.
17.	35692 hrs.	**18.**	743693 secs.
19.	14291 hrs.	**20.**	563410 mins.

EXERCISE II e

Reduce each of the following to inches:

1.	7 yds. 2 ft.	**2.**	12 yds. 1 ft. 3 in.
3.	112 yds. 2 ft. 1 in.	**4.**	72 yds. 5 in.
5.	1 mile 720 yds.	**6.**	14 yds. 1 ft. 1 in.
7.	2 miles 42 yds.	**8.**	1 mile 170 yds. 2 ft.
9.	1 mile 63 yds. 1 ft.	**10.**	2 miles 1050 yds.

Express each of the following in miles, yds., ft., in.:

11.	3765 in.	**12.**	7929 ft.
13.	20000 in.	**14.**	8543 ft.
15.	19000 ft.	**16.**	5943 in.
17.	235764 in.	**18.**	30000 ft.
19.	154360 ft.	**20.**	19757 in.

EXERCISE II f

Reduce each of the following to pints:

1.	15 quarts 1 pint.	**2.**	3 galls. 3 quarts.
3.	5 galls. 1 pint.	**4.**	11 galls. 2 qts. 1 pt.
5.	1 bush. 4 galls.	**6.**	3 bush. 3 quarts.

Express in bushels, gallons, quarts and pints:

7.	423 quarts.	**8.**	1721 pints.
9.	2701 pints.	**10.**	1100 quarts.
11.	10000 pints.	**12.**	1750 quarts.

Compound Addition

EXERCISE II g

Suggestions for oral work

Add together:

1.	8*d.* and 6*d.*	**2.**	1/2 and 11*d.*
3.	3/7 and 5/4.	**4.**	2/1 and 17*d.*

5. 2 ft. 5 in. and 3 ft. 3 in.

6. 6 ft. 11 in. and 4 ft. 9 in.

7. 1 lb. 3 oz. and 2 lbs. 7 oz.

8. 3 lbs. 11 oz. and 2 lbs. 7 oz.

9. 5 mins. 13 secs. and 6 mins. 31 secs.

10. 3 mins. 10 secs. and 14 mins. 53 secs.

Oral practice may be given by starting with a sum of money (say 2/7) or other quantity, and making each member of the class in succession add a fixed amount, say 9*d.*, thus, 2/7, 3/4, 4/1, etc.

EXERCISE II h

Add together:

	£	s.	d.		£	s.	d.		£	s.	d.
1.	4	3	6	**2.**	17	3	2	**3.**	11	0	1
	5	17	2		15	4	7		0	19	3
	6	19	8		3	18	0		6	15	7

	£	s.	d.		£	s.	d.		£	s.	d.
4.	22	17	9	**5.**	2	19	10	**6.**	17	14	$3\frac{1}{2}$
	5	6	$3\frac{1}{2}$		3	17	$6\frac{1}{2}$		18	19	$11\frac{1}{2}$
	4	11	11		4	0	$8\frac{1}{2}$		25	6	$4\frac{1}{2}$

	£	s.	d.		£	s.	d.		£	s.	d.
7.	0	16	$8\frac{1}{2}$	**8.**	0	15	$5\frac{1}{4}$	**9.**	420	0	6
	0	19	$11\frac{1}{2}$		0	17	$2\frac{3}{4}$		921	14	0
	4	17	$9\frac{1}{2}$		0	18	$3\frac{1}{2}$		82	19	6

10. 17/6, 18/2, 14/7.

11. £1. 2*s.* 3*d.*, £4. 11*s.* 6*d.*, £5. 17*s.* 3*d.*

12. £14. 2*s.* 3*d.*, £2. 11*s.* 6*d.*, £18. 7*s.* 6*d.*

13. £25. 6*s.* 8*d.*, 17/6, £2. 11*s.* 3*d.*

14. £18. 2*s.* 3*d.*, £19. 7*s.* $11\frac{1}{2}$*d.*, £6. 0*s.* 2*d.*

15. £9. 3*s.* $8\frac{1}{2}$*d.*, £8. 3*s.* $1\frac{1}{2}$*d.*, 16/$4\frac{1}{2}$*d.*

16. £21. 2*s.* 6*d.*, 5/3, £1. 7*s.* 4*d.*

17. £218. 4*s.* 0*d.*, £512. 16*s.* 0*d.*, £230. 5*s.* 0*d.*

18. £2. 6*s.* $3\frac{1}{4}$*d.*, £4. 1*s.* $5\frac{1}{2}$*d.*, £8. 0*s.* $6\frac{1}{2}$*d.*

19. £173. 4*s.* 7*d.*, £219. 14*s.* 4*d.*, £31. 17*s.* 9*d.*

20. £19. 13*s.* 4*d.*, £71. 10*s.* 10*d.*, £17. 14*s.* 4*d.*

21. £213. 17*s.* 9*d.*, £143. 5*s.* 11*d.*, £230. 11*s.* 11*d.*

22. £43. 11*s.* 3*d.*, £49. 9*s.* 7*d.*, £41. 16*s.* 9*d.*

23. £271. 14*s.* 7*d.*, £117. 17*s.* 4*d.*, £312. 2*s.* 9*d.*

24. £16. 19*s.* 10*d.*, £13. 19*s.* 9*d.*, £20. 16*s.* 4*d.*

25. £14. 13*s*. 6*d*., £16. 1*s*. 8*d*., £13. 15*s*. 7*d*.

26. £91. 2*s*. 7*d*., £94. 7*s*. 11*d*., £103. 9*s*. 5*d*.

27. £63. 5*s*. 1*d*., £46. 14*s*. 4*d*., £25. 7*s*. 3*d*.

28. £491. 13*s*. 11*d*., £133. 8*s*. 6*d*., £72. 11*s*. 9*d*.

29. £941. 16*s*. 5*d*., £421. 6*s*. 8*d*., £83. 6*s*. 5*d*.

30. £249. 7*s*. 6*d*., £157. 19*s*. 8*d*., £142. 11*s*. 10*d*.

EXERCISE II i

Add together:

1.

yds.	ft.	in.
	2	6
	4	10
2	1	10

2.

yds.	ft.	in.
1	2	3
3	2	1
6	2	10

3.

yds.	ft.	in.
7	2	11
2	2	6
3	1	7

4.

yds.	ft.	in.
15	2	6
17	1	7
18	2	8
20	2	10

5.

miles	yds.
2	440
1	880
3	440
2	320

6.

miles	yds.
1	1000
1	500
1	250
1	125

7.

yds.	ft.	in.
	3	4
2	1	6
3	1	11
4	2	7
3	0	10

8.

yds.	ft.	in.
1	0	1
5	1	9
3	2	6
4	1	7
	7	2

9.

miles	yds.
2	300
	700
1	629
2	154
	521

Add together, and express the answer in miles, yards, feet, and inches:

10. 3 ft. 6 in., 216 in., 72 ft., 200 yds., 1000 yds., 2000 yds.

11. 144 in., 501 ft., 3000 yds.

12. 170 in., 73 ft., 400 yds., 1261 in.

EXERCISE II j

Add together:

1.

lbs.	oz.
1	4
3	8
5	12

2.

st.	lbs.
2	8
1	8
3	10

3.

tons	cwts.	qrs.	lbs.
1	3	3	8
	5	3	4
	17	1	16

4.

lbs.	oz.
4	11
5	3
7	13

5.

st.	lbs.
12	3
14	11
13	9

6.

tons	cwts.	st.	lbs.
2	5	2	11
	13	0	3
1	11	3	9

7.

lbs.	oz.
1	13
2	11
4	3
5	7

8.

st.	lbs.
5	6
6	12
5	13
6	1

9.

tons	cwts.	qrs.	lbs.
1	0	0	21
	7	2	11
2	0	3	5
1	13	2	17

10.

lbs.	oz.
2	5
3	13
2	14
1	15
4	6

11.

st.	lbs.
10	11
11	5
10	13
12	14
9	11

12.

tons	cwts.	st.	lbs.
2	14	0	0
1	9	3	5
1	0	1	7
	16	2	4
3	0	1	12

Add together and express the answer in tons, cwts., qrs., lbs.:

13. 56 lbs., 8 stone, 17 cwts.

14. 42 cwts., 50 stone, 1000 lbs.

15. 50 cwts., 80 stone, 1400 lbs.

16. 72 cwts., 22 qrs., 1200 lbs.

17. 43 cwts., 43 qrs., 200 lbs.

EXERCISE II k

Add together:

	hrs.	mins.	secs.
1.	5	11	23
	1	46	29
		14	15

	hrs.	mins.	secs.
2.	3	0	14
		59	21
	1	29	17

	hrs.	mins.	secs.
3.		35	35
	1	47	47
	2	0	13

	days	hrs.	mins.
4.	1	21	14
	3	17	51
		23	43
	1	17	11

	days	hrs.	mins.
5.		20	20
		15	13
	1	2	43
	2	13	31

	days	hrs.	mins.	secs.
6.	1	0	27	0
	1	21	46	17
		17	11	44
	2	0	12	34

EXERCISE II l

Add together:

	galls.	qts.	pts.
1.	3	1	1
		3	0
		1	1

	galls.	qts.	pts.
2.	2	2	0
		3	1
	1	0	1

	galls.	qts.	pts.
3.	5	1	1
	2	0	1
		2	1

	bush.	galls.	qts.	pts.
4.	1	0	3	1
	2	5	3	1
		6	2	0
	1	2	1	1

	bush.	galls.	qts.	pts.
5.	3	3	3	1
	2	2	2	1
	1	6	3	0
		7	3	0

COMPOUND SUBTRACTION

Suggestions for oral exercises:

A. Take 8*d.* from 1/2, £4. 2*s.* 6*d.*, £2. 0*s.* 2*d.*

B. Take 9 in. from 1 ft. 4 in., 1 yard, 2 ft. $7\frac{1}{2}$ in.

C. Take 12 lbs. from 3 stone, 1 cwt. 9 lbs., 8 stone 8 lbs.

D. Take 45 secs. from 1 hour, 2 mins. 25 secs., 1 min. 11 secs.

EXERCISE II m

Subtract the lower quantity from the upper:

	£	s.	d.		£	s.	d.		£	s.	d.
1.	5	4	7	**2.**	5	0	0	**3.**	43	15	0
	4	3	2		2	16	1		13	2	6

	£	s.	d.		£	s.	d.		£	s.	d.
4.	18	17	11	**5.**	19	17	8	**6.**	19	0	4
	16	0	$3\frac{1}{4}$		17	19	$10\frac{1}{2}$			18	$9\frac{3}{4}$

7. Subtract £3. 11*s.* 7*d.* from £4. 5*s.* 6*d.*

8. „ £17. 1*s.* 10*d.* „ £20. 0*s.* 5*d.*

9. From £14. 10*s.* 0*d.* subtract £11. 6*s.* $9\frac{1}{2}$*d.*

10. „ £30. 15*s.* 6*d.* „ £21. 19*s.* 11*d.*

11. „ £20. 5*s.* $9\frac{1}{4}$*d.* „ £7. 8*s.* $9\frac{1}{2}$*d.*

12. „ £12. 8*s.* $8\frac{1}{2}$*d.* „ £5. 11*s.* $10\frac{3}{4}$*d.*

EXERCISE II n

Subtract:

1. 4 yds. 2 ft. 3 in. from 12 yds. 1 ft. 2 in.

2. 3 yds. 1 ft. 7 in. from 5 yds. 0 ft. 9 in.

3. 1 yd. 1 ft. 8 in. from 3 yds. 1 ft. 6 in.

4. 700 yds. from 1 mile 130 yds.

5. 1 mile 329 yds. from 2 miles 295 yds.

6. 5 miles 491 yds. from 7 miles 119 yds.

EXERCISE II o

Subtract:

1. 5 stone 13 lbs. from 13 stone 7 lbs.

2. 3 stone 3 lbs. from 5 stone 1 lb.

3. 2 cwts. 100 lbs. from 4 cwts. 47 lbs.

4. 13 cwts. 3 qrs. from 17 cwts. 2 qrs.

5. 1 ton 19 cwts. 6 lbs. from 3 tons 9 cwts. 4 lbs.

6. 1 ton 12 cwts. 7 st. 11 lbs. from 3 tons 6 st. 9 lbs.

EXERCISE II p

Subtract:

1. 1 hr. 13 mins. 29 secs. from 2 hrs. 0 min. 47 secs.

2. 3 hrs. 12 mins. 43 secs. from 7 hrs. 2 mins. 14 secs.

3. 3 days 14 hrs. 21 mins. from 7 days 9 hrs. 11 mins.

4. 16 days 23 hrs. 50 mins. from 17 days 10 hrs. 10 mins.

5. 7 weeks 3 days 13 hrs. from 10 weeks 1 day 10 hrs.

6. 12 weeks 6 days 22 hrs. from 13 weeks 4 days 4 hrs.

COMPOUND MULTIPLICATION

§ 14. Suggestions for oral exercises:

A. What is 5 times 8*d*., 7 times 6/2?

B. What is 10 times 2 lbs. 2 oz., 3 times 2 cwts. 3 qrs.?

C. What is 3 times 2 ft. 5 in., 4 times 4 yds. 2 ft.?

D. What is 5 times 1 min. 20 secs., 10 times 4 hrs. 29 mins.?

Example. Multiply £64. 3*s*. 7*d*. by 237.

£	*s*.	*d*.		
64	3	7		(*a*)
		10		
641	15	10	= 10 times	(*b*)
		10		
6417	18	4	= 100 times	
		2		
12835	16	8	= 200 times	
1925	7	6	= 30 times, from (*b*) above	
449	5	1	= 7 times, from (*a*) above	
15210	9	3		

Example. Multiply 3 cwts. 2 qrs. 19 lbs. by 323.

In this case multiply 19 lbs. by 323 and express in tons, cwts., qrs. and lbs. Then multiply 2 qrs. by 323 and express in tons, cwts. and qrs. Finally multiply 3 cwts. by 323 and express in tons and cwts. But see Note below.

```
   323                 323                  323
    19                   2                    3
   ---                 ---                  ---
   323              4 |646              20 |969
  2907             20 |161 cwts. 2 qrs.      48 tons 9 cwts.
28)6137(219 qrs.        8 tons 1 cwt. 2 qrs.
   56
   --
    53
    28
   ---
   257
   252
   ---
     5 lbs.
```

```
                                         tons cwts. qrs. lbs.
                       19 lbs. × 323 =    2    14    3    5
 4|219 qrs.             2 qrs. × 323 =    8     1    2    0
20|54 cwts. 3 qrs.     3 cwts. × 323 =   48     9    0    0
    2 tons 14 cwts. 3 qrs.               --------------------
                                         59     5    1    5
```

NOTE.—Compound Multiplication sums are generally more conveniently done by Practice (see p. 45).

EXERCISE II q

1. Multiply 3/6 by 12, 15, 50, 100.
2. Multiply £4. 8*s*. 5*d*. by 6, 10, 6 × 4, 81, 72.
3. Multiply £18. 17*s*. 6½*d*. by 4, 4 × 7, 5 × 9, 72.
4. Multiply £2. 5*s*. 3*d*. by 140 (= 20 × 7).
5. Multiply £53. 9*s*. 7*d*. by 213 (=10×10×2+10+3).
6. Multiply £109. 7*s*. 4*d*. by 93.
7. Multiply £213. 15*s*. 7*d*. by 141.
8. Multiply £61. 17*s*. 2*d*. by 233.
9. Multiply £19. 9*s*. 6½*d*. by 344.
10. Multiply £127. 14*s*. 3¾*d*. by 232.
11. Multiply 3 tons 15 cwts. by 4, 12, 20, 23, 123.

12. Multiply 2 lbs. 10 oz. by 11, 17, 35, 213.

13. Multiply 9 stone 5 lbs. by 10, 50, 53, 153, 253.

14. Multiply 2 tons 14 cwts. 92 lbs. by 52.

15. Multiply 3 cwts. 3 qrs. 20 lbs. by 131.

16. Multiply 2 yds. 2 ft. 7 in. by 2, 12, 23.

17. Multiply 5 miles 240 yds. by 100, 120, 127.

18. Multiply 5 hrs. 24 mins. 13 secs. by 351.

19. Multiply 23 days 17 hrs. 50 mins. by 76.

20. Multiply 17 hrs. 53 mins. 47 secs. by 143.

NOTE.—Further exercises in Compound Multiplication can be found in the section on Practice (p. 46).

COMPOUND DIVISION. PARTITION

§ 15. The following are suggestions for oral work:

A. Divide £5. 5*s*. 0*d*. equally between 3 men.

B. Divide 7 ft. 11 in. into 5 equal parts.

C. Divide 2 stone 8 lbs. into 6 equal parts.

D. Divide 2 hrs. 24 mins. into 8 equal parts.

Example. Divide £93. 5*s*. 5*d*. by 37.

```
      £2
37)£93. 5s. 5d.
    74
    19
     20
37)385s.(10s.
    37
     15
      12
37)185d.(5d.
    185                £2. 10s. 5d.
```

Example. Divide 121 yds. 1 ft. 11 in. by 29.

```
      4 yds.
29)121 yds. 1 ft. 11 in.
   116
     5 yds.
     3
    16 ft.
    12
29)203(7 in.
   203              4 yds. 0 ft. 7 in.
```

Example. Divide 217 tons 13 cwts. 42 lbs. by 141.

```
       1 ton
141)217 tons 13 cwts. 42 lbs.
    141
     76
     20
141)1533 cwts.(10 cwts.
    141
     123
     112
   123
    123
     246
      42
141)13818 lbs.(98 lbs.
    1269
     1128
     1128            1 ton 10 cwts. 98 lbs.
```

Example. Divide 7 days 7 hrs. 35 mins. by 43.

7 days 7 hrs. 35 mins. = 175 hrs. 35 mins.

```
      4 hrs.
43)175 hrs. 35 mins.
   172
     3
    60
43)215 mins.(5 mins.
   215                  4 hrs. 5 mins.
```

EXERCISE II r

1. Divide £15. 15*s*. 0*d*. by 9, 28, 60, 63, 135.
2. Divide £250 by 12, 60, 600, 75, 375.
3. Divide £72 by 48, 80, 90, 96, 144.
4. Divide £4. 11*s*. 8*d*. by 5, 10, 11, 50, 44.
5. Divide £70. 8*s*. 9*d*. by 23.
6. Divide £89. 0*s*. 1*d*. by 41.
7. Divide £597. 2*s*. 10*d*. by 131.
8. Divide £419. 6*s*. 5*d*. by 157.
9. Divide 17 yds. 2 ft. 1 in. by 13.
10. Divide 486 yds. 2 ft. 11 in. by 47.
11. Divide 298 yds. 0 ft. 7 in. by 113.
12. Divide 1441 yds. 2 ft. 6 in. by 246.
13. Divide 220 lbs. 14 oz. by 57.
14. Divide 44 cwts. 2 lbs. by 29.
15. Divide 213 tons 5 cwts. 30 lbs. by 67.
16. Divide 32 tons 13 cwts. 2 qrs. 14 lbs. by 63.
17. Divide 30 tons 4 cwts. 2 qrs. 4 lbs. 3 oz. by 113.
18. Divide 106 tons 19 cwts. 1 qr. 14 lbs. by 163.
19. Divide 325 tons 8 cwts. 14 lbs. by 267.
20. Divide 1622 tons 16 cwts. 2 qrs. by 934.

COMPOUND DIVISION. QUOTITION

§ 16. Suggestions for oral work:

A. How many men can be paid 15*d*. each out of 45*d*., 5/-, 8/9?

B. How many pieces each 10 in. (6 in., 11 in.) long can be cut from a wire 10 ft. 4 in. long, and how much remains?

C. How many sacks containing 100 lbs. can be filled from a ton of coal, and how much remains over?

D. How many periods of $\frac{3}{4}$ hr. are there between 9 a.m. and 1 p.m. on the same day, and how much remains over?

EXERCISE II s

1. How many times is £1. 5*s.* contained in £15?

2. How many times is 17*s.* 4*d.* contained in £13?

3. How many men can be paid 18*s.* 4*d.* out of £21. 1*s.* 8*d.*?

4. How many blocks at 1*s.* 9½*d.* can be bought for £2. 17*s.* 4*d.*?

5. How many shares of 3/- can be paid out of £5, and what sum remains over?

6. How many tickets at 5*s.* 4*d.* each can be bought for £3. 1*s.* 4*d.*, and how much change is there?

7. How many times can 2*s.* 8½*d.* be paid from £2, and what is the change?

8. £50 is spent in buying books at 4*s.* 6*d.* How many can be bought, and what is the change?

9. How many times is 14*s.* 7*d.* contained in £20, and what is the change?

10. How many subscriptions of 17*s.* 6*d.* go to make up £23. 12*s.* 6*d.*?

11. How many times is 18*s.* 4*d.* contained in £15. 3*s.* 7*d.*, and what is the change?

12. How many times is £1. 15*s.* 3*d.* contained in £19. 17*s.* 10*d.*, and what is the change?

13. How many pieces of wood 18 in. long can be cut from a plank 17 feet long? What length remains over?

14. How many pieces of paper 10 in. long can be cut from a roll 22 yds. long, and what length remains?

15. How many hooks can be manufactured from 50 yds. of wire if each hook requires 7 in.? What length remains?

How many times is

16. 11 in. contained in 20 ft.,

17. 1 ft. 5 in. contained in 200 yds.,

18. 29 in. contained in 43 ft.,

and how much remains in each case?

19. How many sacks each holding 160 lbs. can be filled from a load of 15 cwts., and what weight remains?

How many times will A be contained in B and what remains if:

20. $A =$ 31 lbs. and $B =$ 2 cwts. 20 lbs.,

21. $A =$ 2 lbs. 3 oz. and $B =$ 260 lbs.,

22. $A =$ 243 lbs. and $B =$ 3 tons 4 cwts.,

23. $A =$ 24 lbs. 6 oz. and $B =$ 17 cwts. 13 lbs.,

24. $A =$ 15 lbs. 13 oz. and $B =$ 3 qrs. 21 lbs.,

25. $A =$ 2 lbs. 3 oz. and $B =$ 5 stone 10 lbs.?

EXERCISE II t

Problems on the Four Compound Rules

1. The value of a Japanese 20-yen piece is £2. 0*s*. 11$\frac{3}{4}$*d*. What is the value of ten of these coins?

2. I have £54. 14*s.* 6*d.* in the Post Office Savings Bank and £1. 6*s.* 11*d.* interest is to be added to it. How much shall I then have in the bank?

3. There are 5 partners holding equal shares in a business and the profit for the year is £1611. 10*s.* 5*d.* What is each partner's share of the profit?

4. A cart full of coal weighs 1 ton 17 cwts.; the empty cart weighs 14 cwts. 2 qrs. What is the weight of the coal?

5. A library buys 25 copies of a new book, paying 16*s.* 8*d.* for each copy. How much do the 25 cost?

6. A man walks 24 times round his garden every day for exercise. One round measures 430 yards. How far does he walk each day (in miles and yards)?

7. I bought 24 bottles of claret for £1. 12*s.* How much was this per bottle?

8. What weight of tea will be required to make up seven packets, each containing 1 lb. 4 oz. of tea?

9. If a man drinks a pint of beer a day, how long will a 9-gallon cask last him?

10. In January a man had £25. 16*s.* 4*d.* in the savings bank. In February he deposited 18*s.* 6*d.*, and in March a further sum of £1. 5*s.* 10*d.* How much had he then in the bank if he had drawn nothing out in the meantime?

11. What must I pay for 15 railway tickets at 1*s.* 2*d.* each?

12. A man paid £23. 7*s.* for income tax last year, and this year he paid 15*s.* 4*d.* less. How much did he pay this year?

13. A man pays for $6\frac{1}{2}$ railway tickets with a £5 note, and receives back in change £3. 10*s.* 9*d.*; what is the price of a ticket?

14. 1st class fare is $1\frac{3}{4}d.$ a mile, and 3rd class fare is 1*d.* a mile; find the difference between 1st and 3rd class fares for a journey of 236 miles.

15. A father who gives each of his boys 18*d.* a week pocket money finds that this costs him £19. 10*s.* 0*d.* a year (52 weeks); how many sons has he?

16. How much milk (gallons, quarts, etc.) is required for 140 persons, allowing a pint to each person?

17. An ounce bottle of Bovril costs $6\frac{1}{2}d.$, and a pound bottle 4/10. What is the saving in buying a pound bottle rather than the corresponding number of ounce bottles?

18. You may buy stamped postcards at 11 for 6*d.* or plain cards at 25 for one penny and stamp them yourself with halfpenny stamps. What would you gain by buying 1100 cards by the second method instead of by the first?

19. A clerk is engaged from 8.30 a.m. to 1 p.m. on six mornings, from 2 p.m. to 5 p.m. on four afternoons, and from 6.30 p.m. to 8.30 p.m. on two evenings each week. Her wages are 27*s.* 6*d.* a week. Find to the nearest halfpenny the average pay per hour.

20. The wheel of a locomotive is 22 ft. in circumference. How many turns would it make in travelling a mile?

21. In a battle in Poland, the Germans deployed 7 army corps of 22,000 men each on a front of 7 miles. How many men were there to every 2 yards?

22. A soldier takes in a minute 120 paces of 33 inches: how far (miles and yards) does he march in an hour?

PRACTICE

§ 17. The multiplication of compound quantities is more conveniently performed by the method known as Practice. From the example below it will be seen that the essential feature of Practice consists in taking what are known as **Aliquot Parts** of the unit: i.e. fractions such as $\frac{1}{2}, \frac{1}{3}, \frac{1}{4}, \frac{1}{5}$ etc., whose numerators are 1.

The example below is an instance of **Simple Practice.**

Example. Find the cost of $254\frac{1}{2}$ tons at £3. 7*s*. $8\frac{1}{2}d$. per ton.

	£	*s*.	*d*.	
	254	10	0	= cost at £1 per ton.
	763	10	0	= ,, £3 ,,
5/- $= \frac{1}{4}$ of £1	63	12	6	= ,, 5/- ,,
2/6 $= \frac{1}{2}$ of 5/-	31	16	3	= ,, 2/6 ,,
$2\frac{1}{2}d. = \frac{1}{12}$ of 2/6	2	13	$0\frac{1}{4}$	= ,, $2\frac{1}{2}d$. ,,
	861	11	$9\frac{1}{4}$	= cost at £3. 7*s*. $8\frac{1}{2}d$. per ton.

Much labour may be saved by a judicious selection of Aliquot Parts. For instance in the example above £1. 7*s*. $8\frac{1}{2}d$. = £1 + 5/- + 1/- + 1/- + 6*d*. + 2*d*. + $\frac{1}{2}d$., involving the fractions $\frac{1}{4}, \frac{1}{5}, \left(\frac{1}{5}\right), \frac{1}{2}, \frac{1}{3}, \frac{1}{4}$, but this method requires three more lines of work than the calculation above.

EXERCISE II u (oral)

Express each of the sums of money quoted in Exercise II v below by means of a series of convenient aliquot parts, thus:

$$£1.\ 16s.\ 3d. = £1 + 10/\text{-} + 5/\text{-} + 1/3 \left(\frac{1}{2}, \frac{1}{2}, \frac{1}{4}\right),$$

$$17s.\ 9\tfrac{3}{4}d. = 10/\text{-} + 5/\text{-} + 2/6 + 3d. + \tfrac{3}{4}d. \left(\frac{1}{2}, \frac{1}{2}, \frac{1}{10}, \frac{1}{4}\right)$$

and so on.

EXERCISE II v

Find by Practice the cost of:

1. 96 things at £1. 17*s*. 6*d*. each.
2. 54 things at £1. 18*s*. 9*d*. each.
3. 75 things at £1. 12*s*. 8*d*. each.
4. 84 things at £2. 15*s*. 10*d*. each.
5. 23 things at £1. 13*s*. 6*d*. each.
6. 44 things at £1. 6*s*. 3*d*. each.
7. 568 things at £2. 9*s*. 6*d*. each.
8. 139 things at £1. 6*s*. $7\frac{1}{4}d$. each.
9. 137 things at £1. 2*s*. $3\frac{1}{2}d$. each.
10. 223 things at £2. 15*s*. $4\frac{1}{2}d$. each.

Find by Practice the weight of:

11. 220 boxes if each weighs 10 cwts. 1 qr. 14 lbs.
12. 433 boxes if each weighs 10 cwts. 3 qrs. 21 lbs.
13. 491 pianos if each weighs 12 cwts. 1 qr. 23 lbs.
14. 719 motor cars each weighing 1 ton 5 cwts. 48 lbs.
15. 3479 boxes if each weighs 5 cwts. 1 qr. 16 lbs.

Find by Practice the total length of:

16. 100 rails each 2 yds. 1 ft. 3 in. long.
17. 763 rails each 3 yds. 1 ft. 8 in. long.
18. 971 drain pipes each 2 ft. $9\frac{1}{2}$ in. long.

§ 18. When the given quantities are both compound, as in the example below, the process is called **Compound Practice.**

Example. Find the value of 1 cwt. 2 qrs. 12 lbs. at £3. 1*s*. 3*d*. per cwt.

	£	*s*.	*d*.	
	3	1	3	= cost of 1 cwt.
2 qrs. = $\frac{1}{2}$ of 1 cwt.	1	10	$7\frac{1}{2}$	= „ 2 qrs.
8 lbs. = $\frac{1}{7}$ of 2 qrs.		4	$4\frac{1}{2}$	= „ 8 lbs.
4 lbs. = $\frac{1}{2}$ of 8 lbs.		2	$2\frac{1}{4}$	= „ 4 lbs.
	4	18	$5\frac{1}{4}$	= cost of 1 cwt. 2 qrs. 12 lbs.

EXERCISE II w

Find by Compound Practice the cost of:

1. 17 cwts. 2 qrs. at £27. 6*s*. per ton.

2. 3 tons 4 cwts. 2 qrs. at £5. 10*s*. 10*d*. per ton.

3. 6 cwts. 3 qrs. 16 lbs. at £1. 9*s*. 2*d*. per cwt.

4. 3 tons 18 cwts. at £1. 8*s*. 9*d*. per ton.

5. 2 cwts. 1 qr. 21 lbs. at £22. 1*s*. 4*d*. per cwt.

6. 6 cwts. 1 qr. 20 lbs. at £15. 3*s*. 4*d*. per ton.

7. 2 cwts. 3 qrs. 8 lbs. at £56. 11*s*. 8*d*. per ton.

8. 8 yds. 2 ft. 9 in. at £1. 1*s*. 6*d*. per yd.

9. 15 yds. 0 ft. 10 in. at £1. 9*s*. 3*d*. per yd.

10. 5 yds. 2 ft. $10\frac{1}{2}$ in. at 14*s*. 6*d*. per yd.

Find by Practice the tax on:

11. £725. 15*s*. at 1*s*. 8*d*. in the £.

12. £1761. 6*s*. 8*d*. at 2*s*. 9*d*. in the £.

13. £3563. 13*s*. 4*d*. at 3*s*. $7\frac{1}{2}$*d*. in the £.

CHAPTER III

UNITARY METHOD (No Fractions)

§ 19. *Example.* If 12 golf-balls cost 18/-, what do 7 cost?

Since 12 balls cost 18/-,

$\therefore$ 1 ball costs 18/- $\div$ 12 = 1/6,

$\therefore$ 7 balls cost 1/6 $\times$ 7 = 10/6.

Note the assumption that the dozen balls cannot be bought at a cheaper rate than one ball.

EXERCISE III a (oral)

Direct

1. A man earns 3/- a day. How much will he earn in 2 days? 5 days? 12 days?

2. A man earns 21/- in 7 days. How much does he earn in 1 day? 3 days? 6 days?

3. If 1 lb. of tea cost 1/6, how much will 3 lbs. cost?

4. If 6 lbs. of tea cost 12/-, what is the cost of 1 lb.? 3 lbs.? 5 lbs.?

5. A man walks 4 miles in 60 minutes. How long does he take to walk 1 mile? 2 miles? 5 miles?

6. If four boxes of matches cost 1*d.*, what will be the cost of 1 box? 6 boxes? 20 boxes?

7. If six boxes of matches cost 1*d.*, how many can be bought for 2*d.*? $\frac{1}{2}$*d.*? $2\frac{1}{2}$*d.*? 2/6?

8. A man buys 8 pencils for 6*d.* How many can he buy for 1*s.* 3*d.*?

9. A man earns 12/- in 6 days. In how many days will he earn 3 florins? 18/-? £1?

10. If 6 lbs. of tea cost 7/6, how many pounds can I buy with 5 half-crowns? £1? £1. 12*s.* 6*d.*?

EXERCISE III b

Direct

1. If a 24-acre field can be mown in 3 days, how many acres can be mown in 8 days?

2. If 5 tons of coal cost £5. 12*s.* 6*d.*, what will 8 tons cost?

3. If 4 gallons of beer cost 6/-, what will five quarts cost?

4. If a dozen remade golf-balls cost 6/-, how many can I buy with 4/-? 3/6?

5. If 4 bats cost £5, what will 18 cost?

6. If 20 officers are required for 600 men, how many men can 15 officers command?

7. If there are 15 lamps in a street 300 yards long, how many yards will 25 lamps illuminate?

8. If 3 dozen biscuits cost 1/-, what will 5 dozen cost?

9. A quart of methylated spirits weighs 2 lbs. What will 3 pints weigh?

10. If 20 sacks of coal weigh one ton, how many pounds of coal are there in 3 such sacks?

11. If 10 men eat 45 lbs. of bread in 3 days, how much will 12 men eat in 2 days?

12. If 5 hens lay on an average 80 eggs in 4 weeks, how many eggs should 7 hens lay in 3 weeks?

13. If 80 men eat 160 lbs. of bread in one day, how much will 100 men eat in 6 days?

14. If a battalion of six companies is 720 strong, how many men will there be in (1) 4 battalions of 8 companies each, (2) 4 battalions of 6, 5, 4, 5 companies respectively? (Suppose that all the companies are of the same strength.)

§ **20.** *Example.* If 3 pumps can empty the hold of a ship in 15 hours, how long will 5 such pumps take to empty it? (Think whether 1 pump will take a longer or a shorter time than 3 pumps.)

Since 3 pumps take 15 hours to do the work,

$\therefore$ 1 pump will take $15 \times 3 = 45$ hours,

$\therefore$ 5 pumps will take $45 \div 5 = 9$ hours.

EXERCISE III c (oral)

Inverse

1. If six porters can empty the luggage van in 1 minute, how long will it take one porter?

2. If two taps of the same size fill the bath in 10 minutes, how long will it take one tap?

3. If 3 men eat a leg of mutton in two meals, how many meals will it provide for 1 man?

4. If 1 man can bail out a boat in 6 minutes, how long will it take 6 men?

5. If six men take six minutes to load a wagon, how long will it take one man?

6. If one sheep eats a patch of turnips in 4 days, how long will it take 4 sheep to eat them?

7. If one boy takes 45 minutes to clean the boots, in what time ought 3 boys to clean them?

8. If one mouse can eat a pound of cheese in 12 days, how long will it take six mice to eat a pound?

EXERCISE III d

Inverse

1. Last year a farmer carried all his corn in 12 days with 3 carts at work. This year he only has 2 carts. How long will he take to carry the same amount of corn?

2. A contractor has 40 men at work and expects to finish the job in 6 days. If 16 men desert him, how long will the rest take to finish the work?

3. If 15 bricklayers build a wall 15 yards long in one day, how long will it take

(1) 1 bricklayer to build the same length of wall,
(2) 1 bricklayer to build 1 yard of wall,
(3) 1 bricklayer to build 300 yards of wall,
(4) 25 bricklayers to build 300 yards of wall?

4. If sixteen men shear 128 sheep in 16 hours, how long will it take 5 men to shear 40 sheep?

5. A workman carries 2 loads of bricks in 6 minutes. If each load contains 12 bricks, how long will it take (1) 15 men to carry 80 loads? (2) 25 men to carry 1200 bricks?

§ **21.** The method of the foregoing examples is called the **unitary method,** because it involves reduction to the unit (price of *one* golf-ball, time taken by *one* pump, etc.). This reduction forms the second line in the sum; and at this stage *you must think whether the answer will be diminished or increased* by the step you are taking.

EXERCISE III e

Direct and Inverse

1. 364 tons of coal are required to supply 7 furnaces for a month. How many tons would 9 furnaces burn in the same time?

2. If 4 pipes can fill a tank in 5 hours, in what time could 5 pipes fill it?

3. If 22 copies of a book cost £6. 1*s.* 0*d.*, what is the cost of 100 copies?

4. If a garrison of 100 men has food enough for a fortnight, how long would the food last if the garrison is reinforced by 40 more men?

5. If 13 lbs. of sugar cost 2*s.* $8\frac{1}{2}$*d.*, find the cost of 27 lbs. of sugar.

6. 24 cwts. of goods were carried over a certain line of railway for a distance of 336 miles. How far would the same railway company carry 14 cwts. for the same sum of money?

7. If a man can walk to his work in 16 minutes, walking 3 miles an hour, how long will the journey take him if he walks at 4 miles an hour?

8. A 40 horse-power motor car runs 12 miles to a gallon of petrol; what does petrol cost for a 600 mile journey, at 1/9 a gallon?

9. If an army can hold 30 miles of trenches when there are 2 men to a yard, what length of front can they hold when they are 3 to a yard?

10. If eggs cost 9*d.* a dozen, how many for 1/-?

11. Once round a running track is 440 yards. How many times round is 10 miles? If a man runs once round the track in 75 secs., at how many miles an hour is he running?

12. If 60 hens lay 240 eggs in 8 days, how many eggs will 18 hens lay in 12 days?

13. If 4 men in 1 year earn £166. 8*s.* 0*d.*, what do 20 men earn in 1 week? for how many weeks should 20 men work in order to earn £240?

14. If the wages of 29 men for 12 days are £87, how many men should receive £405 for 54 days?

CHAPTER IV

AREA. VOLUME. (No Fractions)

Area of Rectangle

§ **22.** In the figure the rectangles A and B represent two pieces of ground which are to be sown with grass-seed: which of them will need the greater quantity of seed? Notice that A is longer than B, but B is broader than A.

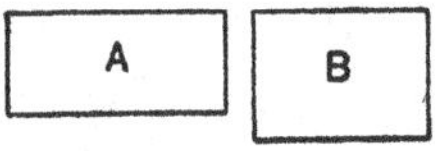

A B

It is not very easy to judge without measurement which piece will require more seed. On measurement it is found that the sides of A are 9 ft. and 4 ft.; and the sides of B are 7 ft. and 5 ft. Let the sides of each rectangle be marked off in feet and let lines be drawn, cutting up the rectangles into small squares.

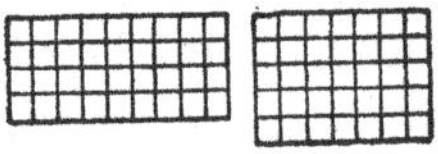

Each small square measures 1 foot by 1 foot; it contains 1 **square foot** of ground.

Now count the number of square feet in each rectangle. The easiest way of counting the squares is as follows. In A, there are 9 squares in each row; there are 4 rows; altogether there are 9×4 squares or 36 squares. Similarly, in B, there are 7×5 squares or 35 squares. It appears, then, that A will require more seed than B.

The **area** of A is said to be 36 square feet; and the area of B is said to be 35 square feet. The **unit of area** used in this case is 1 square foot.

N.B.—*A square foot is not a special kind of foot, but a different thing altogether.* Any convenient area might be used as unit: e.g. what is the area of paper in this book if one of the pages is taken as unit of area? But it is convenient to have a connection between the unit of length and the unit of area.

¶**Ex. 1.** What is the area of a rectangle which measures 3 ft. by 5 ft.? 6 ft. by 10 ft.? 12 ft. by 2 ft.? 9 ft. by 5 ft.?

¶**Ex. 2.** What is the area of a square which measures 6 ft. by 6 ft.? 11 ft. by 11 ft.? Of a square whose side is 4 ft.? 5 ft.? 8 ft.?

¶**Ex. 3.** Given the sides of a rectangle, how is the area calculated?

In finding the area of a sheet of paper, we might measure the side in inches, rather than feet, and express the area in **square inches.**

¶**Ex. 4.** What is a square inch? a square mile? a square yard?

¶**Ex. 5.** In what units should we naturally express the area of a county? a tennis lawn? a handkerchief?

¶**Ex. 6.** What is the area of a rectangle measuring 4 in. by 5 in.? 3 in. by 7 in.? 5 miles by 8 miles? 6 yards by 4 yards?

¶**Ex. 7.** What is the area of a square whose side is 5 inches? 6 miles?

The above exercises lead up to the following rule:

To find the number of units in the area of a rectangle, multiply together the numbers of units in the length and breadth of the rectangle.

This rule is often abbreviated to

area of rectangle = length × breadth,

but this must not be misunderstood. It must not be supposed that 3 ft. × 5 ft. = 15 sq. ft. How can a length of 3 feet be multiplied by a length of 5 feet? We can multiply 3 feet by 5, and the result is 15 feet, but we cannot multiply 3 feet by 5 *feet*. To multiply anything by 5 means to take the thing five times, and add up. But we cannot take a thing *five-feet* times: this is nonsense.

If a rectangle measures 3 ft. by 5 ft.,

it is **wrong** to say.. area = 3 ft. × 5 ft.

It is **right** to say.. area = (3 × 5) sq. ft.

EXERCISE IV a

What is the area of a rectangle whose length and breadth are:

1. 7 ft. and 3 ft., **2.** 10 inches and 6 inches,

3. 13 yds. and 4 yds., **4.** 2 miles and 3 miles?

Calculate the area of:

5. A field 200 yds. long and 120 yds. wide.

6. A field 230 yds. long and 130 yds. wide.

7. A sheet of paper measuring 23 in. by 17 in.

8. A sheet of paper 34 in. long and 29 in. wide.

9. A billiard table 13 feet long and 7 ft. wide.

10. The ceiling of a room 43 ft. by 23 ft.

Find the cost of carpeting the floor of a room:

11. 24 ft. long and 20 ft. wide at 2/- per sq. foot.

12. 16 ft. long and 15 ft. wide at 4/- per sq. foot.

13. 53 ft. long and 47 ft. wide at 3/- per sq. foot.

14. 6 yds. long and 5 yds. wide at 10/- per sq. yd.

15. 11 yds. long and 13 yds. wide at 7/- per sq. yd.

16. 13 yds. long and 13 yds. wide at 11/- per sq. yd.

17. A chess-board 16 inches square contains 64 squares. What is the area of one square? Of all the black squares?

18. If a carpet covers an area of 200 sq. ft. and is 20 ft. long, how wide must it be?

19. If the area of a rectangular sail is known to be 391 sq. ft. and its width is 17 ft., what is its length?

20. If a rectangular field is 430 yds. long and is known to contain 90,300 sq. yds., how wide is it?

21. A rectangular lawn 12 yds. wide and 32 yds. long is to be resown with grass-seed. If the gardener has enough seed for 360 sq. yds. and begins from one end, what length of lawn can he sow?

22. If a rectangle contains 925 sq. inches and is 25 inches long, what is its width?

23. A carpet is made in lengths 4 ft. wide. What length is required to cover a floor 20 ft. long and 12 ft. wide?

24. A wall-paper is made of width 30 inches. What length of paper is required to cover a wall 210 inches long and 120 inches high?

25. A courtyard 27 ft. by 20 ft. is paved with stones each 3 ft. by 2 ft. How many stones are there?

26. A carpet 17 ft. by 13 ft. is laid on a floor 25 ft. by 18 ft. What area is *not* covered?

27. Explain the difference between 2 square feet and 2 feet square.

UNITS OF AREA

§ 23. ¶**Ex. 8.** How many square inches are there in a square whose side is 1 foot?

¶**Ex. 9.** Express in square feet the area of a square of side 1 yard.

A square whose side is 1 foot or 12 inches contains 12^2 sq. ins. or 144 sq. ins.; similarly, a square whose side is 1 yard or 3 ft. contains 3^2 sq. ft. or 9 sq. ft.

EXERCISE IV b

Express:

1. 4 sq. ft. in sq. in. **2.** 57 sq. ft. in sq. in.

3. 288 sq. in. in sq. ft. **4.** 720 sq. in. in sq. ft.

5. 5 sq. yds. in sq. ft. **6.** 4 sq. yds. in sq. in.

7. 135 sq. ft. in sq. yds. **8.** 2592 sq. in. in sq. yds.

Reduce:

9. 1008 sq. in. to sq. ft. **10.** 81 sq. ft. to sq. yds.

11. 16848 sq. in. to sq. yds.

12. 22032 sq. in. to sq. yds.

13. How many bricks 9 in. long by 5 in. broad will be needed for the floor of a stable 12 ft. by 10 ft.?

14. A garden 32 yds. by 20 yds. contains a lawn 80 ft. by 50 ft. What area is occupied by paths and flower-beds?

15. A street 50 ft. wide and 100 yds. long has a foot-path 4 ft. wide on both sides. Find the area of (1) the roadway, (2) the paths.

16. Express in sq. in. the area of a rectangle 2 ft. 3 in. wide and 3 ft. 2 in. long.

17. A roll of cloth measures 30 inches by 12 yds. Find the area of the cloth (1) in sq. ft., (2) in sq. yds.

18. A window contains 8 panes of glass each measuring 16 inches by 9 inches. Find the total glazed surface in (1) sq. in., (2) sq. ft.

Find the cost of a piece of carpet

19. 30 in. by 14 yds. at 1/- per sq. ft.

20. 2 ft. 3 in. by 12 yds. at 5/- per sq. yd.

21. 2 ft. by 15 yds. at 7/6 per sq. yd.

22. If the area of a rectangle is 120 sq. ft. and its length is 5 yds., what is its width in ft.?

23. How many yards of carpet 30 inches wide are required to cover a floor 20 ft. long and 18 ft. wide?

24. What length of planking 9 in. wide will be required to floor a room

(1) 150 sq. ft. in area? (2) 20 ft. by 18 ft.?

25. An oak paling 100 yards long and 5 feet high is made by placing oak planks 5 ft. by 6 in. side by side. What length of planks will be used?

26. A packet of newspaper wrappers contains 120 wrappers, each measuring 12 in. by 5 in. How many square feet of paper is this?

§ 24. Land surveyors measure land with a chain, 22 yards long; this is also the length of a cricket pitch. If a square is marked out with each of its sides 1 chain in length, its area will be 22^2 or 484 square yards. This area might be called a square-chain. Areas of land are not actually expressed in square-chains: but the idea of a square-chain may help you to realise what an **acre** is; 10 square-chains, in fact, make up an acre. Thus

1 acre = 484 × 10 sq. yds. = **4840 sq. yds.**

Notice that a square whose side is 70 yards contains 4900 sq. yds., and therefore a little more than an acre.

The largest unit of area in use is a **square mile;** Hyde Park and Kensington Gardens together are rather more than a square mile in area.

¶**Ex. 10.** How many yards are there in a mile? How many square yards in a square mile? How many square yards in an acre? How many acres in a square mile?

The above exercises lead to the conclusion that

1 square mile = 640 acres.

Hence the following **table of units of area,** sometimes called **square measure:**

12^2 or 144 sq. ins.	= 1 sq. ft.
3^2 or 9 sq. ft.	= 1 sq. yd.
4840 sq. yds.	= 1 acre.
640 acres	= 1 sq. mile.

EXERCISE IV c

How many acres are there in a rectangular field

1. 220 yds. long and 110 yds. wide?
2. 880 yds. long and 330 yds. wide?

How many square miles are there in a rectangle

3. 21 miles long and 13 miles wide?

4. 171 miles long and 233 miles wide?

5. How many chains are there in a sq. mile?

6. In a square field whose side is 110 yds. long?

7. In a rectangular field 240 yds. by 121 yds.?

8. How many cabbages can be grown on an acre if 4 plants require 1 sq. yard?

9. If a fir tree requires 100 sq. feet of ground, how many trees can be planted on 250 acres?

10. How many fir trees are required for a square plantation whose side is 130 yds. long?

11. A rubber estate is planted with 96 trees per acre. Find how many trees there are in a rectangular field measuring $1\frac{1}{2}$ miles by $\frac{3}{4}$ mile.

12. A rectangular 3-acre field is 44 yards broad; how long is it?

13. What is the width of a 20-acre field whose length is a quarter of a mile?

EXERCISE IV d

1. A potato-patch in the form of a rectangle 20 ft. by 30 ft. is surrounded by a gravel walk 1 yd. wide. What is the area of the gravel? (The dotted lines in the fig. show how the figure of the path may be divided up into 4 rectangles. Observe carefully the length of the rectangles *A* and *B*.)

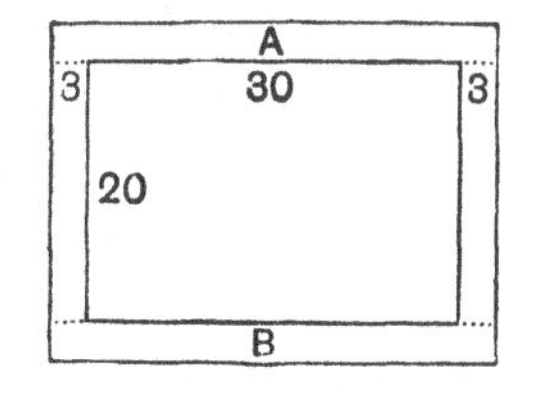

2. Work the above exercise by finding the *difference* between the area of the whole figure and the area of the potato-patch.

3. A carpet 15 ft. by 10 ft. is laid in a room 24 ft. by 18 ft. What area is not covered?

4. Four flower-beds 5 ft. by 5 ft. are separated by paths 1 ft. wide and surrounded by a path 2 ft. wide. Find the total area of the paths.

5. A window frame, external dimensions 2 ft. 6 in. by 1 ft. 9 in., contains four equal panes separated from each other by 1 in. of wood. If the frame is made of wood 2 in. wide, find the area of each pane.

6. For the following mounted pictures, find the area of the mount, and of the picture: (i) Outside measurements of mount 18″ by 10″, width of mount 2″. (ii) Outside measurements of mount 20″ by 12″, width of mount 3″.

7. A reaping machine cuts down a strip of corn a yard wide each time it goes round a rectangular field. If the field is 200 yds. by 120 yds., what is the length, breadth and area of the corn left standing after the machine has been round 10 times? 20 times?

8. A stair-carpet 1 yard wide covers 20 steps 1 foot broad and 6 inches high and projects 1 foot at both the top and bottom of the staircase. What is the length and area of the stair-carpet?

9. The floor of a room 24 ft. by 18 ft. is stained all round within a distance of two feet from the walls. A carpet 21 ft. by 15 ft. is laid symmetrically on the floor. (1) What width, (2) what area, of the staining is covered by the carpet?

10. Find the areas of the shaded and unshaded parts of figs. 1 to 8 below. (Dimensions in feet.)

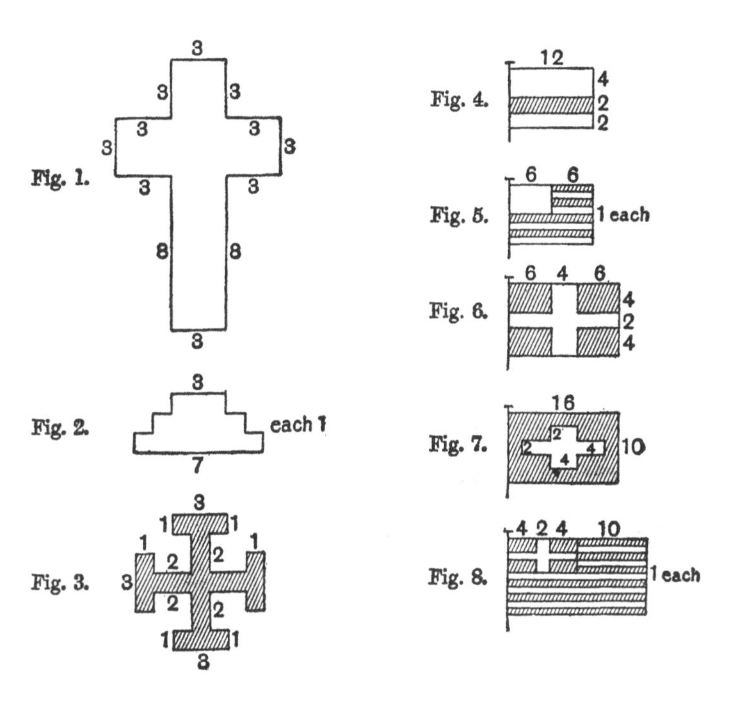

11. A rectangular iron grating is 2 ft. 4 in. wide and 3 ft. 6 in. long and has 12×20 square inch holes in it. What is the weight of the grating if 1 sq. foot of iron of the same thickness weighs 10 lbs.?

12. The panels and the frame of a door are to be painted in two different shades. If the door is 6 ft. 6 in. high and 4 ft. 9 in. wide and has 6 panels 8 inches wide, 3 being 3 ft. 3 in. long and the other 3 being 24 in. long, find the area to be painted in each shade. (One side of the door to be painted.)

13. What is the cost of painting a venetian blind made of 24 thin laths each 4 ft. 6 in. long and 3 inches wide at 2*d*. per sq. foot?

SURFACE OF CUBOID

§ **25.** A thing shaped like a brick or an ordinary box is called a **rectangular block** or a **cuboid.** A cuboid may be solid like a brick, or hollow like a box. A cuboid is bounded by rectangles, called its **faces**: how many faces has a cuboid?

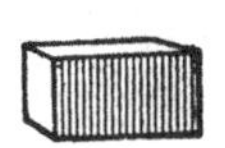

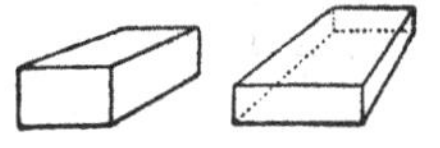

A cuboid has length and breadth and thickness (or length, breadth and depth); these three measurements are called its three **dimensions.**

If the three dimensions of a cuboid happen to be equal, it is a special kind of cuboid called a **cube** (e.g. dice, cube sugar).

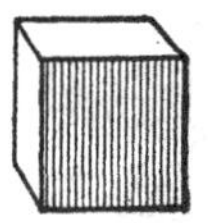

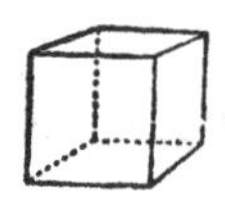

NOTE.—A simple method of drawing figures of cuboids, etc. has been used in most of the figures in this book. It consists in drawing one face to scale, the opposite face somewhat smaller and rather to one side, and joining the corresponding corners. Another method of producing the same effect is suggested by the fact that these joins, when produced, meet at a point.

§ 26. To paint the outside of a box, a man has to paint 6 faces: namely, the 4 sides, the top, and the bottom. The amount of paint needed for each face depends on two things: (1) how thickly the paint is laid on, and (2) the area of the face. The sum of the areas of the 6 faces is called the **surface** of the cuboid.

Ex. 11. Find the surface of a cuboid which measures 6 inches by 7 inches by 3 inches. Notice that opposite faces are equal.

Example. A closed cistern is made of thin sheet iron; the length is 10 ft., the breadth 5 ft., the depth 3 ft. If sheet iron costs $1\frac{1}{2}d.$ per sq. ft., what is the cost of the metal?

Area of top and bottom together = **twice** 50 sq. ft. = 100 sq. ft.

Area of two long sides together = **twice** 30 sq. ft. = 60 sq. ft.

Area of two short sides together = **twice** 15 sq. ft. = 30 sq. ft.

Total surface = 190 sq. ft.

Cost = $1\frac{1}{2}d. \times 190 = (190 + 95)d. = 285d. = \underline{£1.\ 3s.\ 9d.}$

Example. An open box 1 foot high, 2 ft. 6 in. long, 1 ft. 6 in. broad, is to be lined with sheet lead. How many square inches are required?

Here there are only 5 rectangles to be considered.

Area of bottom = 30×18 sq. in. = 540 sq. in.

Area of two long sides = twice 12×30 sq. in. = 720 sq. in.

Area of two short sides = twice 12×18 sq. in. = 432 sq. in.

Total area of lead required = <u>1692 sq. in.</u>

EXERCISE IV e

Find in sq. ft. the total surface of a solid cuboid whose dimensions are

1. 3 ft., 5 ft., 7 ft. **2.** 6 ft., 10 ft., 14 ft.

3. 7 ft., 13 ft., 17 ft. **4.** 14 ft., 26 ft., 34 ft.

5. 60 ft., 12 ft., 2 ft. **6.** 30 ft., 6 ft., 1 ft.

7. 12 in., 18 in., 24 in. **8.** 8 in., 8 in., 14 in.

9. 2 yds., 2 ft. 4 in., 18 in.

10. 3 ft. 6 in., 3 yds., 1 ft.

11. Find how much cardboard is needed to make open boxes of the following dimensions:

	i	ii	iii	iv
Length	4″	6″	1 foot	1′ 6″
Breadth	3″	4″	3 in.	9″
Depth	2″	2″	3 in.	5″

12. Repeat Ex. 11, doubling all the dimensions.

13. Find the total area of the four walls of a room, given that the length, breadth and height of the room are (1) 24 ft., 18 ft., 12 ft., (2) 20 ft., 15 ft., 12 ft., (3) 20 ft., 18 ft., 12 ft. 6 in.

14. What is the area of the four walls of a room, the length, breadth and height being l ft., b ft., h ft.?

VOLUME OF CUBOID

§ **27.** A cube each of whose dimensions is 1 inch is called a **cubic inch.** If you have a supply of wooden cubic inches, you should do the following practical exercises.

Ex. 12. Build up with cubic inches a cube each of whose dimensions is 2 inches. How many cubic inches are used?

Ex. 13. Build up, with cubic inches, a cuboid having the following dimensions: 2 in. by 3 in. by 4 in. How many cubes are used? How is this number connected with 2, 3, 4?

§ **28.** The number of cubic inches in a cuboid is called its **volume.** You will have found that the volume of the cuboid you made is 24 cubic inches; but note the following way of counting the cubes.

The cuboid shown in the figure consists of 4 layers of cubes; each layer corresponds to an inch of height: there are as many layers as there are inches in the height.

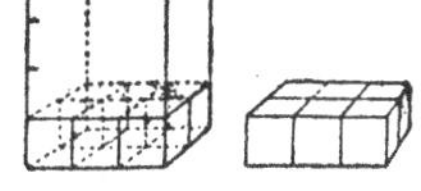

How many cubes in each layer? Clearly 6. And how many *square* inches in the base of the cuboid? Again 6. These two numbers must always be the same, for on each square inch of the base stands one cube of the bottom layer. Thus

Number of cubes in one layer = number of sq. in. in base;
Number of layers = number of inches in height.

∴ Total number of cubic inches in cuboid = number of sq. in. in base × number of inches in height.

This result is often abbreviated to

volume of cuboid = **base** × **height.**

(But remember that this is rather slipshod wording. It is absurd to speak of multiplying by a height: all that you can do is to multiply by the number of units in the height.)

Lastly, consider how the area of the base is connected with the length (3 in.) and the breadth (2 in.).

$$\text{Area of base} = \text{length} \times \text{breadth},$$

$$\therefore\ \textbf{volume} \text{ of cuboid} = \textbf{length} \times \textbf{breadth} \times \textbf{height},$$

or, to be quite accurate, **To find the number of cubic inches in a cuboid, multiply together the numbers of inches in the length, breadth and height.**

§ 29. **Solid things and hollow things.** You have seen how to find the volume of a solid cuboid. The same methods apply in finding the volume of the interior of a hollow cuboid, e.g. the interior of a box. To ask what is the volume of the interior is the same as to ask how much the box will hold. This point could be settled by filling the box with cubic inches; counting them would lead to the same rule as that already found. Of course, in this case the inside measurements of the box must be used.

The terms **cubical content** and **capacity** are often used in the sense of "volume contained."

§ 30. *Example.* An open box is made of wood 1 inch thick; the outside of the box is 12 in. long, 10 in. wide, 6 in. deep. Find the volume of the wood used and the cubical content of the box.

Imagine that the box has been hollowed out of a solid block of wood measuring 12 in. by 10 in. by 6 in. The volume of wood in the box could then be calculated by subtracting the volume removed from the volume of the block. The volume removed is equal to the *inside* volume of the box: the volume of the block may be called the *outside* volume.

Outside volume $= 12 \times 10 \times 6$ cu. in. $= 720$ cu. in.

As regards the inside volume,

Inside length $= (12 - 2)$ in. $= 10$ in.

Inside width $= (10 - 2)$ in. $= 8$ in.

Inside depth $= (\ 6 - 1)$ in. $= 5$ in.

$\therefore$ Inside volume $= 10 \times 8 \times 5$ cu. in. $= 400$ cu. in.

$\therefore$ Volume of wood $= (720 - 400)$ cu. in. $= \underline{320 \text{ cu. in.}}$

The cubical content $=$ inside volume $= \underline{400 \text{ cu. in.}}$

EXERCISE IV f

1. How many inch cubes can be put inside the following boxes, whose internal dimensions are:

(1) 6 in., 4 in., 3 in. (2) 1 ft., 6 in., 6 in.
(3) 1 ft., 1 ft. 6 in., 6 in. (4) 6 ft., 5 in., 3 in.
(5) $6\frac{1}{2}$ ft., 4 in., $2\frac{1}{2}$ ft. (6) $1\frac{3}{4}$ ft., 9 in., $2\frac{1}{4}$ ft.?

2. Find the volume of each of the following cuboids, the length, breadth and height being:

(1) 6 ft., 5 ft., 4 ft. (2) 21 in., 11 in., 7 in.
(3) 2 in., 1 in., $\frac{1}{2}$ ft. (4) 12 in., 2 in., $2\frac{1}{2}$ ft.

3. There are two closed water tanks, the first measuring 6 ft. by 9 ft. by 4 ft., and the second 10 ft. by 8 ft. by 2 ft. Prove that the tank which requires the larger amount of sheet iron contains the smaller quantity of water.

4. Find the number of cubic feet of water contained by open tanks of the given length, breadth and depth; also the number of square feet of sheet iron that will be needed to make them:

(1) 8 ft. by 8 ft. by 8 ft. (2) 16 ft., 8 ft., 4 ft.
(3) 32 ft., 4 ft., 4 ft. (4) 64 ft., 4 ft., 2 ft.

5. The internal dimensions of a tank are 4 ft. by 3 ft., by 2 ft. deep, and it contains water to a depth of 1 ft. 6 in. How many bricks 9 in. by 6 in. by 4 in. must be put into the tank so that the water may just overflow?

6. A closed box is 16 in. long, 12 in. wide and 15 in. high outside. What total space in cubic in.

does it occupy? If the wood is 1 inch thick, what are the inside dimensions and the cubic content? How many cubic in. of wood are there in it?

7. Find in cubic in. the cubic content and the amount of wood in the following closed boxes:

	(1)	(2)	(3)	(4)
Outside length	24 in.	18 in.	4 ft. 6 in.	2 ft.
Outside width	18 in.	13 in.	7 in.	1 ft. 6 in.
Outside height	10 in.	14 in.	4 in.	1 ft. 4 in.
Thickness of wood	1 in.	$\frac{1}{2}$ in.	$\frac{1}{2}$ in.	$1\frac{1}{2}$ in.

8. Find in cubic in. the cubic content and the amount of wood in the sides and bottom of the following open rectangular troughs:

	(1)	(2)	(3)
Inside length	23 in.	3 ft. 6 in.	10 ft.
Inside width	11 in.	2 ft. 4 in.	6 in.
Inside depth	5 in.	11 in.	3 in.
Thickness of wood	1 in.	1 in.	1 in.

9. Find the volumes of the solids represented by Figs. 1, 2, 3 below.

10. Repeat Ex. 8, doubling all the dimensions.

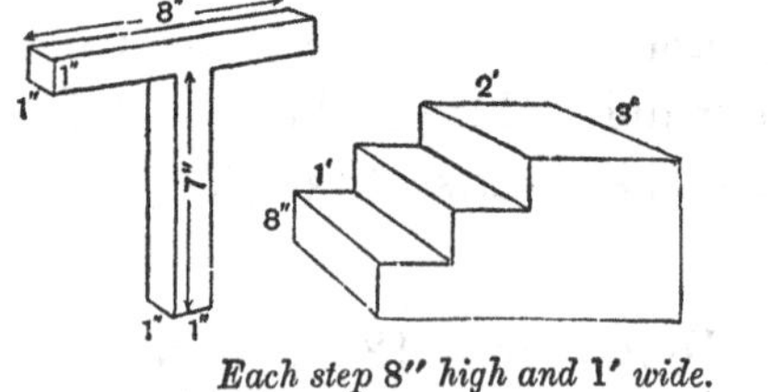

Each step 8" high and 1' wide.

Fig. 1. Fig. 2.

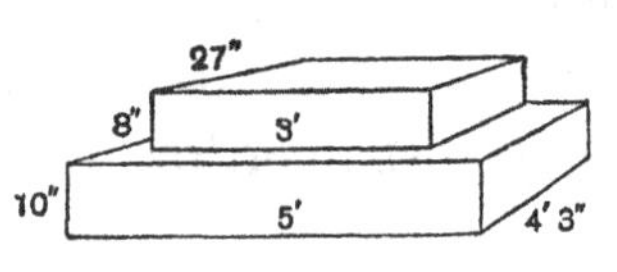

Fig. 3.

11. Find the total surface in Figs. 1 and 2.

12. Find (1) the total vertical surface, (2) the total exposed horizontal surface in Fig. 3.

Units of Volume

§ 31. Since 1 foot = 12 inches, 1 cubic foot contains 12^3, or **1728** cubic inches. A cubic foot is something like the volume of an ordinary cubical biscuit-tin.

Since 1 yard = 3 feet, 1 cubic yard contains 3^3, or **27** cubic feet. A cart is usually reckoned to hold something like a cubic yard of material.

Other British units of volume are the pint, the quart, the gallon, the bushel. These are generally called units of **capacity**; but capacity is only another word for volume.

12^3, or 1728 cu. in. = 1 cu. ft.
3^3, or 27 cu. ft. = 1 cu. yd.

2 pints = 1 quart
4 quarts = 1 gallon
8 gallons = 1 bushel

A gallon contains about 277 cu. in., but this number need not be remembered.

EXERCISE IV g

1. How many cubic inches are there in a 2-inch, a 3-inch, a 4-inch cube?

2. How many 6-inch, 4-inch, 3-inch, 2-inch cubes are there in a cubic foot?

3. How many 18-inch, 9-inch, 6-inch, 3-inch cubes are there in a cubic yard?

4. If there is a cubic inch of copper in 55 inches of a certain telephone wire, how many cubic feet of copper are there in 6 miles of it?

5. A cubic foot of lead weighs 720 lbs. How many ounces do 3 cubic inches weigh?

CHAPTER V

PRIMES. H.C.F. L.C.M. ROOTS

PRIME FACTORS

§ 32. If a number* is exactly divisible by another number the second is said to be a **factor** of the first: e.g. 3 is a factor of 15 and also of 24.

Some numbers can be expressed as the product of a *pair* of factors in several ways.

Examples. $12 = 1 \times 12$ or 2×6 or 3×4.
$56 = 1 \times 56$ or 2×28 or 4×14 or 7×8.

All possible ways should be discovered by testing in succession whether any of the numbers 1, 2, 3, 4, etc. are factors.

¶**Ex. 1.** Express each of the following as the product of two factors in as many ways as possible: 6, 8, 10, 11, 15, 19, 27, 47, 49, 38.

§ 33. If a number has no factor other than itself or 1 it is called a **prime number.** Thus 7, 13, 23, 31 are prime numbers.

¶**Ex. 2.** Which of the following are prime numbers: 5, 27, 42, 43, 51, 52, 53, 71, 72, 73?

§ 34. Any number can be expressed as the product of factors which are prime numbers, i.e. in **prime factors.**

Examples. $72 = 8 \times 9 = 2 \times 2 \times 2 \times 3 \times 3 = 2^3 \times 3^2$.
$182 = 2 \times 91 = 2 \times 7 \times 13$.

* The word "number" is used in this chapter in the sense of "whole number" or "integer."

A pair of factors may be recognizable at once, as in the first example. If not, test in succession whether any of the prime numbers 2, 3, 5, 7, 11, etc. are factors.

¶**Ex. 3.** Express as the products of prime factors: 6, 9, 12, 13, 15, 19, 21, 24, 37, 49, 93.

EXERCISE V a

Express as the products of prime factors:

1. 32.	**2.** 60.	**3.** 64.	**4.** 71.
5. 85.	**6.** 92.	**7.** 168.	**8.** 192.
9. 211.	**10.** 224.	**11.** 251.	**12.** 256.
13. 315.	**14.** 343.	**15.** 360.	**16.** 484.
17. 571.	**18.** 693.	**19.** 9261.	**20.** 74088.
21. 28413.	**22.** 11025.	**23.** 89712.	**24.** 111.
25. 1111.	**26.** 14641.	**27.** 111111.	**28.** 25088.

Greatest Common Measure

§ 35. Suppose that we have to find a *common measure* for two rods 4 ft. 6 in. and 3 ft. long; i.e. a length which may be used to measure both exactly, so that each rod contains the length an exact number of times.

If we choose as unit a length of 1 inch, the rods contain 54 and 36 of these units.

If we choose as unit a 2-inch length, the rods contain 27 and 18 of these units.

Proceeding in this way we find that a length of 1 inch, 2 inches, 3 inches, 6 inches, 9 inches or 18 inches may be chosen as a common measure of the lengths of the two rods, but we shall not find any unit greater than 18 inches which will exactly measure *both* rods.

The **greatest common measure** (G.C.M.) is therefore 18 inches.

¶**Ex. 4.** What is the G.C.M. of:

1. 18 inches and 48 inches? **2.** 24 pints and 56 pints?

3. 1 foot, 2 ft. 6 in. and 3 ft. 6 in.?

4. 45 lbs. and 72 lbs.? **5.** 54 ft. and 81 ft.?

HIGHEST COMMON FACTOR

§ 36. 3 is a common factor of the numbers 15, 24 and 60, for $15 = 3 \times 5$, $24 = 3 \times 8$ and $60 = 3 \times 20$.

It is easy to see that 3 is the **highest common factor** (H.C.F.) of 15, 24 and 60. But 3 is *not* the highest common factor of 24 and 60, for 6 is a factor of both.

Observe that the G.C.M. of the *lengths* 54 and 36 inches is 18 inches (see § 35), and the H.C.F. of the *numbers* 54 and 36 is 18. To find the G.C.M. of a set of quantities expressed in terms of the same unit, we must find the H.C.F. of the set of corresponding numbers.

EXERCISE V b (oral)

What is the H.C.F. of:

1. 4 and 6. **2.** 6 and 9.
3. 14 and 21. **4.** 8 and 32.
5. 17 and 34. **6.** 15 and 21.
7. 8 and 15. **8.** 10 and 24.
9. 16 and 24. **10.** 21 and 42.
11. 6, 8 and 14. **12.** 4, 16 and 24.
13. 5, 15 and 9. **14.** 14, 21 and 42.
15. 50, 15 and 20. **16.** 16, 36 and 44?

§ 37. To find the H.C.F. of two or more numbers, each should be expressed in prime factors.

Example. To find the H.C.F. of 180, 168 and 132:

$$180 = 4 \times 45 = 2^2 \times 3^2 \times 5.$$
$$168 = 8 \times 21 = 2^3 \times 3 \times 7.$$
$$132 = 4 \times 33 = 2^2 \times 3 \times 11.$$

It is clear that 2^2 is a common factor, and that 3 is a common factor. There is no prime factor other than 2 or 3 common to all three numbers, and no higher powers than 2^2 and 3 are common factors of all three. Therefore the H.C.F. is $2^2 \times 3$ or 12. This may be seen clearly if we write: $180 = \mathbf{12} \times 15$, $168 = \mathbf{12} \times 14$ and $132 = \mathbf{12} \times 11$.

H.C.F. BY RULE*

§38. Sometimes it is difficult to find the prime factors of a set of numbers. The method of finding the H.C.F. adopted in this case is shown in the example below; the proof of the rule is too difficult to be given at this stage.

Example. To find the H.C.F. of 9991 and 9506. Divide the greater (9991) by the smaller (9506) and find the remainder (485). Divide the first divisor (9506) by this remainder (485) and find a new remainder (291). Divide the last divisor (485) by this new remainder (291) and continue the process until the division is exact. The last divisor is the H.C.F.

The work may be arranged in either of the following ways:

```
9506)9991(1
     9506
      485)9506(19
          485
          4656
          4365
           291)485(1
               291
               194)291(1
                   194
                    97)194(2
                       194
```

19	9506	9991	1
	485	9506	
	4656	485	1
	4365	291	
1	291	194	
	194	194	2
	97	—	

The H.C.F. of 9506 and 9991 is 97.

By division, $9506 = 97 \times 98$ and $9991 = 97 \times 103$.

EXERCISE V c

Find the H.C.F. of:

¶1. 2×3 and $2^2 \times 7$.

¶2. $2 \times 3^2 \times 5$ and $2^2 \times 3^2 \times 7$.

¶3. $2^2 \times 3$; $2^4 \times 3^3$ and $2^2 \times 3^2 \times 11$.

¶4. 2×3; 3×5 and 5×7.

¶5. $2^2 \times 3 \times 37$; $3^2 \times 5 \times 13$ and $2^2 \times 13 \times 37$.

* May be postponed.

Find the prime factors and hence the H.C.F. of:

6. 48 and 72.
7. 45 and 100.
8. 128 and 112.
9. 720 and 960.
10. 343 and 98.
11. 242 and 1111.
12. 370 and 555.
13. 24, 42 and 112.
14. 14, 112 and 686.
15. 32, 56, 112 and 2056.
16. 1024, 216, 256 and 728.
17. 125, 625, 600, 1200 and 275.
18. 48, 72, 480 and 216.
19. 45, 81, 72, 108 and 99.
20. 16650, 407, 555 and 1147.

EXERCISE V d*

Find the H.C.F. of:

1. 391 and 1081.
2. 312 and 552.
3. 1517 and 1681.
4. 6351 and 6873.
5. 957 and 1023.
6. 6893 and 6441.
7. 6549 and 4307.
8. 12319 and 14605.
9. 30592 and 60945.
10. 145673 and 103683.
11. 199279 and 250669.
12. 284924 and 544502.

Least Common Multiple

§ 39. Along one side of a road is a row of lamps 20 yards apart, and on the other side is a row of trees 12 yards apart. At a certain point a tree is exactly opposite to a lamp. How far down the road must a man go to find the next tree which is exactly opposite to a lamp?

The required distance must contain 20 yards an exact number of times; it must also contain 12 yards an exact number of times. It is therefore clear that at a distance of

* May be postponed.

240 yards (= 20 × 12 yds.) a tree is exactly opposite to a lamp. But by trial (or by drawing a diagram to scale) we find that 120 yards away the same is true, and also at a distance of 60 yards. Now there is no number less than 60 which contains 20 and also 12 an exact number of times. Therefore the man must go at least 60 yards to find the next tree exactly opposite to a lamp.

A number is said to be a **multiple** of each of its factors. Therefore 240, 120, 60 are multiples of 20. They are also multiples of 12; and are said to be **common multiples** of 20 and 12.

The **least common multiple** (L.C.M.) of 20 and 12 is 60.

EXERCISE V e (oral)

Find the L.C.M. of:

1. 4 and 6. **2.** 8 and 12.
3. 3 and 12. **4.** 9 and 18.
5. 18 and 27. **6.** 12 and 10.
7. 6 and 8. **8.** 7 and 14.
9. 14 and 21. **10.** 9 and 12.
11. 24 and 36. **12.** 1, 2 and 3.
13. 2, 3 and 4. **14.** 3, 4 and 5.
15. 4, 5 and 6. **16.** 2, 6 and 8.
17. 6, 8 and 12. **18.** 2, 4, 6 and 8.
19. 3, 6 and 15. **20.** 2, 12, 6 and 18.
21. 3, 4, 12 and 18. **22.** 10, 15, 25 and 30.

§ **40.** To find the L.C.M. of two or more numbers it is necessary to express each of them in factors—generally in prime factors.

Example. Find the L.C.M. of 24, 36 and 54:

$$24 = 8 \times 3 = 2^3 \times 3.$$
$$36 = 4 \times 9 = 2^2 \times 3^2.$$
$$54 = 2 \times 27 = 2 \times 3^3.$$

The L.C.M. must be the product of a power of 2 and a power of 3. The power of 2 must not be less than 2^3, in order that 24 may be a factor of the L.C.M. The power of 3 must not be less than 3^3, in order that 54 may be a factor of the L.C.M.

Thus the L.C.M. must be $2^3 \times 3^3$ or 216. To verify that 24, 36, 54 are all factors, notice that $216 = \mathbf{24} \times 9$, $216 = \mathbf{36} \times 6$ and $216 = \mathbf{54} \times 4$.

EXERCISE V f

Find in prime factors the L.C.M. of:

1. $3^2 \times 5$ and $3^3 \times 5^2$.

2. $2^2 \times 5 \times 7$ and $2^3 \times 5 \times 13$.

3. $2 \times 3^2 \times 5$ and 7×13.

4. $2^2 \times 3^2 \times 5$; $3^2 \times 7^2 \times 5^2$ and $7^2 \times 5^2 \times 3$.

5. $2^2 \times 3 \times 7^2$; $2^3 \times 3 \times 7^2$ and $2^3 \times 3^3 \times 7^3$.

Find the prime factors and hence the L.C.M. of:

6. 24 and 60.
7. 48 and 60.

8. 12, 48 and 60.
9. 14, 21 and 84.

10. 32, 96, 144, 128.
11. 3, 4, 5, 6, 9, 10.

12. 64, 144, 576, 512.
13. 1, 11, 111, 1111.

14. 7, 14, 98, 343.
15. 5, 10, 15, 20, 25, 125.

In the following leave the L.C.M. in its prime factors:

16. 2, 4, 6, 8, 10, 12, 14, 16, 18, 20.

17. 18, 42, 49, 98, 294.
18. 42, 44, 77, 462.

19. 81, 1029, 9261.
20. 111, 555, 101, 407.

21. 240, 360, 144, 576, 112.

22. 81, 27, 324, 1296, 1584.

23. 2, 4, 8, 16, 32, 64, 128, 256.

24. 28, 63, 49, 798, 508.

25. 110, 1155, 1111, 1470.
26. 1274, 455, 1092.

EXERCISE V g

Problems on H.C.F. and L.C.M.

1. Find the greatest number which divides 32 and 48 exactly.

2. Find the greatest number which leaves a remainder 5 when divided into 37 and 53.

3. Find the greatest number which leaves remainders 5 and 7 when divided into 37 and 55 respectively.

4. What is the least number which is exactly divisible by 6, 9, and 12?

5. What is the least number which leaves a remainder 3 when divided by 6, 9, or 12?

6. A farmer takes 9 pints of milk from a milk can, and puts in 12 pints of water. How many pints should his measure contain, if he fills it every time, and wishes to carry out the two operations as rapidly as possible?

7. Two companies of 96 and 120 men respectively are to be told off into squads of equal strength. What is the greatest possible number in each squad?

8. One side of a road is planted with trees 24 yards apart. On the other side are lamps 32 yards apart. If the first tree is opposite the first lamp, how far off is the next tree that has a lamp opposite?

9. The circumferences of the wheels of my bicycle are 84 and 90 inches respectively. If I start with the valves of both wheels at the bottom of the wheels, how far shall I have gone when next they are in the same position at the same moment?

10. Two clocks begin to strike twelve together. They strike at intervals of 2 seconds and 3 seconds respectively. When do they strike together again? How many has the slower struck when the quicker one finishes?

11. The flagstones on a pavement are 36 inches wide. A man who strides 30 inches starts with his right toe on the crack between two stones. After how many steps will (1) his right toe, (2) his left toe be on the line between two stones?

12. The *Teutonic* and *Majestic* used to leave Liverpool and New York alternately every 12 and 18 days respectively. If they leave Liverpool together on a certain day, when will they next leave (1) Liverpool, (2) New York together?

SQUARE ROOT AND CUBE ROOT BY PRIME FACTORS

§ 41. In Chapter I we learned that 25 is called the *square* of 5. Inversely 5 is called the **square root** of 25, and this statement is written $\sqrt{25} = 5$.

¶**Ex. 5.** What is the square root of 4, 9, 49, 64, 16, 36, 81, 100, 169, 144?

If the square root of a given number is known to be a whole number, it may be determined by finding the prime factors.

Example. Find $\sqrt{5184}$.

$$5184 = 2 \times 2 \times 2 \times 2 \times 2 \times 2 \times 3 \times 3 \times 3 \times 3$$
$$= (2 \times 2 \times 2 \times 3 \times 3) \times (2 \times 2 \times 2 \times 3 \times 3)$$
(rearranged in two identical groups),
$$= 72 \times 72,$$
$$\therefore \sqrt{5184} = \underline{72}.$$

EXERCISE V h

1. Find the square root of 2^4, 3^6, 2^8, 2^{16}, $7^2 \times 2^2$, $3^2 \times 11^2$.

What is the value of:

2. $\sqrt{7^4 \times 2^2}$.　　**3.** $\sqrt{2^2 \times 3^2 \times 5^2}$.

4. $\sqrt{3^4 \times 2^6}$.　　**5.** $\sqrt{3^4 \times 11^2}$.

6. $\sqrt{7^2 \times 5^4}$.　　**7.** $\sqrt{2^4 \times 3^2 \times 5^4}$?

Find the prime factors and hence the square root of:

8. 144.　　**9.** 1024.　　**10.** 900.

11. 1764.　　**12.** 7056.　　**13.** 11025.

14. 17424.　　**15.** 3025.　　**16.** 122500.

17. 432×27.　　**18.** $12 \times 48 \times 72 \times 32$.

19. $17 \times 51 \times 27$.　　**20.** $8 \times 27 \times 216$.

21. $20 \times 30 \times 40 \times 60$.

Find by prime factors:

22. $\sqrt{225}$.　　**23.** $\sqrt{25600}$.　　**24.** $\sqrt{324}$.

25. $\sqrt{5929}$.　　**26.** $\sqrt{409600}$.　　**27.** $\sqrt{3136}$.

28. $\sqrt{893025}$.　　**29.** $\sqrt{384160000}$.　　**30.** $\sqrt{184041}$.

§ **42.** 2^3 or 8 is called the cube of 2. (Note that 8 cu. in. is the volume of a cube whose edge is 2 in.)

2 is called the **cube root** of 8, and this statement may by written $\sqrt[3]{8} = 2$.

¶**Ex. 6.** What is the value of $7 \times 7 \times 7$? What is the cube root of 343? of 27, 64, 1, 125, 216?

Prime factors may be used to find the cube root of a number, if the cube root is known to be a whole number.

EXERCISE V i

1. Find the cube root of 2^3, 3^6, 5^3, $2^3 \times 3^3$, $2^3 \times 3^6$, $7^6 \times 2^3$.

2. $5^3 \times 2^6 \times 3^3$.

3. $11^3 \times 7^3$.

4. $11^3 \times 2^3 \times 3^3$.

5. $3^3 \times 2^3 \times 5^6$.

6. $11^6 \times 7^6 \times 2^3$.

7. $3^6 \times 5^6 \times 2^{12}$.

Find the prime factors and hence the cube root of:

8. 1728. **9.** 2744. **10.** 216000. **11.** 125000.

12. 3375. **13.** 74088. **14.** 144×96. **15.** 81×72.

Find by prime factors the value of:

16. $\sqrt[3]{512}$.

17. $\sqrt[3]{91125}$.

18. $\sqrt[3]{35937000}$.

19. $\sqrt[3]{373248}$.

REVISION PAPERS

Paper 1 (*on Chaps. I–III*)

1. Divide 282282 by 987.

2. Simplify: (1) $3 + 4 - 5 + 6 - 7$.

(2) $3 + 4 - (5 + 6 - 7)$.

(3) $3 + (4 - 5) + (6 - 7)$.

3. A case weighs 3 cwts. 47 lbs. How many pounds is that?

4. How many days are there in 5 years, one of which is a leap year?

5. How many half-pint mugs can be filled from a gallon jar?

6. What would be the cost of 4850 tons of coal at £1. 3*s*. 6*d*. per ton?

7. The hours of work at a certain factory are: Monday to Friday (both days inclusive) 6 a.m. to 8.15 a.m., 9 a.m. to 1 p.m., 2 p.m. to 5.30 p.m., and on Saturday 6 a.m. to 8.15 a.m., and 9 a.m. to 12 noon. Find the weekly wages of a man who earns 10*d*. an hour, if he works full time.

Paper 2 (*on Chaps. I–III*)

1. Multiply 18357 by 17245.

2. Four quarts are drawn from a 9-gallon cask; how many quarts are left? If 50 pints are next drawn off, how many pints are left?

3. A boy has collected 278 pennies. What is their value in pounds, shillings and pence?

4. Day and Martin's blacking costs 9*d.* per bottle. What do I save by buying 12 bottles for 8/9?

5. A railway company issued (*a*) 13,247 tickets at 11*d.*; (*b*) 120,347 at 1*s.* 1*d.*; (*c*) 1375 at 7*d.*; (*d*) 20,735 at 2*s.* 3*d.* Calculate in pence the total amount received for each class of ticket.

6. How many weeks are there in (1) 100 days, (2) between Jan. 1st and Sept. 15th, 1901 (including one of the dates)?

7. Find the cost of 73 things at £2. 17*s.* 11*d.*

Paper 3 (*on Chaps. I–III*)

1. Divide 18423618 by 347.

2. Simplify: (i) $9 + 12 \div 3 - 2$.

(ii) $9 + (12 \div 3) - 2$.

(iii) $(9 + 12) \div (3 - 2)$.

(iv) $9 \times 12 - 10 \div 2$.

3. Express 4800 lbs. in tons, cwts., qrs., and lbs.

4. What shall I save in a year by taking a half-penny daily paper, instead of a penny one? (No paper on Sunday, 52 Sundays.)

5. A certain blend of tea costs 1/4 per lb., but if 20 lbs. or more are bought, the price is reduced by $\frac{1}{2}d.$ per lb. What is the cost of 23 lbs.?

6. A contractor agrees to cart away 19 loads of earth for £2. 15*s.* Find how much this is per load, to the nearest penny.

7. Find the cost of 703 things at £1. 15*s.* 10*d.*

Paper 4 (*on Chaps. I–III*)

1. Show that the following short methods of multiplication are correct:

(i) $79 \times \mathbf{25} = 7900 \div 4 = 1975$.
(ii) $69 \times \mathbf{125} = 69000 \div 8 = 8625$.
(iii) $54 \times \mathbf{99} = 5400 - 54 = 5346$.
(iv) $27 \times \mathbf{98} = 27 \times (100 - 2) = 2700 - 54 = 2646$.

2. A set of Jubilee coins consisted of—A five-pound piece, a two-pound piece, a guinea, a sovereign, half-a-sovereign, a crown, half-a-crown, a florin, a shilling, a sixpence, a threepenny piece, a penny, and a halfpenny. What was the total value in £ *s.* *d.*?

3. Express 10101 pence as £ *s.* *d.*

4. Gingernuts cost 7*d.* a lb. loose, but a 9-lb. tin costs 5/11. What is paid for the empty tin?

5. Divide £825. 1*s.* $10\frac{3}{4}d.$ by 19.

[*See next page*

6. A coal merchant has 90 sacks which hold 120 lbs. of coal each and others which hold 50 lbs. each. How many sacks must he use for 5 tons of coal?

7. Find the value of 250 articles at £2. 17*s*. $8\frac{1}{2}$*d*.

Paper 5 (*on Chaps. I–III*)

1. Multiply 12345 by 997.

2. Simplify the following expressions:

(i) $5(14+4)-6(2+1)$.

(ii) $5\times 14+4-6\times 2+1$.

(iii) $5\{14+4-6(2+1)\}$.

3. The weight required to break a rope is found to be 4 tons 6 cwts. How many lbs. is this?

4. The following amounts were the totals expended in providing dinner for 31 children on 16 occasions: Baker 12*s*. 7*d*.; butcher £5. 4*s*. 7*d*.; dairyman 13*s*. 7*d*.; fishmonger 18*s*. 5*d*.; greengrocer £1. 7*s*. 9*d*.; grocer 19*s*. 6*d*.; provision merchant £2. 19*s*. 9*d*. What was the total profit if $7\frac{1}{2}$*d*. was charged for each dinner?

5. Multiply 4 tons 17 cwts. 3 qrs. 24 lbs. by 43.

6. The following servants are employed in a household: a cook at £30 a year, a kitchen-maid at £14 a year, a house-maid at £22 a year, and a nurse at £24 a year. How much a month in all is paid?

If 6*d*. a week is paid for each of them under the Insurance Act, how much will be paid for them all in a year (52 weeks)?

7. Find the value of 512 things at 16*s*. $0\frac{3}{4}$*d*.

Paper 6 (*on Chaps. I–III*)

1. Multiply 1234 by 999.

2. How much travelling money shall I need if my ticket costs 17/2, cab fares 3/6, tips 9*d.*, lunch 2/-? How much will be left if I was allowed £1. 5*s.*?

3. How many sixpences are there in $5\frac{1}{2}$ guineas?

4. What is the total cost of 29 books at 12*s.* $4\frac{1}{2}d.$?

5. The velocity of sound is about 1100 feet per second. How many miles away is the storm if I hear the thunder 24 seconds after seeing the flash?

6. A boy buys 100 boxes of matches at 5 for 1*d.* and sells them at 4 for 1*d.* What is his profit?

7. Find the cost of 21 tons of lead at £21. 7*s.* 6*d.* a ton.

Paper 7 (*on Chaps. I–III*)

1. Divide 526926439416 by 1794.

2. Find the value (i) of $(7 + 3 - 2) \div (7 - 3)$, and (ii) of $(8 + 6) \div (5 + 2) + (4 - 0)$.

3. A deep-sea sounding gave the depth as 10,800 feet. Express this depth in miles and yards.

4. Find the cost of tea and cake for 1793 children at $3\frac{1}{2}d.$ a head.

5. Divide £38208. 2*s.* 8*d.* by 239.

6. Find the cost of 350 yards of stair carpet at 2*s.* $9\frac{1}{2}d.$ per yard.

[*See next page*

7. Verify by trial with different numbers that (i) a number is divisible by 4 if the number formed by the last two digits is divisible by 4; (ii) a number is divisible by 8 if the number formed by the last three digits is divisible by 8; (iii) a number is divisible by 3 or 9 if the sum of the digits is divisible by 3 or 9.

Paper 8 (*on Chaps. I–III*)

1. Multiply, in the best way you can, 348769 (i) by 125, (ii) by 999.

2. I go shopping with £5 in my purse, and buy the following articles: 3 shirts at 7/6 each; 4 pairs of socks at 2/6; 2 ties 2/-, and 1/6; one pair of boots, 25/-. How much will be left in my purse?

3. To what height would a pile of 1,000,000 sheets of paper reach, if a book of 400 *pages* is 1 inch thick? (Answer in yds., ft., in.)

4. A certain kind of oil can be bought in 10-gallon drums costing 30*s.* and in quart tins costing 1*s.* 5*d.* Find what is saved by buying 10 gallons in the drum instead of in quart tins. What is the difference per pint in the two costs?

5. Divide 25 miles 176 yards 1 inch by 7.

6. Find the cost of 2397 things at £10/15/10.

7. £139 is to be divided as equally as possible among 29 people, each person's share to be an exact number of shillings. What is each share and how much money remains undistributed?

Paper 9 (*on Chaps. I–III*)

1. Divide two thousand six hundred millions five hundred and nine thousand and fifty, by two thousand two hundred and fourteen.

2. Calculate all the different numbers which can be made out of $36 - 24 \div 3 + 1$ by putting in brackets in various ways, the signs and the order of the figures being left unchanged.

3. An oblong field is 18 chains long and 12 chains wide. How many feet of fencing will be required to go round the field? (A chain = 22 yards.)

4. If jams are 2*d.* per lb. cheaper in London than where I live, what shall I save or lose by ordering (1) 1 lb., (2) 2 lbs., (3) 3 lbs., (4) 56 lbs. from London? *Parcel rates:* 1 lb. 3*d.*, 2 lbs. 4*d.*, 3 lbs. 6*d.* *Railway rates:* 8*d.* per stone.

5. A 36-gallon barrel of beer costs £2. 17*s.* 6*d.* Find the price per pint to the nearest farthing.

6. Find the cost of 4 pieces of cloth, each 80 yards long, at 6*s.* $8\frac{1}{2}$*d.* per yard.

7. In quick time a soldier steps 33 in., and 120 paces are taken in a minute; how far would he march in an hour?

Paper 10 (*on Chaps. I–V*)

1. How often does the sum of 2064 and 1892 contain their difference?

2. How many days were there between the following dates? Include one of the dates in each case:

(1) Jan. 1st, 1902, and Oct. 2nd, 1902,

(2) February 2nd, 1904, and Aug. 3rd, 1904.

3. A woman bought 12 yards of flannel at 1*s*. 3*d*. a yard, 9 yards of calico at $10\frac{1}{2}d.$ a yard, 3 pounds of wool at $2\frac{1}{2}d.$ an ounce, 9 reels of cotton at $2\frac{1}{2}d.$ a reel, and 54 buttons at 1*d*. a dozen. How much change did she get out of two sovereigns?

4. The following prices are quoted per cwt.: biscuits £1. 5*s*. 11*d*.; sugar 14*s*. 10*d*.; coffee £5. 11*s*. 11*d*.; chocolate £2. 18*s*. 3*d*.; mustard £2. 18*s*. 6*d*.; pepper £3. 16*s*. In each case find the price per lb. to the nearest farthing.

5. Find the cost of 5 tons 4 cwts. 2 qrs. at £4 a ton.

6. Find, in square yards, the area of a football ground 115 yards long and 83 yards wide.

7. Find the H.C.F. and L.C.M. of 24, 50, 64, 15.

Paper 11 (*on Chaps. I–V*)

1. How many days were there (1) between noon on Feb. 3rd, 1900, and noon on March 25, 1900, and (2) between 6 a.m. on Oct. 28, 1900, and 6 a.m. on Jan. 15th, 1901?

2. A box full of books weighs 1 cwt. 8 lbs. If the box alone weighs 14 lbs., what is the weight of the books?

3. If 5 tons of coal were put in the cellar 100 days ago, and since then 4 stones have been burnt per day, how much coal is there still in the cellar?

4. The return railway ticket from a suburb to London costs 2*s.* 2*d.*, and the annual ticket costs £12. 5*s.* 0*d.* How many journeys in the year must a man make to save money by taking an annual ticket, instead of paying for each journey?

5. Find the cost of 4 tons 11 cwts. 3 qrs. 14 lbs. at 6 shillings per cwt.

6. How many acres are there in a square field whose edge is a quarter of a mile long?

7. If a cubic foot of wood were cut up into 2-inch cubes, and these cubes were placed on top of one another in a column, what would be the height of the column?

8. Find the H.C.F. of 7875 and 21560.

Paper 12 (*on Chaps. I–V*)

1. Add together 12 cwts. 3 qrs. 17 lbs. 11 oz.; 13 cwts. 1 qr. 4 lbs.; 1 qr. 25 lbs. 9 oz.

2. 120 sacks, each holding 100 lbs. of corn, are put into a truck weighing 4 tons. What is the weight of the truck and load combined?

3. The sum allowed for feeding a boy at a certain school is 1/9 per day, or £22. 16*s*. 9*d*. per year. How many days are reckoned in the year?

4. Find the cost of 3 tons 10 cwts. of coal at 25*s*. the ton.

5.	Population	Area in sq. miles
Japan	51,591,361	162,700
British Isles	46,035,570	121,400

Calculate the number of inhabitants to the square mile in Japan and the British Isles respectively (to the nearest whole number).

6. The area of a rectangular field is 7 acres. One side is 84 yards long. Find the length of the other side.

7. How many inch cubes can be packed in a wooden box 1 foot 5 inches long, 1 foot 1 inch wide and 11 inches deep—all *inside* measurements?

If the box has a cover and the wood is half an inch thick, what are the *outside* measurements? Calculate the volume of wood used.

8. Find the H.C.F. of

$2^3 \times 3^2 \times 5;\ 2^5 \times 3^4 \times 5^2;\ 2^7 \times 3 \times 5^3 \times 7.$

Paper 13 (*on Chaps. I–V*)

1. Three packages weigh 2 cwts. 1 qr. 10 lbs., 2 qrs. 14 lbs., and 1 cwt. 1 qr. 13 lbs. respectively. Find the weight of the three together.

2. How many times is £37. 15*s.* $7\frac{3}{4}$*d.* contained in £2871. 9*s.* 1*d.*?

3. How many steps of 1 yard must a man take in a minute in order to walk 3 miles in an hour?

4. What is the cost of 9 yards 27 inches of cloth at 3*s.* 9*d.* a yard?

5. How many square tiles, each edge 9 inches, will be required to cover a rectangular floor 19 ft. 6 in. long, and 15 ft. 9 in. wide?

6. Find the H.C.F. of 8575 and 67375.

7. Find the prime factors of (1) 1024, (2) 74088. Hence find $\sqrt{1024}$ and $\sqrt[3]{74088}$.

Paper 14 (*on Chaps. I–V*)

1. A family uses 5 pints of milk a day. What will be the weekly cost at 1*s.* 4*d.* a gallon?

2. Find correct to the nearest shilling the price of 127 tons 16 cwts. at 24/6 per ton.

3. The numbers 3454, 79871 are divisible by 11. Now $3+5=8=4+4$ and $7+8+1=16=9+7$. See if this is true for other numbers divisible by 11.

4. A courtyard consists of a border of pavement surrounding a rectangle of asphalt. If the courtyard is 30 yards long and 20 yards broad, and the pavement is 10 ft. wide, find the area of asphalt.

5. From each corner of a rectangular piece of cardboard 16 in. by 10 in. a square of side 2 in. is cut away. The remainder is then folded to form an open box. Find its volume.

6. A closed rectangular tin box measures 4 ft. by 2 ft. by 2 ft. Find its surface in square feet and its volume in cubic feet.

7. Find by prime factors $\sqrt{1764}$ and $\sqrt[3]{216}$.

Paper 15 (*on Chaps. I–V*)

1. A society has 26367 members, each paying 6*d.* a week. How much does the society receive in a year (52 weeks)?

2. Find to the nearest £ the cost of 586 tons 5 cwts. of copper at £74. 12*s.* 6*d.* per ton.

3. A lawn is 74 yards long and 56 yards wide: find how many yards a mowing machine, 2 feet wide, travels in cutting it.

4. In the centre of a room, 21 feet square, there is a carpet, 18 feet square, worth 16*s.* a square yard, which is surrounded by oilcloth worth 3*s.* 4*d.* a square yard. What is the total value of the carpet and oilcloth together?

5. An open cistern, made of zinc, is 3 feet deep, 10 feet long, and 8 feet wide. Find (1) how many cubic feet of water it will hold; (2) how many square feet of zinc are used in making it. Also find the weight of the water in pounds, given that 1 cubic foot of water weighs 1000 ounces.

6. A box, with a lid, is made of wood 1/2″ thick. The external dimensions of the box are 12″ by 11″ by 10″. Find the weight of the box, given that 1 cubic inch of wood weighs 1/2 oz.

7. Find the H.C.F. and L.C.M. of 315 and 294.

Paper 16 (*on Chaps. I–V*)

1. A hurdle consists of two upright stakes 4 feet long joined by 3 horizontal pieces 6 feet long, strengthened by a vertical piece 3 feet long. What length of wood is required for (1) one hurdle, (2) 100 yards of hurdles?

2. Find the value of 2 tons 11 cwts. 17 lbs. at £10. 4*s.* 9*d.* a ton, to the nearest shilling.

[*See next page*

3. What would be the price per ton of bell metal (4 copper to 1 tin) when copper is £73. 12*s.* 6*d.* a ton, and tin £215. 10*s.* per ton?

4. How many boards each 12 ft. 6 in. long, 5 in. wide, will cover a floor 25 ft. long and 15 ft. wide?

5. Two rooms contain equal quantities of air. The area of the floor of one is 340 square feet, and its height is 12 feet. Find the area of the floor of the second whose height is 17 feet.

6. Find the H.C.F. and L.C.M. of 385 and 3003.

7. Find the value of $\sqrt{2304}$ and of $\sqrt[3]{91125}$.

Paper 17 (*on Chaps. I–V*)

1. If income tax is charged at the rate of 1*s.* 2*d.* on every £, find the tax on an income of £1387.

2. Find the cost of 9 cwts. 3 qrs. 18 lbs. at £3. 15*s.* 4*d.* per cwt. (to the nearest penny).

3. A cricket field is 300 yards square, and within it eight pieces of ground, each measuring 12 yards by 30 yards, are specially prepared for pitches at a cost of 2*s.* 6*d.* per square yard. The preparation of the rest of the field costs £20 per acre. Find the total cost of preparing the field.

4. Find the number of square chains in 1 acre.

5. The cost of building a room is 9*d.* per cubic foot of its internal volume. What will it cost to build a room 30 ft. long, 17 ft. wide and 11 ft. high?

It is desired to reduce the estimated cost to £150. The height is reduced to 9 feet, and the width to 15 feet. Find the length to the nearest inch.

6. A cubic foot of granite weighs 2600 oz. What is the weight of a cubic yard? Give the answer in tons and cwts. to the nearest cwt.

7. Find the L.C.M. of 12, 18, 48, 90.

Paper 18 (*on Chaps. I–V*)

1. Find the cost of (i) 1 stone 8 lbs. of sugar at $2\frac{1}{2}d.$ per lb.; (ii) 6 lbs. of raisins at $7d.$ per lb.; (iii) 1 cwt. of sugar at $2d.$ per lb.; (iv) 1 ton of rice at 2/- per stone.

2. Find by Practice the cost of 17 gallons 3 quarts 1 pint at £26. 7*s.* 4*d.* per gallon.

3. A closed box made of wood 1 inch thick, whose inside measurements are 3 ft. 10 in. long, 2 ft. 4 in. wide, 1 ft. 6 in. deep, is covered inside and outside with gold leaf costing $2d.$ per square inch. Find the cost of doing this.

4. How many panels measuring 2 ft. by 1 ft. will be required to cover the walls of a room 25 ft. long and 18 ft. broad to a height of 10 ft.?

5. A hall is 100 feet long, 50 feet broad, and 40 feet high; at one end there is a platform 5 feet high and 20 feet deep. How many people can exist in it for 2 hours, if each person requires 100 cubic feet of air per hour?

6. Find the H.C.F. and L.C.M. of 48, 100, 256.

7. Find (i) the square root of 4356, (ii) $\sqrt[3]{10648}$.

Paper 19 (*on Chaps. I–V*)

1. A household uses three quarts of lamp oil a week. If the oil costs 1*s.* 1*d.* a gallon, what will be the total cost of the oil used from Nov. 1 to the following Feb. 20, both days inclusive?

2. A profit of 2*s.* $7\frac{1}{2}$*d.* in the pound is made on £2340. 6*s.* 8*d.* Find the total profit.

3. Find the external and the internal surfaces of a wooden box, the external dimensions being 12 inches long, 10 inches broad, and 8 inches high, the wood being 1 inch thick. (No lid.)

4. Find to the nearest penny the rent of 5 acres 120 sq. yards of land at £3 per acre.

5. A rectangular tank has a base area of 8 sq. yds. How many cubic feet of water will be required to fill it to a depth of 8 yds.? Find also the weight of the water to the nearest cwt., taking the weight of a cubic foot of water as 1000 oz.

6. An open lead trough measures 12 in. by 10 in. by 6 in. outside, and 10 in. by 9 in. by 4 in. inside. Find the volume of the lead.

The lead weighs 720 lbs. a cubic foot. Find the weight of the trough in pounds.

7. Find the L.C.M. of 16, 24, 36, 45, 120.

Paper 20 (*on Chaps. I–V*)

1. There are employed in a certain office a correspondence clerk at £160 a year, a typewriting clerk at £79 a year, one junior clerk at 18*s.* a week, one at 12*s.* 6*d.* a week, and two at 7*s.* 6*d.* a week. Find the total wages bill for a year (52 weeks).

2. A debtor is able to pay only 14*s.* 9*d.* out of every pound he owes. What will a creditor receive to whom he owes £93. 10*s.*?

3. In a room 18 ft. by 22 ft., there is a Turkey carpet worth 15*s.* a sq. yd., with a space 2 ft. wide all round. What is the value of the carpet?

4. How many cubic inches of wood are required to make a closed box one inch thick, whose external measurements are 5 ft. by 2 ft. 6 in. by 4 ft.?

5. A cubic foot of water weighs 1000 oz. What is the weight of a cubic yard (in cwts. and lbs.)?

6. How often is each of the numbers 420, 378, 231 contained in their L.C.M.?

7. Find $\sqrt{11664}$. By what number less than 10 must 11664 be multiplied to produce a perfect cube?

MISCELLANEOUS EXERCISES

On Chaps. I–V

1. Starting with $2^4 = 16$, $2^8 = (2^4)^2 = 256$, calculate 2^{16} and 2^{32}. Verify that $2^{32} + 1$ is divisible by 641.

2. How many zeros are there in $10^8 \times 10^4$? Also in $(10^8)^4$?

What is $\sqrt{4^4}$? $\sqrt[3]{8^8}$?

Between what powers of 10 does 123456 lie?

3. What is the least number which, when divided by either 10, 11, or 12, gives in each case remainder 6?

4. What is the least number which leaves a remainder 3 when divided by either 6, 9, or 12?

5. Find the least multiple of 10 which gives 1 for remainder when it is divided by 7, 9, or 11.

6. Three bells begin tolling together at 6 o'clock. One (A) tolls every 6 minutes; another (B) tolls every 4 minutes; the third (C) tolls every 3 minutes.

When will (1) A and B next toll together?

(2) B and C ,, ,,

(3) C and A ,, ,,

(4) A, B and C ,, ,,

7. What is the smallest sum of money which is an exact multiple of each of the following amounts:

1*s*. 4*d*., 2*s*., 1*s*. 8*d*. and 1*s*. 3*d*.?

8. A man with a barrel organ makes on an average a penny every five minutes. How much will he earn in a week if he works 8 hours a day and takes a rest on Sundays?

9. I buy 100 eggs at 9*d*. a doz. 9 are broken and I sell the rest at 13 for 1/-. How much do I gain?

10. Find all the factors which are common to the three numbers 24, 60 and 72; and the common multiples of these three numbers which lie between 1000 and 2000.

11. Prove that the sum of all whole numbers which are less than 50 and have no factor in common with 50 is divisible by 50.

12. Strawberry jam costs $6\frac{1}{2}$*d*. per lb. jar, $1/4\frac{1}{2}$ for a 3 lb. jar and 3/3 for a 7 lb. jar. What is the cheapest way of buying 15 lbs., and the cost?

13. The price of a lady's ticket for a ball is 10*s*. 6*d*. and a gentleman's 15*s*.; the supper and refreshments are contracted for at 7*s*. a head, and the miscellaneous fixed expenditure is £57. 10*s*. 0*d*. Find the smallest number that must be present, on the supposition that there are as many gentlemen as ladies, in order that the expenses may be covered.

14. Seven families occupy the different flats of a seven-storied house, and they have to pay £9. 2*s*. 0*d*. per annum to keep the common staircase clean; how should this expense be divided between them?

15. A sum of £1407. 7*s*. is to be divided among 56 people; if 8 of them receive their shares, and 6 of the others give up their shares to the rest, how much more than the first will each of the last receive (to the nearest penny)?

16. A boy has been doing a multiplication sum on a slate and accidentally rubbed it out. All that remains is 47 in the product, there having been two figures preceding these and one after; however he remembers that the multiplicand was 792; find for him the multiplier.

17. Reproduce a multiplication sum which, after being worked correctly, is partially rubbed out, and left thus:

```
      4 - -
        3 -
  ---------
  - - 7 -
  3 6 - -
  ---------
  - - 3 - -
```

18. Three smiths are making horse-shoes. They begin hammering together, and strike 100, 120, 116 blows per minute respectively. How long is it before all three again strike together?

19. £x. y*s*. 0*d*. divided by 2 becomes £y. x*s*. 0*d*. What must x and y be?

20. A man intends to write a cheque for £$x/y/z$. By mistake he writes it £$y/x/z$. After cashing the cheque and paying away 2/6 he finds he has twice as much as he intended to draw from the bank. What must x, y, and z be?

21. A room is 18 ft. long by 13 ft. 6 in. wide. Find the cost of laying Indian matting round the edge of the floor against the walls, the matting being 2 ft. 3 in. wide, and the price of it 1*s.* 2*d.* per yard.

22. A square courtyard of 80 sq. ft. is to be paved with stones each 1 ft. square. How many stones will be needed? How many can be put in whole?

23. There are four square courtyards; the first and second are 32 feet square; the third and fourth 32 square feet in area. They are paved with square stones; in the first and third courtyards each stone is 2 feet square; in the second and fourth each stone is 2 square feet in area. Find the number of stones in each yard, and determine in which of them the stones could be laid unbroken.

24. A rectangular block of square section is 20 feet high. Its volume is 980 cubic feet. What is the length of the other edges?

25. An open box, external dimensions 4 ft. by 3 ft. by 2 ft., is to be made of wood 1 in. thick. The pieces are to be fastened together without dove-tailing. What must be the dimensions of the ends and bottom if the sides are made (i) 4′ by 2′? (ii) 4′ by 1′ 11″? (iii) 3′ 10″ by 2′? (iv) 3′ 10″ by 1′ 11″?

How many sq. ft. of wood will be required in each case?

26. A man walks 1000 miles in 1000 consecutive hours, beginning at 8 p.m. on June 9; when must he have completed his undertaking?

27. A firm that conveyed goods from one town to another found that the total cost of wagon and horses (including wages, etc.) was £5. 8*s.* 7*d.* per week for a three-horse wagon; each wagon carried 15 tons in the course of a week. A steam wagon was later put into service, and carried 30 tons a week at a total cost of £6. 11*s.* 8½*d.* Find to the nearest penny the saving in cost per ton carried.

28. On March 31st, 1912, there were in force in the United Kingdom 889,783 old-age pensions of 5*s.* a week, 19,805 of 4*s.* a week, 19,351 of 3*s.* a week, 8867 of 2*s.* a week, and 4354 of 1*s.* a week. Taking a year as 365 days, find what sum of money is required to pay these pensions for a year.

29. A gramophone with 5 records costs £7; the same gramophone with 20 records costs £9. How much should be paid for the gramophone with 50 records?

30. At a shop which grants a discount of 2*d.* in the shilling, I buy a parcel of books for £7. 6*s.* 4*d.*, but this includes two books, together worth 38*s.*, on which no discount at all is allowed. What would the same books cost at a shop which grants no discount on any book?

31. Postal Orders are issued for every 6*d.* from 6*d.* to 20*s.*, and for 21*s.* The postal charge for an order is ½*d.* for orders from 6*d.* to 2*s.* 6*d.*, 1*d.* for orders from 3*s.* to 15*s.*, and 1½*d.* for orders from

15*s*. 6*d*. to 21*s*. What is the lowest charge at which you could send orders for (1) £1. 3*s*. 6*d*., (2) £3. 18*s*.? Give two possible choices of orders in each case.

32. In an office which employs 22 clerks, all of whom write at equal speeds, it is required to make 300 hand-written copies of a document which takes 5 minutes a copy. The first clerk copies the document, hands his copy to a second to copy, and makes another copy himself; each hands his newly-made copy to another clerk to copy, and so on until all the clerks are employed. If the work begins at 2.45, at what time will the 300 copies be completed? How many will be in the first clerk's handwriting?

33. Find the length of thread used in weaving a piece of silk 10 yards long by 22 inches broad, if it is woven with 120 threads to the inch each way.

34. Which three months begin on the same day of the week (1) in an ordinary year, (2) in a leap year?

35. Take a sum of money (£ *s*. *d*.) less than £12. Under this write the sum obtained by interchanging the pounds and pence. Subtract the smaller of these from the greater. Interchange the pounds and pence in this difference and add the two together.

	£	*s*.	*d*.	
E.g.	11	7	4	
	4	7	11	Try this with different sums.
	6	19	5	
	5	19	6	Do you always get the same
	£12	18	11	answer?

Try the same operations with (1) yds., ft. and in., (2) cwts., stones, lbs.

PART II

FRACTIONS

CHAPTER VI

VULGAR FRACTIONS

MEANING OF A FRACTION

§ 43. You already know the meaning of the words "half" and "quarter." You also know that one-half of a penny is written $\frac{1}{2}$ penny or $\frac{1}{2}d.$: one-quarter of a penny is written $\frac{1}{4}$ penny or $\frac{1}{4}d.$

One-half and one-quarter are called **fractions** (Latin *fractus* = broken); and are written $\frac{1}{2}$, $\frac{1}{4}$.

Suppose that some object, e.g. the circle in the figure, is divided into a number of equal parts, say 6. Each of these parts is called **one-sixth** of the circle: or, in figures, $\frac{1}{6}$ of the circle. The number 6 *below* the line shows how many equal parts there are: it gives the *name* to the fraction, and is therefore called the **denominator** (Latin *nomen* = name).

We might take 5 of these parts; these 5 sixths would make up the circle with the shaded slice removed. 5 sixths is written $\frac{5}{6}$*. If we compare 1 sixth of a circle with 5 sixths of a circle, we have to do with sixths in each case, but different

* $\frac{5}{6}$ is often printed 5/6, to save space. 5*s.* 6*d.* is also written 5/6, but there should be no danger of confusing the two meanings.

numbers of sixths: the 1 and the 5 are therefore called the **numerators** (Latin *numerus* = number) of the fractions $\frac{1}{6}$ and $\frac{5}{6}$: the numerator is the figure *above* the line.

§ 44. Fractions form a new class of number. Hitherto we have been concerned mainly with **whole numbers** such as 1, 2, 3, 27, etc.: whole numbers are sometimes called **integers** (Latin *integer* = whole, or unbroken).

EXERCISE VI a (oral)

1. Which is the greater, one-third of a loaf or one-fifth of a loaf? $\frac{1}{4}$ of an hour or $\frac{1}{6}$ of an hour? What do you learn from these two questions?

Give the value of:

2. One-sixth of a foot. **3.** One-fifth of an hour.
4. 1 third of a shilling. **5.** 1 eighth of a pound.
6. $\frac{1}{3}$ of a yard. **7.** $\frac{1}{4}$ of a foot. **8.** $\frac{1}{8}$ of a mile.
9. $\frac{1}{3}$ of a £. **10.** $\frac{1}{6}$ of a £. **11.** $\frac{1}{8}$ of a £.
12. $\frac{1}{10}$ of a £. **13.** $\frac{1}{7}$ of a stone. **14.** $\frac{1}{4}$ of a cwt.
15. 1/5 of a ton. **16.** 1/8 of a gallon. **17.** 1/10 of a min.
18. 1/15 of an hour. **19.** 1/12 of a day. **20.** 1/5 of a year.
21. $\frac{1}{4}$ of £3. **22.** $\frac{1}{30}$ of 5 mins. **23.** $\frac{1}{7}$ of 2 stone.
24. $\frac{1}{5}$ of 2/6. **25.** $\frac{1}{6}$ of 2 ft. **26.** $\frac{1}{4}$ of a yard.

What fraction is:

27. 6*d*. of 1/-? **28.** 3*d*. of 1/-?
29. 1*d*. of 1/-? **30.** 1/- of £1?
31. 5/- of £1? **32.** 2 oz. of 1 lb.?
33. 2 lbs. of 1 stone? **34.** 12 secs. of 1 minute?
35. 4 pints of 1 gallon? **36.** 2/6 of £1?
37. 2/- of £1? **38.** 6/8 of £1?
39. 3/4 of £1? **40.** 7/- of one guinea?
41. 9 in. of 1 yard? **42.** 11 ft. of 1 chain?

What is the value of:

43. 1 seventh of a week?

44. 3 sevenths of a week?

45. $\frac{1}{12}$, $\frac{5}{12}$, $\frac{11}{12}$ of a shilling?

46. $\frac{5}{8}$ of 1 lb.?

47. $\frac{2}{3}$ of a yard?

48. $\frac{3}{8}$, $\frac{5}{8}$, $\frac{7}{8}$ of £1?

49. $\frac{3}{10}$, $\frac{9}{10}$ of £1?

50. $\frac{3}{7}$ of a stone?

51. 3/4 of 1 cwt.?

52. 4/5 of 1 ton?

53. 3/8 of 1 gallon?

54. 7/10, 4/15 of 1 minute?

55. $\frac{11}{12}$ of a day?

56. $\frac{3}{4}$ of £3?

57. $\frac{2}{5}$ of 2/6?

58. $\frac{13}{30}$ of 5 minutes?

What fraction is:

59. 1*d.*, 5*d.*, 7*d.* of 1/-?

60. 3/-, 7/- of £1?

61. 3 oz. of 1 lb.?

62. 5 lbs. of 1 stone?

63. 7 secs. of 1 minute?

64. 3 pints of 1 gallon?

65. 5/- of 1 guinea?

66. 7 cwts. of 1 ton?

67. 5 in. of 1 foot?

68. 21 in. of 1 yard?

§ 45. *Example.* What is $\frac{5}{8}$ of £3/4/-?

$$\frac{1}{8} \text{ of £3/4/-} = \frac{1}{8} \text{ of 64/-} = \text{8/-},$$

$$\therefore \frac{5}{8} \text{ of £3/4/-} = \text{8/-} \times 5 = \text{40/-} = \text{£2}.$$

Example. What fraction of £2 is 7/-?

$$\text{1/- is } \frac{1}{40} \text{ of £2}, \qquad \therefore \text{7/- is } \frac{7}{40} \text{ of £2}.$$

EXERCISE VI b

Find the value of:

1. $\frac{3}{7}$ of 11/8.

2. $\frac{5}{9}$ of 10/6.

3. $\frac{7}{11}$ of 12/10.

4. $\frac{4}{5}$ of £1/2/11.

5. $\frac{7}{10}$ of £1/11/8.

6. $\frac{11}{14}$ of a guinea.

7. 4/7 of 3 yds. 1 ft. 6 in.

8. 7/9 of 16 yds. 9 in.

9. 3/11 of 1 mile 440 yds.

10. 5/6 of 7 st. 11 lbs. 14 oz.

11. 3/8 of 2 qrs. 11 lbs. 8 oz.
12. 7/11 of 19 cwts. 1 qr.
13. 5/12 of 22 gallons 2 qts.
14. 3/5 of 13 gallons 1 pint.

What fraction is:

15. £1/11/- of £10?
16. 3*s*. 7*d*. of £3?
17. 4*s*. 11*d*. of £5?
18. 5 ft. 1 in. of 10 ft.?
19. 3 yds. 2 ft. of 12 yds.?
20. 5 yds. 2 ft. of 10 yds.?
21. 1 stone 9 lbs. of 1 qr.?
22. 6 cwts. 3 qrs. of 1 ton?
23. 3 stone 9 lbs. of 5 stone?
24. 53 yds. of 3 miles?

Equal Fractions with Unequal Denominators

EXERCISE VI c (oral)

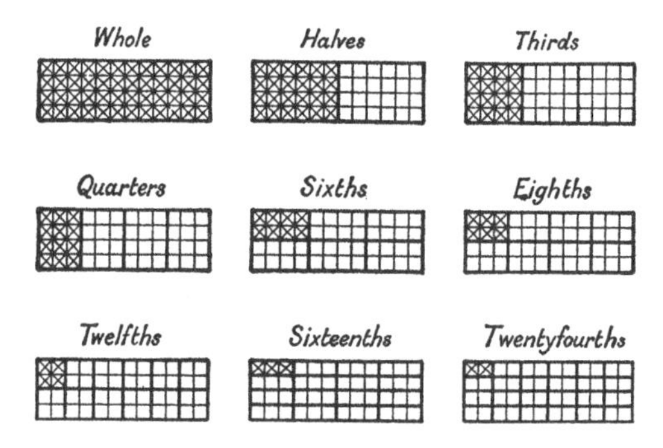

§ 46. Collect the results of the following exercises and attempt to discover the underlying principle.

1. Using the above figure, express 1/2 as quarters, sixths, eighths, 12ths, 16ths, 24ths.

2. Express 2/3 as fractions with denominators 6, 12, 24.

3. Express 3/4 as fractions with denominators 8, 16.

4. Express 3/8 as a fraction with denominator 16.

5. Express 8/24 as thirds, sixths, twelfths.

6. Noticing that the rim of your watch dial is divided into twelfths and sixtieths, find fractions equal to $\frac{5}{60}, \frac{25}{60}, \frac{35}{60}, \frac{15}{60}, \frac{45}{60}, \frac{30}{60}$.

§ 47. In the adjoining figure, **OAB** is $\frac{2}{5}$ of the circle. Now divide each of the fifths, **OAC**, **OCB**, into 3 equal parts. Of the small parts the circle would contain 5×3, and **OAB** contains 2×3,

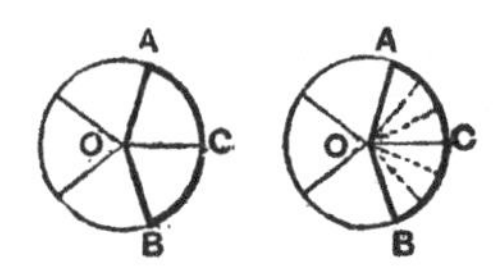

$$\therefore \frac{2}{5} = \frac{2 \times 3}{5 \times 3} = \frac{6}{15}.$$

This suggests the following general rule:

The value of a fraction is unaltered if numerator and denominator are multiplied by the same number.

Again, the fact that $\frac{6}{15} = \frac{2}{5}$ will now suggest to you that

The value of a fraction is unaltered if numerator and denominator are divided by the same number.

These rules may be illustrated as follows. One inch is $\frac{1}{36}$ of a yard; 24 inches $= \frac{24}{36}$ of a yard. But 24 inches $= 2$ feet $= \frac{2}{3}$ of a yard, which shows that $\frac{2}{3}$ and $\frac{24}{36}$ are equal.

§ 48. If we find a fraction equal to a given fraction, but with smaller numerator and denominator, we are said to have **reduced** the fraction **to lower terms.** To achieve this we must divide numerator and denominator by a factor common to both. If this factor is the Highest Common Factor of numerator and denominator, the resulting fraction is in **its lowest terms**: no further reduction is possible.

In reducing a fraction to its lowest terms we shall use the H.C.F. at once if we can see at a glance what number this is: otherwise, it will as a rule be sufficient to divide by such factors as are obvious, and to continue this process until numerator and denominator are prime to one another.

WARNING. In dividing numerator and denominator of a fraction by a number, *we are not dividing the fraction* by this number: the value of the fraction remains unaltered.

EXERCISE VI d (oral)

Reduce each of the following to its lowest terms:

1. $\frac{2}{4}$. **2.** $\frac{2}{6}$. **3.** $\frac{3}{6}$. **4.** $\frac{2}{8}$. **5.** $\frac{4}{8}$.

6. $\frac{6}{8}$. **7.** $\frac{3}{9}$. **8.** $\frac{6}{9}$. **9.** $\frac{2}{10}$. **10.** $\frac{4}{10}$.

11. $\frac{6}{10}$. **12.** $\frac{8}{10}$. **13.** $\frac{5}{15}$. **14.** $\frac{10}{15}$. **15.** $\frac{12}{16}$.

16. $\frac{12}{18}$. **17.** $\frac{15}{20}$. **18.** $\frac{16}{20}$. **19.** $\frac{14}{21}$. **20.** $\frac{16}{24}$.

21. $\frac{10}{25}$. **22.** $\frac{39}{26}$. **23.** $\frac{18}{27}$. **24.** $\frac{12}{27}$. **25.** $\frac{35}{28}$.

26. $\frac{12}{32}$. **27.** $\frac{51}{34}$. **28.** $\frac{14}{36}$. **29.** $\frac{24}{40}$. **30.** $\frac{35}{42}$.

31. $\frac{12}{42}$. **32.** $\frac{28}{49}$. **33.** $\frac{22}{55}$. **34.** $\frac{27}{63}$. **35.** $\frac{63}{81}$.

Fill up the gaps in the following:

36. $\frac{1}{2} = \frac{?}{16}$. **37.** $\frac{2}{3} = \frac{?}{12}$. **38.** $\frac{2}{5} = \frac{?}{25}$.

39. $\frac{3}{8} = \frac{12}{?}$. **40.** $\frac{4}{5} = \frac{20}{?}$. **41.** $\frac{7}{8} = \frac{35}{?}$.

42. $\frac{3}{7} = \frac{?}{21} = \frac{?}{49} = \frac{9}{?}$. **43.** $\frac{7}{10} = \frac{?}{20} = \frac{?}{100}$.

44. $10 = \frac{?}{3} = \frac{100}{?}$. **45.** $2 = \frac{?}{4} = \frac{?}{7} = \frac{12}{?}$.

Find what value x must have in each of the following:

46. $\frac{1}{3} = \frac{x}{9}$. **47.** $\frac{2}{5} = \frac{x}{30}$. **48.** $\frac{3}{7} = \frac{21}{x}$. **49.** $\frac{5}{10} = \frac{x}{2}$.

50. $\frac{1}{9} = \frac{3}{x}$. **51.** $\frac{x}{3} = \frac{24}{36}$. **52.** $\frac{5}{x} = \frac{25}{30}$. **53.** $\frac{7}{x} = \frac{35}{40}$.

§ 49. *Example.* Reduce $\frac{210}{315}$ to its lowest terms.

$$\frac{210}{315} = \frac{42}{63} = \frac{2}{3}.$$

(In the step $\frac{210}{315} = \frac{42}{63}$, numerator and denominator were divided by 5, and in the next step by 21.)

The process is often shown as follows: $\frac{\overset{\overset{2}{\cancel{42}}}{\cancel{210}}}{\underset{\underset{3}{\cancel{63}}}{\cancel{315}}} = \frac{2}{3}.$

EXERCISE VI e

Reduce each of the following fractions to its lowest terms:

1. $\frac{24}{80}$.	**2.** $\frac{114}{150}$.	**3.** $\frac{72}{180}$.	**4.** $\frac{30}{54}$.
5. $\frac{48}{72}$.	**6.** $\frac{45}{100}$.	**7.** $\frac{90}{135}$.	**8.** $\frac{144}{960}$.
9. $\frac{112}{128}$.	**10.** $\frac{98}{343}$.	**11.** $\frac{242}{1111}$.	**12.** $\frac{370}{555}$.
13. $\frac{728}{1024}$.	**14.** $\frac{275}{625}$.	**15.** $\frac{216}{480}$.	**16.** $\frac{126}{294}$.
17. $\frac{352}{880}$.	**18.** $\frac{216}{243}$.	**19.** $\frac{220}{924}$.	**20.** $\frac{465}{705}$.
21. $\frac{729}{945}$.	**22.** $\frac{336}{728}$.	**23.** $\frac{429}{462}$.	**24.** $\frac{572}{1012}$.
25. $\frac{1617}{1815}$.	**26.** $\frac{264}{1760}$.	**27.** $\frac{1008}{1728}$.	**28.** $\frac{39}{169}$.

§ 50. *Example.* Which is the greater: $\frac{4}{5}$ or $\frac{7}{9}$?

The least common multiple of 5 and 9 is 45.

$$\frac{4}{5} = \frac{4 \times 9}{5 \times 9} = \frac{36}{45} \quad \text{and} \quad \frac{7}{9} = \frac{7 \times 5}{9 \times 5} = \frac{35}{45}.$$

But $$\frac{36}{45} > \frac{35}{45},$$

$$\therefore \frac{4}{5} > \frac{7}{9}.$$

Example. Arrange the following in descending order of magnitude (i.e. the greatest first and the least last):

$$\frac{1}{4}, \frac{1}{3} \text{ and } \frac{5}{16}.$$

The **least common denominator** is 48.

$$\frac{1}{4} = \frac{12}{48}, \ \frac{1}{3} = \frac{16}{48}, \text{ and } \frac{5}{16} = \frac{15}{48},$$

$$\therefore \ \frac{1}{3}, \frac{5}{16}, \frac{1}{4} \text{ are in descending order.}$$

EXERCISE VI f

Which is the greater of the following pairs?

1. $\frac{2}{3}, \frac{5}{6}$. **2.** $\frac{3}{4}, \frac{2}{3}$. **3.** $\frac{7}{12}, \frac{11}{15}$.

4. 4/9, 5/12. **5.** 3/8, 4/9. **6.** 7/10, 10/13.

7. 5/24, 7/36. **8.** 11/14, 17/21. **9.** 23/48, 31/64.

10. 8/27, 11/36. **11.** 13/50, 19/75. **12.** 22/7, 355/113.

Arrange each of the following groups in descending order:

13. $\frac{1}{2}, \frac{2}{3}, \frac{3}{4}$. **14.** $\frac{4}{5}, \frac{3}{4}, \frac{2}{3}$. **15.** $\frac{3}{10}, \frac{2}{5}, \frac{1}{3}$.

16. $\frac{7}{12}, \frac{7}{10}, \frac{2}{3}$. **17.** $\frac{3}{4}, \frac{5}{6}, \frac{7}{9}$. **18.** $\frac{7}{12}, \frac{5}{9}, \frac{9}{16}$.

§ 51. **Improper fractions and mixed numbers.** If we divide a week into 7 equal parts and take 7 of them we have $\frac{7}{7}$ of a week: at the same time we obviously have 1 week. In fact $\frac{7}{7} = 1$; thus, if the numerator and denominator of a fraction are equal, the fraction is equal to 1.

Again, since one seventh of a week is a day, 21 sevenths of a week = 21 days = 3 weeks. Thus $\frac{21}{7} = 3$. Similarly $\frac{28}{7} = 4$, $\frac{35}{7} = 5$, etc.

There is no need for a special rule to meet these cases. If we reduce $\frac{28}{7}$ in the ordinary way, dividing numerator and denominator by 7, we have $\frac{4}{1}$. This means that we are to divide a week into *one* part and then take 4 of these parts. But to divide anything into one part is the same as to leave it undivided. Thus $\frac{4}{1}$ of a week = 4 weeks, and

$$\frac{28}{7} = \frac{4}{1} = 4.$$

Incidentally, we see that **a fraction whose denominator is 1 is equal to a whole number, namely the numerator.**

This fact is often overlooked by beginners, who are apt to leave their answer in such a form as $\frac{12}{1}$, instead of taking the final step and writing 12.

§ 52. A fraction such as $\frac{30}{7}$, whose numerator is greater than its denominator, is called an **improper** fraction. On the other hand, if the numerator of a fraction is less than the denominator, as in $\frac{3}{5}$, the fraction is called **proper.**

If we take a week as unit,

$$\frac{30}{7} \text{ of a week} = 30 \text{ days} = 4 \text{ weeks} + 2 \text{ days}$$
$$= 4 \text{ weeks} + \frac{2}{7} \text{ of a week.}$$

This is written $4\frac{2}{7}$ weeks, and is read thus: "four and two-sevenths" weeks.

A number like $4\frac{2}{7}$ $\left(\text{short for } 4 + \frac{2}{7}\right)$ which consists of a whole number with a proper fraction added, is sometimes called a **mixed number***.

§ 53. Let us examine more closely how we might obtain the result $\frac{30}{7} = 4\frac{2}{7}$. The steps are

$$\frac{30}{7} = \frac{28+2}{7} = \frac{28}{7} + \frac{2}{7} = 4 + \frac{2}{7} = 4\frac{2}{7}.$$

* Note that $4\frac{2}{7}$ means $4 + \frac{2}{7}$; but $4 \,.\, \frac{2}{7}$ means $4 \times \frac{2}{7}$.

The **4** and the **2** are obtained as in dividing 30 by 7.

$$\begin{array}{r} 7)30 \\ \hline 4 \end{array} \text{ remainder } 2.$$

Example. Express $4\frac{6}{17}$ as an improper fraction.

$$4 + \frac{6}{17} = \frac{68}{17} + \frac{6}{17} = \frac{74}{17}.$$

EXERCISE VI g (oral)

Improper Fractions. Mixed Numbers

Express each of the following mixed numbers as an improper fraction:

1. $1\frac{1}{2}$.	**2.** $1\frac{1}{4}$.	**3.** $1\frac{3}{4}$.	**4.** $5\frac{1}{4}$.
5. $6\frac{3}{4}$.	**6.** $4\frac{2}{5}$.	**7.** $8\frac{1}{7}$.	**8.** $3\frac{3}{10}$.
9. $12\frac{1}{7}$.	**10.** $3\frac{5}{8}$.	**11.** $4\frac{2}{9}$.	**12.** $6\frac{4}{5}$.
13. $7\frac{3}{10}$.	**14.** $9\frac{7}{10}$.	**15.** $4\frac{3}{100}$.	**16.** $5\frac{7}{100}$.
17. $2\frac{27}{100}$.	**18.** $4\frac{57}{100}$.	**19.** $3\frac{5}{16}$.	**20.** $4\frac{7}{22}$.

Express each of the following as a mixed number:

21. $\frac{3}{2}$.	**22.** $\frac{7}{4}$.	**23.** $\frac{8}{3}$.	**24.** $\frac{6}{5}$.
25. $\frac{9}{2}$.	**26.** $\frac{14}{5}$.	**27.** $\frac{17}{12}$.	**28.** $\frac{22}{7}$.
29. $\frac{17}{10}$.	**30.** $\frac{29}{10}$.	**31.** $\frac{143}{100}$.	**32.** $\frac{719}{100}$.
33. $\frac{709}{100}$.	**34.** $\frac{507}{100}$.	**35.** $\frac{5063}{100}$.	**36.** $\frac{5063}{1000}$.

EXERCISE VI h

Express each of the following as an improper fraction:

1. $13\frac{2}{7}$.	**2.** $25\frac{6}{13}$.	**3.** $41\frac{3}{11}$.	**4.** $72\frac{7}{9}$.
5. $19\frac{4}{17}$.	**6.** $12\frac{7}{18}$.	**7.** $22\frac{9}{14}$.	**8.** $13\frac{2}{15}$.

9. $17\frac{3}{8}$. **10.** $14\frac{17}{23}$. **11.** $27\frac{17}{24}$. **12.** $19\frac{5}{43}$.

13. $23\frac{5}{13}$. **14.** $13\frac{4}{19}$. **15.** $23\frac{9}{34}$. **16.** $41\frac{11}{32}$.

17. $25\frac{2}{25}$. **18.** $17\frac{7}{31}$. **19.** $32\frac{5}{64}$. **20.** $51\frac{29}{75}$.

Express each of the following as a mixed number:

21. $\frac{960}{41}$. **22.** $\frac{237}{17}$. **23.** $\frac{633}{43}$. **24.** $\frac{725}{36}$.

25. $\frac{1143}{112}$. **26.** $\frac{2569}{53}$. **27.** $\frac{1790}{101}$. **28.** $\frac{3764}{97}$.

29. $\frac{637}{128}$. **30.** $\frac{1273}{77}$. **31.** $\frac{11111}{202}$. **32.** $\frac{4335}{143}$.

33. $\frac{7441}{89}$. **34.** $\frac{9250}{157}$. **35.** $\frac{4439}{47}$. **36.** $\frac{3479}{95}$.

§ 54. The connection, shown at the end of the last paragraph, between a fraction and the process of division may seem rather unexpected. To find the value of $\frac{30}{7}$ of a week involves dividing a *week* into 7 equal parts and taking 30 such parts. What connection has this with dividing 30 by 7?

Take a simple case; e.g. $\frac{4}{7}$ of a week.

(i) 4 sevenths of a week = 4 days.

(ii) 4 weeks ÷ 7 = 28 days ÷ 7 = 4 days.

Thus $\frac{4}{7}$ of a week is the same as 4 weeks ÷ 7.

§ 55. Hitherto we have not given any meaning to an expression such as $\frac{\text{4 weeks}}{7}$, in which a *quantity* takes the place of the numerator, while the denominator is still a *number*. We now define this as equivalent to 4 weeks ÷ 7. In fact, $\frac{\text{quantity } A}{\text{number } b}$ means "quantity A ÷ number b." This is the kind of division known as "partition" (cf. § 9).

The result obtained in § 54 may now be written

$$\frac{4}{7} \text{ of a week} = \frac{4 \text{ weeks}}{7}.$$

In general, it is true that

$$\frac{a}{b} \text{ of a unit} = \frac{a \text{ units}}{b}$$

EXERCISE VI i (oral)

Find the value of:

1. $\frac{6 \text{ in.}}{3}$.	**2.** $\frac{15 \text{ pence}}{5}$.	**3.** $\frac{£45}{9}$.
4. $\frac{28 \text{ lbs.}}{7}$.	**5.** $\frac{3 \text{ weeks}}{7}$.	**6.** $\frac{11 \text{ yards}}{3}$.
7. $\frac{5 \text{ lbs.}}{8}$.	**8.** $\frac{£2}{10}$.	**9.** $\frac{7/-}{4}$.
10. $\frac{2 \text{ ft.}}{6}$.	**11.** $\frac{3 \text{ lbs.}}{16}$.	**12.** $\frac{2/6}{5}$.
13 $\frac{8/4}{8}$.	**14.** $\frac{5 \text{ gallons}}{8}$.	**15.** $\frac{5 \text{ ft. } 3 \text{ in.}}{9}$.

***§ 56.** We have explained the meaning of $\frac{a}{b}$ and of $\frac{a \text{ units}}{b}$, where a, b are numbers. We have not said anything about the meaning of $\frac{a \text{ units}}{b \text{ units}}$. This we now define as "the number of times that b units are contained in a units"........(i).

Thus $\frac{10 \text{ miles}}{5 \text{ miles}}$ is the same as 10 miles ÷ 5 miles, and equal to 2. This is an example of "quotition."

Consider $\frac{13/-}{5/-}$. In dividing 13/- by 5/- (i.e. a crown) we have hitherto taken the quotient to be 2, with a remainder of 3/-. But these 3/- are $\frac{3}{5}$ of a crown; thus 13/- contains 2 crowns *plus* $\frac{3}{5}$ of a crown: in all $2\frac{3}{5}$ crowns. We may therefore say that 5/- is contained $2\frac{3}{5}$ times in 13/-, or

$$\frac{13/-}{5/-} = 2\frac{3}{5}.$$

* May be omitted till the chapter on ratio is read.

Similarly $\frac{3/\text{-}}{5/\text{-}} = \frac{3}{5}$, for 3/- contains $\frac{3}{5}$ of a crown.

We may, in fact, quote the definition (i) above in the following form: "To simplify $\frac{a \text{ units}}{b \text{ units}}$ is to find what fraction *a* units is of *b* units."

*EXERCISE VI j (oral)

What is the value of.

1. $\frac{9d.}{3d.}$?	**2.** $\frac{\text{1 foot}}{\text{1 inch}}$?	**3.** $\frac{\text{5 gallons}}{\text{1 quart}}$?
4. $\frac{\text{2 stone}}{\text{7 lbs.}}$?	**5.** $\frac{£1/10/\text{-}}{2/6}$?	**6.** $\frac{15/\text{-}}{1/6}$?
7. $\frac{9d.}{6d.}$?	**8.** $\frac{\text{1 foot}}{\text{9 in.}}$?	**9.** $\frac{6d.}{9d.}$?
10. $\frac{\text{8 in.}}{\text{1 foot}}$?	**11.** $\frac{\text{1 ft. 6 in.}}{\text{1 yard}}$?	**12.** $\frac{\text{7 cwts.}}{\text{1 ton}}$?

What is the value of:

13. $\frac{£8}{£4}$?	**14.** $\frac{£8}{4}$?	**15.** $\frac{\text{7 lbs.}}{\text{14 lbs.}}$?
16. $\frac{\text{7 lbs.}}{14}$?	**17.** $\frac{2/6}{5/\text{-}}$?	**18.** $\frac{2/6}{5}$?
19. $\frac{\text{4 yds.}}{\text{6 yds.}}$?	**20.** $\frac{\text{4 yds.}}{6}$?	**21.** $\frac{\text{7 tons}}{20}$?
22. $\frac{\text{7 tons}}{\text{20 tons}}$?	**23.** $\frac{\text{1 yd. 2 ft.}}{\text{10 ft.}}$?	**24.** $\frac{\text{1 yd. 2 ft.}}{10}$?

ADDITION OF FRACTIONS

§ 57. **Fractions with the same denominator.** Suppose that it is required to find the sum of 3 tenths of an inch and 5 tenths of an inch. Since a tenth of an inch is a definite thing, we have to find the sum of 3 things and 5 things of the same kind; the result is 8 things of this kind; in other words, 8 tenths of an inch.

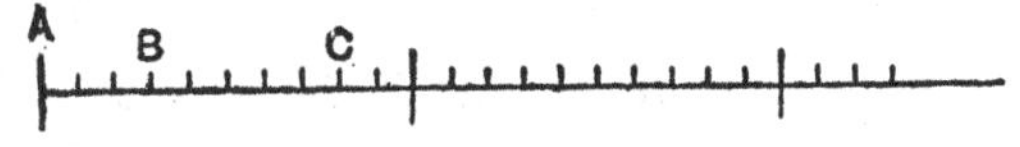

In the figure, AB $= \frac{3}{10}$ of an inch, BC $= \frac{5}{10}$ of an inch,

AB and BC = AC $= \frac{8}{10}$ of an inch.

* May be omitted till the chapter on ratio is read.

Whether the unit is an inch or anything else,

$$\frac{3}{10} \text{ of a unit} + \frac{5}{10} \text{ of a unit} = \frac{8}{10} \text{ of a unit.}$$

That is a statement as to the addition of fractional *quantities*; the corresponding statement for fractional *numbers* is

$$\frac{3}{10} + \frac{5}{10} = \frac{8}{10}.$$

If the answer is to be expressed as a fraction in its lowest terms, we should say

$$\frac{3}{10} + \frac{5}{10} = \frac{8}{10} = \frac{4}{5}.$$

Take the example:

$$\frac{7}{10} \text{ of an inch} + \frac{9}{10} \text{ of an inch.}$$

This equals $\frac{16}{10}$ of an inch $= 1\frac{6}{10}$ of an inch $= 1\frac{3}{5}$ of an inch.

If you follow this in the figure, you will see why the answer is a mixed number.

EXERCISE VI k (oral)

Simplify:

1. 1 ninth + 4 ninths. **2.** 3 sevenths + 2 sevenths.

3. $\frac{3}{7} + \frac{4}{7}$. **4.** $\frac{3}{7} + \frac{5}{7}$. **5.** $\frac{3}{8} + \frac{5}{8}$.

6. $\frac{1}{10} + \frac{7}{10}$. **7.** $\frac{1}{4} + \frac{1}{4}$. **8.** $\frac{3}{10} + \frac{1}{10}$.

9. $\frac{1}{12} + \frac{5}{12}$. **10.** $\frac{17}{24} + \frac{1}{24}$. **11.** $\frac{8}{15} + \frac{8}{15}$.

12. $\frac{4}{9} + \frac{2}{9}$. **13.** $\frac{5}{11} + \frac{10}{11}$. **14.** $\frac{5}{18} + \frac{7}{18}$.

15. $\frac{1}{5} + \frac{2}{5} + \frac{1}{5}$. **16.** $\frac{2}{7} + \frac{2}{7} + \frac{2}{7}$. **17.** $\frac{7}{10} + \frac{7}{10} + \frac{7}{10}$.

18. $\frac{4}{9} + \frac{4}{9} + \frac{4}{9}$. **19.** $\frac{4}{15} + \frac{4}{15} + \frac{4}{15}$. **20.** $\frac{7}{30} + \frac{7}{30} + \frac{7}{30}$.

§ 58. **Fractions with unequal denominators.** Let us find the sum of $\frac{1}{2}$ an inch and $\frac{2}{5}$ of an inch.

In the figure, $\mathsf{PQ} = \frac{1}{2}$ an inch. What length is $\frac{2}{5}$ of an inch? $\frac{1}{5}$ of an inch contains 2 tenths; $\frac{2}{5}$ of an inch contains 4 tenths; thus $\mathsf{QR} = \frac{2}{5}$ of an inch.

$$\therefore \ \frac{1}{2} \text{ an inch} + \frac{2}{5} \text{ of an inch} = \mathsf{PQ} + \mathsf{QR} = \mathsf{PR}$$
$$= \frac{9}{10} \text{ of an inch.}$$

We have found the answer through being able to express both fractions as tenths. We will now express the operation as a sum of *numbers*.

$$\frac{1}{2} + \frac{2}{5} = 5 \text{ tenths} + 4 \text{ tenths} = 9 \text{ tenths} = \frac{9}{10}.$$

More shortly still $\frac{1}{2} + \frac{2}{5} = \frac{5}{10} + \frac{4}{10} = \frac{9}{10}$.

§ 59. Now let us add together $\frac{3}{5}$ and $\frac{4}{7}$.

The previous example will suggest that both must be expressed as fractions with the same denominator.

Now $\frac{3}{5} = \frac{6}{10} = \frac{9}{15} = \frac{12}{20} = \ldots$ etc.

and $\frac{4}{7} = \frac{8}{14} = \frac{12}{21} = \frac{16}{28} = \ldots$ etc.

So far we have not succeeded in expressing both as a fraction with the same denominator.

The successive denominators in the first case are:

5, 10, 15, 20, etc., i.e. multiples of 5.

In the second case the successive denominators are multiples of 7.

Thus we want to find a **common multiple** of 5 and 7, and the **least common multiple** 35 is the simplest (cf. § 50).

Now $$\frac{3}{5} = \frac{21}{35}, \quad \frac{4}{7} = \frac{20}{35}$$

and $$\frac{3}{5} + \frac{4}{7} = \frac{21}{35} + \frac{20}{35} = \frac{41}{35} = 1\frac{6}{35}.$$

§ 60. The process for addition of fractions is therefore:

First, find the L.C.M. of the denominators.

Secondly, express each fraction as a fraction with this L.C.M. as denominator. This must be done by multiplying numerator and denominator by *some* number—the same number for numerator and denominator. This number must be found by common sense.

The remaining steps should be obvious.

Example 1. Find the sum of $\frac{5}{12}, \frac{7}{18}, \frac{3}{10}$.

$$12 = 2^2 \times 3, \quad 18 = 2 \times 3^2, \quad 10 = 2 \times 5.$$

$\therefore$ the L.C.M. of the denominators $= 2^2 \times 3^2 \times 5 = 180.$

$$\frac{5}{12} + \frac{7}{18} + \frac{3}{10} = \frac{75}{180} + \frac{70}{180} + \frac{54}{180}$$

$$= \frac{199}{180} = \underline{1\frac{19}{180}}.$$

Example 2. Simplify $2\frac{1}{4} + 3\frac{5}{6} + 1\frac{3}{8}$.

$$2\frac{1}{4} + 3\frac{5}{6} + 1\frac{3}{8} = 6 + \frac{1}{4} + \frac{5}{6} + \frac{3}{3}$$

$$= 6 + \frac{6}{24} + \frac{20}{24} + \frac{9}{24}$$

$$= 6 + \frac{35}{24}$$

$$= 6 + 1\frac{11}{24} = \underline{7\frac{11}{24}}.$$

Example 3. Simplify $\frac{36}{25}+\frac{17}{10}+\frac{4}{5}$.

First, express the improper fractions as mixed numbers.

$$\frac{36}{25}+\frac{17}{10}+\frac{4}{5}=1\frac{11}{25}+1\frac{7}{10}+\frac{4}{5}$$
$$=2+\frac{22}{50}+\frac{35}{50}+\frac{40}{50}$$
$$=2+\frac{97}{50}=\underline{3\frac{47}{50}}.$$

EXERCISE VI l

Simplify:

1. $\frac{1}{2}+\frac{1}{4}$. **2.** $\frac{2}{5}+\frac{3}{10}$. **3.** $\frac{5}{14}+\frac{2}{7}$.

4. $\frac{3}{10}+\frac{4}{5}$. **5.** $\frac{3}{14}+\frac{1}{21}$. **6.** $\frac{1}{6}+\frac{2}{9}$.

7. $\frac{4}{15}+\frac{3}{10}$. **8.** $\frac{5}{12}+\frac{4}{15}$. **9.** $\frac{3}{10}+\frac{7}{100}$.

10. $\frac{4}{21}+\frac{3}{35}+\frac{1}{7}$. **11.** $2\frac{1}{2}+3\frac{1}{3}$. **12.** $4\frac{1}{4}+5\frac{1}{2}$.

13. $6\frac{2}{3}+1\frac{1}{6}$. **14.** $2\frac{1}{8}+3\frac{3}{4}$. **15.** $5\frac{1}{7}+4\frac{3}{14}$.

16. $3\frac{4}{5}+2\frac{1}{15}$. **17.** $1\frac{2}{9}+3\frac{1}{3}$. **18.** $3\frac{3}{14}+2\frac{5}{11}$.

19. $2\frac{3}{4}+1\frac{1}{2}$. **20.** $\frac{4}{5}+3\frac{3}{10}$. **21.** $\frac{5}{6}+1\frac{5}{12}$.

22. $\frac{7}{9}+3\frac{5}{12}$. **23.** $4\frac{3}{8}+3\frac{5}{12}$. **24.** $2\frac{15}{22}+3\frac{20}{33}$.

25. $2\frac{5}{8}+3\frac{1}{6}+1\frac{2}{3}$. **26.** $1\frac{2}{5}+3\frac{1}{4}+\frac{7}{10}$. **27.** $7\frac{6}{15}+4\frac{11}{12}+3\frac{7}{30}$.

28. $3\frac{5}{8}+1\frac{7}{12}+6\frac{2}{3}$. **29.** $\frac{17}{5}+2\frac{3}{5}$. **30.** $6\frac{4}{7}+\frac{22}{14}$.

31. $\frac{4}{3}+1\frac{1}{9}+\frac{17}{6}$. **32.** $\frac{5}{9}+\frac{5}{9}+\frac{5}{9}+\frac{5}{9}$.

33. $\frac{8}{9}+2\frac{5}{6}+1\frac{11}{12}$. **34.** $\frac{17}{15}+\frac{39}{20}+1\frac{2}{3}$.

35. $\frac{11}{24}+\frac{15}{8}+1\frac{7}{10}$. **36.** $\frac{21}{10}+\frac{1}{42}+\frac{107}{105}$.

37. $\frac{17}{15}+2\frac{5}{12}+\frac{47}{20}$. **38.** $\frac{1}{56}+\frac{1}{679}+\frac{1}{776}$.

Subtraction of Fractions

§ 61. *Example* 1. Simplify $\frac{17}{36} - \frac{5}{24}$.

The L.C.M. of the denominators is 72.

$$\frac{17}{36} - \frac{5}{24} = \frac{34}{72} - \frac{15}{72} = \underline{\frac{19}{72}}.$$

Example 2. Simplify $3\frac{9}{14} - 1\frac{1}{12}$.

(If we had to simplify 3*s*. 7*d*. – 2*s*. 1*d*. we should subtract the 2*s*. from the 3*s*., and the 1*d*. from the 7*d*. Similarly we may subtract 1 from 3, and $\frac{1}{12}$ from $\frac{9}{14}$.)

$$3\frac{9}{14} - 1\frac{1}{12} = 3 - 1 + \frac{9}{14} - \frac{1}{12}$$
$$= 2 + \frac{54}{84} - \frac{7}{84}$$
$$= 2 + \frac{47}{84} = \underline{2\frac{47}{84}}.$$

Example 3. Simplify $4\frac{3}{10} - 2\frac{11}{15}$.

$$4\frac{3}{10} - 2\frac{11}{15} = 2 + \frac{3}{10} - \frac{11}{15}$$
$$= 2 + \frac{9}{30} - \frac{22}{30}.$$

(But we cannot subtract 22 from 9. We therefore borrow 1 or $\frac{30}{30}$ from the 2, and proceed as follows:)

$$= 1 + \frac{39}{30} - \frac{22}{30}$$
$$= 1 + \frac{17}{30} = \underline{1\frac{17}{30}}.$$

EXERCISE VI m

Simplify:

1. $\frac{3}{4} - \frac{1}{4}$. **2.** $\frac{5}{6} - \frac{1}{6}$. **3.** $\frac{1}{2} - \frac{1}{4}$.

4. $\frac{2}{3} - \frac{1}{6}$. **5.** $\frac{1}{2} - \frac{1}{3}$. **6.** $\frac{1}{3} - \frac{1}{7}$.

7. $\frac{1}{4}-\frac{1}{5}$.
8. $\frac{3}{4}-\frac{1}{2}$.
9. $\frac{3}{10}-\frac{1}{5}$.
10. $\frac{2}{3}-\frac{1}{2}$.
11. $\frac{3}{4}-\frac{1}{6}$.
12. $\frac{3}{8}-\frac{1}{3}$.
13. $\frac{5}{12}-\frac{5}{18}$.
14. $\frac{5}{14}-\frac{4}{21}$.
15. $\frac{7}{15}-\frac{3}{10}$.
16. $\frac{8}{9}-\frac{5}{6}$.
17. $1\frac{3}{8}-\frac{1}{16}$.
18. $2\frac{1}{2}-1\frac{1}{4}$.
19. $4\frac{2}{3}-1\frac{1}{6}$.
20. $3\frac{1}{4}-1\frac{1}{8}$.
21. $5\frac{3}{4}-2\frac{7}{12}$.
22. $4\frac{5}{6}-2\frac{1}{3}$.
23. $6\frac{5}{8}-1\frac{1}{2}$.
24. $5\frac{1}{7}-2\frac{1}{14}$.
25. $2\frac{1}{4}-1\frac{1}{2}$.
26. $3\frac{1}{6}-2\frac{1}{3}$.
27. $4\frac{1}{8}-3\frac{1}{4}$.
28. $5\frac{1}{12}-4\frac{1}{6}$.
29. $4\frac{5}{12}-1\frac{1}{2}$.
30. $3\frac{5}{8}-1\frac{3}{4}$.
31. $6\frac{3}{4}-2\frac{7}{8}$.
32. $5\frac{1}{6}-2\frac{2}{3}$.
33. $6\frac{1}{4}-1\frac{7}{8}$.
34. $1-\frac{1}{4}$.
35. $1-\frac{3}{7}$.
36. $1-\frac{4}{5}$.
37. $1-\frac{5}{8}$.
38. $1-\frac{7}{11}$.
39. $1-\frac{5}{12}$.
40. $2-\frac{3}{8}$.
41. $4-\frac{6}{7}$.
42. $10-\frac{4}{9}$.
43. $4\frac{5}{8}-2\frac{7}{12}$.
44. $4\frac{5}{8}-2\frac{11}{12}$.
45. $6-4\frac{1}{5}$.
46. $4\frac{5}{12}-2\frac{7}{15}$.
47. $7\frac{3}{5}-4\frac{1}{2}$.
48. $6\frac{3}{4}-4\frac{5}{6}$.
49. $2\frac{3}{14}-1\frac{5}{21}$.
50. $7\frac{4}{7}-5\frac{3}{5}$.
51. $8\frac{10}{21}-4\frac{5}{42}$.
52. $4\frac{7}{20}-2\frac{8}{15}$.
53. $5\frac{6}{7}-4\frac{1}{2}$.
54. $4\frac{5}{42}-2\frac{17}{36}$.

§ 62. *Example.* Simplify $4\frac{5}{12}-3\frac{7}{8}+1\frac{5}{6}$.

It is convenient to rearrange the terms so that the fractions preceded by *minus* signs come last: this may do away with the necessity of "borrowing." In rearranging, remember that the sign and the number that follows it must be moved in one piece.

$$\text{Sum required} = 4\tfrac{5}{12} + 1\tfrac{5}{6} - 3\tfrac{7}{8}$$
$$= 2 + \frac{10}{24} + \frac{20}{24} - \frac{21}{24}$$
$$= 2 + \frac{30}{24} - \frac{21}{24}$$
$$= 2 + \frac{9}{24} = \underline{2\tfrac{3}{8}}.$$

EXERCISE VI n

Simplify:

1. $4\tfrac{1}{2} + 3\tfrac{1}{6} - 2\tfrac{1}{3}$.

2. $5\tfrac{3}{4} + 2\tfrac{1}{4} - 1\tfrac{1}{2}$.

3. $1\tfrac{3}{5} - \tfrac{7}{10} + 2\tfrac{1}{5}$.

4. $4\tfrac{1}{8} - 3\tfrac{1}{4} + 4\tfrac{3}{4}$.

5. $5\tfrac{2}{3} - 5\tfrac{1}{6} + 2\tfrac{5}{6}$.

6. $3\tfrac{1}{8} + 4\tfrac{1}{12} - 3\tfrac{1}{6}$.

7. $2\tfrac{1}{6} - 1\tfrac{5}{6} + 3\tfrac{1}{3}$.

8. $7\tfrac{7}{12} - 2\tfrac{17}{18} + 3\tfrac{1}{6}$.

9. $5\tfrac{1}{2} - 3\tfrac{1}{6} + 2\tfrac{1}{12} - 3\tfrac{3}{4}$.

10. $5\tfrac{3}{8} - 4\tfrac{1}{2} - \tfrac{5}{8} + 10$.

11. $3\tfrac{5}{24} - 1\tfrac{7}{36} + \tfrac{11}{16}$.

12. $2\tfrac{7}{15} - 1\tfrac{13}{20} + \tfrac{17}{25}$.

13. $14\tfrac{11}{21} - 2\tfrac{12}{49} + \tfrac{1}{14} - 1\tfrac{1}{7}$.

14. $5\tfrac{3}{22} - 2\tfrac{17}{33} + 1\tfrac{5}{6} - 1\tfrac{1}{2}$.

In the following exercises, the contents of the brackets should be simplified first.

15. $4\tfrac{1}{8} - \left(1\tfrac{1}{4} + 2\tfrac{1}{2}\right)$.

16. $3\tfrac{5}{12} - \left(\tfrac{1}{6} + 2\tfrac{1}{4}\right)$.

17. $6\tfrac{1}{5} - \left(4\tfrac{3}{10} - 1\tfrac{1}{2}\right)$.

18. $5\tfrac{11}{14} - \left(2\tfrac{1}{7} - \tfrac{1}{2}\right)$.

19. $3\tfrac{5}{9} - \left(2\tfrac{1}{3} - 1\tfrac{5}{12}\right) + 1\tfrac{5}{6}$.

20. $6\tfrac{3}{20} - \left(5\tfrac{7}{15} - 1\tfrac{4}{5}\right) + 2\tfrac{7}{10}$.

21. Add 2/3 to the sum of 1/2 and 1/3.

22. Subtract 3/5 from the sum of 1/2 and 1/5.

23. Add 3/4 to the difference of 1/2 and 1/3.

24. Subtract $\tfrac{1}{2}$ from the difference of $1\tfrac{1}{2}$ and $\tfrac{1}{3}$.

25. Subtract from 1 the sum of 3/4 and 1/8.

26. Subtract from 1 the difference of $1\tfrac{1}{2}$ and $1\tfrac{1}{3}$.

EXERCISE VI o

Problems on Addition and Subtraction of Fractions

1. What length must be cut from a rod $2\frac{1}{4}$ inches long in order that the remaining length may be $1\frac{7}{16}$ inches?

2. A draper sells $4\frac{2}{3}$ yds. and $6\frac{2}{3}$ yds. of calico; and has $5\frac{1}{2}$ yds. left. How much had he at first?

3. Of a mushroom 19/20 is water; what fraction of it is solid matter?

4. A man has to dig a garden in 3 days. The first day he digs 1/2 of it and the second day 3/8 of it. Find what fraction he has dug in two days together: and what fraction remains.

5. What must I add to the sum of $\frac{3}{5}$ and $\frac{4}{15}$ to make $2\frac{1}{2}$?

6. A person reads 1/5 of a book on Monday, 1/3 on Tuesday, and 1/4 on Wednesday; how much remains to be read?

7. If I lose a purse containing $\frac{2}{3}$ of my money, what fraction have I left? If this is 10/-, how much had I to begin with; and how much was in the purse?

8. During a campaign an army loses 1/10 of its numbers in battle, and 1/5 by sickness; what fraction survives?

9. To reach a certain place 9/10 of the journey is performed by rail, 2/25 by motor-car and the rest by mule. What fraction of the whole is the mule-journey?

10. I have to cycle a certain distance in the course of a day. I ride 1/8 of the whole journey before breakfast, 2/5 between breakfast and lunch, 3/10 between lunch and tea; what fraction of the journey remains? If the whole distance is 80 miles, how much do I ride in each interval?

11. *A* can mow a field in 3 days; what fraction of it can he mow in 1 day? *B* can mow the same field in 4 days; what fraction can *B* mow in 1 day? If *A* and *B* work together what fraction do they mow in 1 day?

12. Of a certain magazine 11/23 consists of advertisements and 1/8 of illustrations; what fraction is readable matter?

13. The surface of a tidal river first rises $14\frac{1}{2}$ ft. and then falls $15\frac{5}{6}$ ft. How much lower is it than at first?

14. A ruler 1 ft. long is divided into 12 equal parts by red lines, and into 13 equal parts by blue lines. What is the distance between the first red line and the first blue line?

MULTIPLICATION OF FRACTIONS

§ 63. We will begin this subject with a reminder that 3×2 means "3 multiplied by 2," *not* "2 multiplied by 3." As $3 \times 2 = 2 \times 3$ it may no longer be thought essential to know which is multiplier and which is multiplicand. But if you meet with the expression $\frac{3}{4} \times 2$, you may hope to be able to multiply $\frac{3}{4}$ by 2; whereas to multiply 2 by $\frac{3}{4}$ may seem at first sight a much more difficult undertaking.

§ 64. **When the multiplier is a whole number.** To multiply a fractional quantity—say $\frac{5}{16}$ of a mile—by a whole number —say 3, remember that to multiply anything by 3 means simply to take the thing 3 times and add up.

$$\text{Thus}\quad \frac{5}{16}\text{ mile} \times 3 = \frac{5}{16}\text{ mile} + \frac{5}{16}\text{ mile} + \frac{5}{16}\text{ mile}$$
$$= \frac{15}{16}\text{ mile.}$$

As this is equally true whatever unit we take, we have the purely numerical fact that

$$\frac{5}{16} \times 3 = \frac{5}{16} + \frac{5}{16} + \frac{5}{16} = \frac{15}{16}.$$

Or we may put the matter thus. A sixteenth (of some unit) is a thing; we want to multiply 5 of these things by 3, and the result is 15 of these things—namely, "sixteenths."

No rule will be given for multiplying by a whole number, but after working through the following set of exercises you ought to be able to frame a rule for yourself.

EXERCISE VI p (oral)

Simplify:

1. $\frac{1}{5}+\frac{1}{5}+\frac{1}{5}+\frac{1}{5}$.
2. $\frac{1}{5}\times 4$.
3. $\frac{2}{7}+\frac{2}{7}+\frac{2}{7}$.
4. $\frac{2}{7}\times 3$.
5. $\frac{3}{8}+\frac{3}{8}+\frac{3}{8}+\frac{3}{8}+\frac{3}{8}$.
6. $\frac{3}{8}\times 5$.
7. $\frac{3}{5}+\frac{3}{5}+\frac{3}{5}+\frac{3}{5}$.
8. $\frac{3}{5}\times 4$.
9. $\frac{1}{11}\times 3$.
10. $\frac{2}{5}\times 2$.
11. $\frac{3}{7}\times 2$.
12. $\frac{3}{7}\times 3$.
13. $\frac{5}{8}\times 3$.
14. $\frac{3}{4}\times 7$.
15. $\frac{1}{9}\times 11$.
16. $\frac{3}{11}\times 10$.
17. $\frac{1}{3}\times 5$.
18. $\frac{3}{5}\times 8$.

§ 65. *Example.* Simplify $4\frac{7}{15}\times 24$.

$$\text{The product} = 4\times 24+\frac{7}{15}\times 24$$

$$= 96+\frac{7\times \overset{8}{\cancel{24}}}{\underset{5}{\cancel{15}}}$$

$$= 96+\frac{56}{5}$$

$$= 96+11\tfrac{1}{5} = \underline{107\tfrac{1}{5}}.$$

EXERCISE VI q

Simplify:

1. $\frac{3}{4}\times 8$.
2. $\frac{5}{16}\times 4$.
3. $\frac{7}{10}\times 15$.
4. $\frac{2}{3}\times 6$.
5. $\frac{3}{5}\times 15$.
6. $\frac{3}{8}\times 8$.
7. $\frac{5}{7}\times 21$.
8. $\frac{4}{5}\times 5$.
9. $\frac{2}{3}\times 3$.
10. $\frac{1}{6}\times 3$.
11. $\frac{5}{12}\times 2$.
12. $\frac{4}{15}\times 5$.
13. $\frac{1}{15}\times 10$.
14. $\frac{1}{8}\times 6$.
15. $\frac{1}{12}\times 8$.
16. $\frac{3}{12}\times 18$.
17. $\frac{5}{6}\times 9$.
18. $\frac{3}{10}\times 15$.

19. $\frac{5}{14} \times 21$. **20.** $\frac{5}{8} \times 12$. **21.** $\frac{5}{12} \times 8$.

22. $\frac{2}{27} \times 18$. **23.** $1\frac{1}{4} \times 3$. **24.** $2\frac{1}{8} \times 5$.

25. $2\frac{1}{2} \times 4$. **26.** $3\frac{2}{3} \times 6$. **27.** $1\frac{1}{3} \times 5$.

28. $2\frac{3}{4} \times 3$. **29.** $3\frac{5}{6} \times 5$. **30.** $2\frac{3}{8} \times 4$.

31. $4\frac{1}{6} \times 3$. **32.** $3\frac{5}{6} \times 4$. **33.** $2\frac{1}{10} \times 15$.

34. $1\frac{1}{6} \times 8$. **35.** $3\frac{5}{12} \times 8$. **36.** $4\frac{3}{14} \times 35$.

§ 66. **When the multiplier is a fraction.** What is the meaning of "6 feet multiplied by $\frac{3}{5}$," i.e. 6 feet $\times \frac{3}{5}$?

6×4 means $6 + 6 + 6 + 6$; but we cannot interpret $6 \times \frac{3}{5}$ on these lines. We must try to find some other reasonable meaning for "multiplied by $\frac{3}{5}$."

Suppose that we have a number of sticks, each 6 ft. long. The expression 6 feet $\times$ 4 suggests that we are to take 4 of these sticks. It is reasonable therefore to say that 6 feet $\times \frac{3}{5}$ means that we are to take $\frac{3}{5}$ of a stick; i.e. that 6 feet $\times \frac{3}{5}$ means $\frac{3}{5}$ of 6 feet.

Now we have already seen (§ 55) that $\frac{1}{5}$ of 6 feet $= \frac{6}{5}$ feet.

$$\therefore \ \frac{3}{5} \text{ of 6 feet} = \left(\frac{1}{5} \text{ of 6 feet}\right) \times 3$$

$$= \frac{6}{5} \text{ ft.} \times 3$$

$$= \frac{18}{5} \text{ ft}$$

$$\therefore \ 6 \times \frac{3}{5} = \frac{18}{5}.$$

¶**Ex.** What is the meaning of 7 in. $\times \frac{2}{9}$? (What is $\frac{1}{9}$ of 7 in.? $\frac{2}{9}$ of 7 in.? 7 in. $\times \frac{2}{9}$?)

EXERCISE VI r (oral)

What is

1. $\frac{1}{2}$ of 6*d.*?	**2.** 6*d.* $\times \frac{1}{2}$?	**3.** $\frac{1}{3}$ of 6*s.*?
4. 6*s.* $\times \frac{2}{3}$?	**5.** $\frac{1}{7}$ of 10′?	**6.** 10′ $\times \frac{3}{7}$?
7. 5″ $\times \frac{3}{8}$?	**8.** 4′ $\times \frac{7}{8}$?	**9.** 2*d.* $\times \frac{4}{5}$?
10. 3/- $\times \frac{5}{7}$?	**11.** 2″ $\times \frac{2}{9}$?	**12.** 6 lbs. $\times \frac{3}{5}$?
13. 5 oz. $\times \frac{2}{11}$?	**14.** 7 galls. $\times \frac{6}{11}$?	**15.** £5 $\times \frac{3}{7}$?
16. 5*s.* $\times \frac{6}{13}$?	**17.** 11′ $\times \frac{3}{10}$?	**18.** 13″ $\times \frac{3}{20}$?
19. 7*d.* $\times \frac{4}{13}$?	**20.** 6 yds. $\times \frac{5}{11}$?	**21.** 8*d.* $\times \frac{4}{7}$?

§ 67. **How can multiplying diminish?** It is worth while to stop at this point for a moment, and explain a difficulty that has always troubled beginners.

We have seen that $6 \times \frac{1}{2}$ means one-half of 6, i.e. 3. Multiplying anything by $\frac{1}{2}$ apparently makes it less. Now we have been accustomed to think of multiplication as a process of "making more," and yet we find that multiplication by a proper fraction is a process of "making less."

A modern steamship has no sails; yet we speak of a steamship "sailing" from a port. The old word "sailing" is applied to the new conditions—steamships. In the same way the old word "multiply" is applied to the new conditions—fractions. Steamships do not sail in the literal sense: a fractional multiplier does not MULT-iply in the literal and Latin sense. We widen the meaning of "multiply" when we define $6 \times \frac{1}{2}$ as $\frac{1}{2}$ of 6, just as we widen the meaning of "sail" when we say that a steamer sails.

Multiplication: general case.

§ 68. We will now multiply a fractional quantity by a fractional number. We will find what is the value of $\frac{4}{5}$ of a unit $\times \frac{1}{3}$. Take the unit to be the circle in the figures.

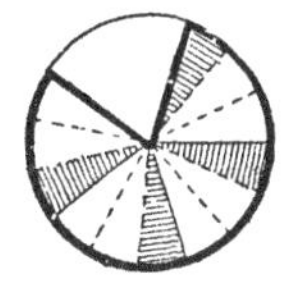

$\frac{4}{5}$ of the circle is the part of the circle with a thick edging.

$\frac{4}{5}$ of the circle $\times \frac{1}{3}$ means $\frac{1}{3}$ of this thick-edged part.

The broken lines divide each fifth into 3 equal parts; these are fifteenths of the circle. We can make up 1/3 of the thick-edged part by picking out the 4 shaded bits: these taken together amount to 1/3 of the thick-edged part. (If there are 4 baskets each containing 3 apples, I can take 1/3 of the whole lot by picking out 1 apple from each basket.)

The 4 shaded bits together make up $\frac{4}{15}$ of the circle.

Thus
$$\frac{1}{3} \text{ of } \frac{4}{5} = \frac{4}{15},$$
$$\therefore \frac{4}{5} \times \frac{1}{3} = \frac{4}{15}.$$

§ 69. Next, what is the value of $\frac{4}{5} \times \frac{2}{3}$, i.e. of $\frac{2}{3}$ of $\frac{4}{5}$?

$$\frac{2}{3} \text{ of } \frac{4}{5} = \text{twice } \frac{1}{3} \text{ of } \frac{4}{5}$$
$$= \text{twice } \frac{4}{15} = \frac{8}{15},$$
$$\therefore \frac{4}{5} \times \frac{2}{3} = \frac{8}{15}.$$

Note that in this result the 15 arose when we divided each of the 5 parts into 3 parts, making $5 \times 3 = 15$ parts. The 8 arose when we took "twice $\frac{4}{15}$"; the 8 was 4×2.

In fact
$$\frac{4}{5} \times \frac{2}{3} = \frac{4 \times 2}{5 \times 3},$$
and the same rule would hold good with any other numbers.

The product of two fractions has for numerator the product of the numerators, and for denominator the product of the denominators.

EXERCISE VI s (oral)

Simplify: $\frac{2}{3} \times \frac{4}{5}$, $\frac{4}{5} \times \frac{2}{3}$, $\frac{3}{7} \times \frac{2}{5}$, $\frac{2}{5} \times \frac{3}{7}$,

$\frac{4}{11} \times \frac{3}{7}$, $\frac{3}{7} \times \frac{4}{11}$, $\frac{1}{2} \times \frac{5}{8}$, $\frac{4}{7} \times \frac{2}{5}$, $\frac{3}{4} \times \frac{3}{4}$, $\left(\frac{2}{5}\right)^2$,

$\left(\frac{3}{8}\right)^2$, $\left(\frac{7}{10}\right)^2$, $\frac{4}{9} \times \frac{3}{7}$, $\left(\frac{4}{5}\right)^2$, $\frac{3}{8} \times \frac{5}{11}$, $\left(\frac{6}{11}\right)^2$.

§ 70. *Example.* Find the product of $1\frac{9}{16} \times 4\frac{4}{5}$.

$$\text{Product} = \frac{\overset{5}{\cancel{25}}}{\underset{2}{\cancel{16}}} \times \frac{\overset{3}{\cancel{24}}}{\cancel{5}} = \frac{15}{2} = 7\frac{1}{2}.$$

N.B. Before *multiplying* fractions it is generally convenient to convert mixed numbers into improper fractions.

Before *adding or subtracting* fractions, convert improper fractions into mixed numbers.

EXERCISE VI t

Simplify:

1. $\frac{2}{3} \times \frac{9}{11}$. **2.** $\frac{5}{6} \times \frac{12}{13}$. **3.** $\frac{7}{12} \times \frac{3}{14}$. **4.** $\frac{5}{9} \times \frac{12}{25}$.

Multiply:

5. $\frac{3}{4}$ by $\frac{8}{9}$, $\frac{2}{15}$, $\frac{10}{21}$, $1\frac{1}{3}$. **6.** $\frac{9}{16}$ by $\frac{8}{15}$, $\frac{10}{27}$, $1\frac{7}{9}$, $3\frac{5}{9}$.

7. $\frac{8}{21}$ by $\frac{7}{8}$, $\frac{35}{64}$, $2\frac{5}{8}$, $1\frac{1}{6}$. **8.** $\frac{6}{25}$ by $\frac{10}{21}$, $4\frac{1}{6}$, $8\frac{1}{3}$, 25.

9. $1\frac{3}{5}$ by $\frac{5}{8}$, $1\frac{1}{4}$, $\frac{15}{16}$, $3\frac{3}{4}$.

10. $1\frac{11}{25}$ by $\frac{25}{11}$, $\frac{5}{27}$, $1\frac{19}{36}$, $\frac{125}{144}$.

Simplify:

11. $\left(1\frac{1}{2}\right)^2$. **12.** $\left(1\frac{1}{3}\right)^2$. **13.** $\left(1\frac{1}{4}\right)^2$.

14. $\left(2\frac{1}{2}\right)^2$. **15.** $\left(1\frac{3}{4}\right)^2$. **16.** $\frac{1}{2} \times \frac{1}{3} \times \frac{1}{4}$.

17. $\frac{1}{2} \times \frac{2}{3} \times \frac{3}{4}$. **18.** $1\frac{1}{2} \times \frac{2}{5} \times 2\frac{1}{2}$. **19.** $\left(\frac{1}{2}\right)^2 \times \frac{1}{3}$.

20. $\left(\frac{3}{4}\right)^2 \times 1\frac{7}{9}$. **21.** $\left(2\frac{1}{2}\right)^3$. **22.** $\left(3\frac{1}{3}\right)^2$.

23. $\left(\frac{1}{2}\right)^4$. **24.** $\left(\frac{1}{10}\right)^4$. **25.** $\left(\frac{1}{2}\right)^5$.

What is

26. $2\frac{1}{2}$ of $3\frac{1}{5}$? **27.** $4\frac{1}{4}$ of $\frac{5}{34}$? **28.** $5\frac{1}{5}$ of $\frac{10}{13}$?

29. $3\frac{3}{4}$ of $2\frac{6}{25}$? **30.** $1\frac{5}{8}$ of $\frac{7}{26}$? **31.** $2\frac{5}{12}$ of $1\frac{7}{29}$?

Division of Fractions

§ 71. **When the divisor is a whole number.** To divide a thing by 5 is the same as to take one-fifth of the thing; thus

$$\frac{3}{8} \text{ mile} \div 5 = \frac{1}{5} \text{ of } \frac{3}{8} \text{ mile}$$

$$= \frac{3}{40} \text{ mile (see § 66).}$$

Notice that $$\frac{3}{8} \div 5 = \frac{3}{8} \times \frac{1}{5}.$$

EXERCISE VI u (oral)

1. Divide 1/2 by 3. **2.** Divide 1/4 by 3.

3. Divide 1/5 by 3. **4.** Divide 3/10 by 3.

5. Divide 3/7 by 2. **6.** Divide 4/5 by 5.

7. Divide 3/8 by 7. **8.** Divide 3/8 by 3.

9. Divide 12/17 by 3, 4, 6, 8, 12, 24.

10. Divide 6/7 by 2, 3, 4, 5, 6, 12, 60.

11. Divide 15/16 by 2, 3, 5, 10, 15, 30.

12. Divide $2\frac{2}{5}$ by 2, 3, 4, 6, 8, 12, 24.

13. Divide $3\frac{3}{7}$ by 2, 3, 4, 6, 8, 12, 18, 24.

14. Divide $3\frac{3}{5}$ by 2, 3, 4, 6, 9, 12, 18.

15. Divide $4\frac{2}{7}$ by 3, 4, 5, 6, 7, 10, 15, 24, 30.

§ 72. **When the divisor is a fraction.** Suppose that there are 2 cakes, and that each boy is to have $\frac{1}{5}$ of a cake. How many boys can be supplied? In other words, how many times does 2 contain $\frac{1}{5}$? or what is the value of $2 \div \frac{1}{5}$? Clearly, **10.**

§ 73. We will next inquire how many times £$\frac{2}{5}$ is contained in £$\frac{3}{4}$. Now £$\frac{2}{5}$ = 8/- and £$\frac{3}{4}$ = 15/-, and 8 shillings is contained in 15 shillings $\frac{15}{8}$ times, or $1\frac{7}{8}$ times (more than once, and not quite twice).

Thus
$$\frac{3}{4} \div \frac{2}{5} = \frac{15}{8}.$$

Look at the same question in another way.

In reducing to shillings we really reduced to twentieths of a pound: in fact, we reduced dividend and divisor to a **common denominator.**

$$£\frac{3}{4} = £\frac{15}{20} \text{ and } £\frac{2}{5} = £\frac{8}{20}.$$

But 15 twentieths contains 8 twentieths $\frac{15}{8}$ times.

$$\therefore \frac{3}{4} \div \frac{2}{5} = \frac{15}{20} \div \frac{8}{20} = \frac{15}{8} = 1\frac{7}{8}.$$

EXERCISE VI v (oral)

1. How many times does 9 tenths contain 3 tenths? 9 tenths contain 5 tenths? 9/10 contain 7/10? 22/10 contain 7/10? 3/5 contain 4/5? 4/7 contain 3/7? 5/9 contain 1/9? 5/9 contain 2/9? etc.

2. How many times is 3/11 contained in 9/11? 2/5 contained in 4/5? 3/13 contained in 7/13? 4/15 contained in 13/15? 2/7 contained in 9/7?

3. What is the value of $\frac{4}{5} \div \frac{2}{5}$, of $\frac{9}{17} \div \frac{4}{17}$, of $\frac{7}{20} \div \frac{3}{20}$, of $\frac{15}{31} \div \frac{7}{31}$, of $\frac{7}{12} \div \frac{5}{12}$, of $\frac{9}{17} \div \frac{5}{17}$?

EXERCISE VI w

By reducing dividend and divisor to a common denominator, simplify:

1. $\frac{4}{5} \div \frac{5}{7}$.	**2.** $\frac{5}{7} \div \frac{4}{5}$.	**3.** $\frac{3}{8} \div \frac{2}{5}$.
4. $\frac{2}{5} \div \frac{3}{8}$.	**5.** $\frac{4}{15} \div \frac{2}{3}$.	**6.** $\frac{2}{3} \div \frac{4}{15}$.
7. $\frac{8}{21} \div \frac{4}{7}$.	**8.** $\frac{4}{7} \div \frac{8}{21}$.	**9.** $\frac{6}{25} \div \frac{9}{10}$.
10. $\frac{9}{10} \div \frac{6}{25}$.	**11.** $\frac{25}{6} \div \frac{10}{9}$.	**12.** $4\frac{1}{6} \div 1\frac{1}{9}$.
13. $3\frac{3}{4} \div 2\frac{1}{2}$.	**14.** $2\frac{1}{2} \div 3\frac{3}{4}$.	**15.** $5\frac{2}{5} \div \frac{18}{25}$.
16. $\frac{18}{25} \div 5\frac{2}{5}$.	**17.** $1 \div \frac{1}{3}$.	**18.** $1 \div \frac{3}{5}$.
19. $1 \div \frac{7}{8}$.	**20.** $1 \div \frac{3}{10}$.	**21.** $1 \div 2\frac{1}{2}$.
22. $1 \div 3\frac{3}{4}$.	**23.** $1 \div \frac{17}{25}$.	**24.** $\frac{14}{45} \div \frac{21}{35}$.
25. $\frac{36}{49} \div 3\frac{3}{14}$.	**26.** $\frac{25}{42} \div \frac{35}{36}$.	**27.** $10\frac{9}{16} \div 1\frac{43}{48}$.

COMPLEX FRACTIONS

§ 74. We may now regard the fraction notation as entirely equivalent to the ÷ notation; in other words $17 \div 11$ and $\frac{17}{11}$ may be taken to have the same meaning.

Accordingly we may write $\frac{3}{4} \div \frac{2}{5}$ thus $\frac{\frac{3}{4}}{\frac{2}{5}}$.

This is called a **complex fraction:** it has a fractional numerator $\frac{3}{4}$ and a fractional denominator $\frac{2}{5}$. The line separating numerator and denominator is often printed slanting to save space; thus $\frac{3}{4}\Big/\frac{2}{5}$.

§ 75. We shall assume that the value of a complex fraction is unaltered when numerator and denominator are multiplied (or divided) by the same number. This is the familiar rule for ordinary fractions.

This rule enables us to reduce any complex fraction to an ordinary fraction. By what number must $\frac{3}{4}$ and $\frac{2}{5}$ be multiplied in order to make both products whole numbers? Clearly by the L.C.M. of 4 and 5, namely 20.

Thus
$$\frac{\frac{3}{4}}{\frac{2}{5}} = \frac{\frac{3}{4} \times 20}{\frac{2}{5} \times 20} = \frac{15}{8}.$$

This of course is the same result as we obtained for $\frac{3}{4} \div \frac{2}{5}$ in § 73. Thus you can divide one fraction by another without learning any new rule.

EXERCISE VI x

Simplify:

1. $\dfrac{\frac{1}{4}}{\frac{3}{4}}$. 2. $\dfrac{\frac{5}{12}}{\frac{7}{12}}$. 3. $\dfrac{\frac{1}{20}}{\frac{17}{20}}$. 4. $\dfrac{\frac{2}{3}}{\frac{5}{7}}$.

5. $\frac{5}{7}\Big/\frac{2}{3}$. 6. $1\frac{1}{4}\Big/2\frac{1}{2}$. 7. $2\frac{1}{2}\Big/1\frac{1}{4}$. 8. $7\frac{1}{8}\Big/2\frac{3}{8}$.

9. $2\frac{3}{8}\Big/7\frac{1}{8}$. 10. $4\frac{1}{2}\Big/5\frac{1}{4}$. 11. $\frac{5}{6}\Big/\frac{10}{9}$. 12. $\frac{14}{15}\Big/4\frac{9}{10}$.

Express the following as complex fractions, and simplify:

13. $4\frac{3}{4} \div 2\frac{3}{8}$. 14. $2\frac{5}{12} \div 3\frac{2}{9}$. 15. $5\frac{1}{5} \div \frac{8}{25}$.

16. $3\frac{13}{21} \div 1\frac{5}{14}$. 17. $4\frac{2}{7} \div 5\frac{5}{14}$. 18. $7\frac{1}{8} \div 5\frac{7}{10}$.

19. $6\frac{1}{15} \div 3\frac{3}{20}$. 20. $3\frac{1}{15} \div 2\frac{5}{9}$. 21. $1 \div \frac{2}{5}$.

22. $1\Big/\frac{4}{7}$. 23. $1 \div \frac{7}{12}$. 24. $1\Big/\frac{9}{13}$.

25. $1 \div 3\frac{1}{2}$. 26. $1\Big/3\frac{1}{4}$. 27. $1 \div \frac{5}{6}$.

28. $1\Big/\frac{7}{9}$. 29. $1 \div 4\frac{1}{4}$. 30. $5\Big/5\frac{1}{5}$.

EXERCISE VI y

Use of Brackets

Simplify:

1. $3 + 5 \times 2$.
2. $(3 + 5) \times 2$.
3. $3 + (5 \times 2)$.
4. $3 + 5 \div 2$.
5. $(3 + 5) \div 2$.
6. $3 + (5 \div 2)$.
7. $(3 \times 5) \div 2$.
8. $3 \times (5 \div 2)$.
9. $3 \div (5 \times 2)$.
10. $(3 \div 5) \times 2$.
11. $\frac{1}{2} + \frac{1}{3} \times \frac{1}{4}$.
12. $\left(\frac{1}{2} + \frac{1}{3}\right) \times \frac{1}{4}$.
13. $\frac{1}{2} + \left(\frac{1}{3} \times \frac{1}{4}\right)$.
14. $\left(\frac{1}{2} + \frac{1}{3} \times \frac{1}{4}\right)$.
15. $\frac{1}{2} + \frac{1}{3} \div \frac{1}{4}$.
16. $\left(\frac{1}{2} + \frac{1}{3}\right) \div \frac{1}{4}$.
17. $\frac{1}{2} + \left(\frac{1}{3} \div \frac{1}{4}\right)$.
18. $\left(\frac{1}{2} \times \frac{1}{3}\right) \div \frac{1}{4}$.
19. $\frac{1}{2} \times \left(\frac{1}{3} \div \frac{1}{4}\right)$.
20. $\left(\frac{1}{2} \div \frac{1}{3}\right) \times \frac{1}{4}$.
21. $\frac{1}{2} \div \left(\frac{1}{3} \times \frac{1}{4}\right)$.
22. $\frac{5}{12} + \frac{7}{24} \div \frac{5}{8} + \frac{3}{8}$.
23. $\left(\frac{5}{12} + \frac{7}{24}\right) \div \frac{5}{8} + \frac{3}{8}$.
24. $\frac{5}{12} + \frac{7}{24} \div \left(\frac{5}{8} + \frac{3}{8}\right)$.
25. $\left(\frac{5}{12} + \frac{7}{24}\right) \div \left(\frac{5}{8} + \frac{3}{8}\right)$.
26. $\frac{5}{12} + \left(\frac{7}{24} \div \frac{5}{8}\right) + \frac{3}{8}$.
27. $\left(\frac{5}{12} + \frac{7}{24} \div \frac{5}{8}\right) + \frac{3}{8}$.
28. $\frac{5}{12} + \left(\frac{7}{24} \div \frac{5}{8} + \frac{3}{8}\right)$.
29. $\frac{3}{7} \div \frac{1}{14} + \frac{13}{14} \times \frac{7}{13}$.
30. $\left(\frac{3}{7} \div \frac{1}{14}\right) + \frac{13}{14} \times \frac{7}{13}$.
31. $\frac{3}{7} \div \left(\frac{1}{14} + \frac{13}{14} \times \frac{7}{13}\right)$.
32. $\frac{3}{7} \div \left\{\left(\frac{1}{14} + \frac{13}{14}\right) \times \frac{7}{13}\right\}$.
33. $\left(\frac{3}{7} \div \frac{1}{14} + \frac{13}{14}\right) \times \frac{7}{13}$.
34. $\left\{\frac{3}{7} \div \left(\frac{1}{14} + \frac{13}{14}\right)\right\} \times \frac{7}{13}$.

PROBLEMS ON VULGAR FRACTIONS

§ 76. An important principle is illustrated in Method 2 of the following example.

Example. If 5/8 gallons of milk is the daily supply of a household of 4, for how many people would 1 gallon a day suffice, at the same rate of consumption?

Method 1. $\frac{5}{8}$ galls. are enough for 4 persons,

$\therefore$ 5 galls. are enough for 4×8 persons,

$\therefore$ 1 gall. is enough for $\frac{4 \times 8}{5}$ persons.

This is $6\frac{2}{5}$ persons: in fact, the supply is more than enough for 6 and not enough for 7.

Method 2. If we were told that **2** galls. are enough for 4 persons, we shall conclude that 1 gall. is enough for $4 \div \mathbf{2}$ persons.

In the actual example, $\frac{5}{8}$ takes the place of 2. But we may still argue in the same way, thus

$$\mathbf{\frac{5}{8}} \text{ galls. supply 4 persons,}$$

$$\therefore \text{ 1 gall. supplies } 4 \div \mathbf{\frac{5}{8}} \text{ persons} = 4 \times \frac{8}{5} \text{ persons.}$$

This gives the same result as the more elementary Method 1. It is important to notice that, in wide classes of problems, we are justified in operating with fractional data on the same lines as if the fractions were replaced by whole numbers.

EXERCISE VI z

1. A boy earns 12*s*. 8*d*. a week and spends 9*s*. 6*d*. Express his expenses as a fraction of his earnings.

2. The gross receipts in a business are £5745; the net receipts £920. What fraction are the net receipts of the gross receipts?

3. A train (including the engine) weighs 400 tons, and the engine weighs 2/9 of the whole. Find the weight of the engine. Give its weight in tons and a fraction of a ton.

4. A pint of water weighs a pound and a quarter, what is the weight in lbs. of a quart, 5 pints, 3 gallons, $3\frac{1}{2}$ pints, $5\frac{3}{4}$ pints, $6\frac{1}{4}$ gallons, 25 gallons of water?

5. What is the cost of $3\frac{3}{4}$ lbs. of coffee at 1*s*. 8*d*. a pound?

6. If my stride is $2\frac{3}{4}$ ft., how many steps shall I take in 100 ft.? in 100 yds.? in 1 mile? in $1\frac{3}{8}$ miles?

7. If a cup holds 1/2 pint, how many people could have a cup of milk from a gallon can? How many if each cup was only 3/4 full?

8. A railway company issues return tickets at a single fare and a quarter; what would the single fare be when the return ticket costs 6*s*. 3*d*.?

9. A map is drawn to the scale of 12 miles to an inch. What is the distance between two towns that are $7\frac{3}{4}$ inches apart on the map?

10. Of a total weight of 78 cwts. there remains, after filling 9 equal cases, enough to fill half of a tenth case. What weight in cwts. does each hold?

11. If 3/4 of a pound of tobacco cost 3*s*., what is the cost of 1/8 lb.? of 1 lb.? of 5 lbs.?

12. If $\frac{7}{10}$ of a ton of coal cost 14*s*., what is the cost of 1 ton? of 1 cwt.? of $2\frac{1}{2}$ tons? of 2 tons 5 cwts.? of $14\frac{3}{4}$ tons?

13. Given that $1\frac{1}{4}$ lbs. of water occupy 1 pint, what fraction of a pint is filled by 1/4 lb., by 1 lb., by 10 lbs., by $2\frac{1}{2}$ lbs., by $3\frac{1}{3}$ lbs., by 5 oz.? How many gallons are filled by 1 cwt. of water?

14. If 4/5 of a drove of pigs are sold for £12, what should the whole drove realize?

15. After completing 3/7 of his journey, a man finds that he has still 21 miles to go. How far has he gone already?

16. If I can do 3/16 of a job in a day, how long will it take me to do the whole job?

17. A father's age is $2\frac{3}{4}$ times that of his son. What fraction is the son's age of his father's?

18. Working $9\frac{1}{2}$ hours a day, a man can finish a piece of work in 6 days. What fraction of the whole can he do in 1 hour? in a day of 8 hours? In how many days of 8 hours can he finish the job?

19. Of the cultivated area of Great Britain 7/32 produces corn and 1/4 other crops (including hay and green crops); the rest is permanent pasture: what fraction is the pasture-land of the whole cultivated area? Given that 17 million acres are pasture-land, find how many acres are corn-land.

20. An iron bridge crosses a road 29 feet wide, 1/6 of the girder resting on the bank on one side of the road, and 1/7 resting on the other side; how long is the girder?

21. A owns 3/7 of a piece of land, B owns 2/5 of it, and C owns the remainder. If C's share is worth £72, what are A's and B's shares each worth?

22. A, B, and C shared a sum of money so that A had $\frac{1}{3}$, B had $\frac{1}{5}$, and C had the remainder which came to £1. 15*s*. What was the sum of money?

23. In setting up a post, 1/3 of its length is buried in the ground, and 1/5 of the remainder is tarred, the rest being painted green. What fraction of (i) the whole, (ii) the part above ground, is green?

24. A boy spends $\frac{1}{2}$ of his pocket money in the first week of term, and in the second week $\frac{2}{3}$ of what remains; what fraction is left?

25. A man loses $\frac{1}{2}$ of his money, then $\frac{2}{5}$ of the remainder, and finally $\frac{3}{7}$ of the rest. If he has now a shilling, how much had he at first?

26. A man leaves 3/4 of his property to his eldest son, the rest to be equally divided among the other children. The total property was £28,000, and each of the younger children received £1750. How many children were there?

27. "Tell me, illustrious Pythagoras, how many pupils are in thy school?" "One-half," replied the philosopher, "study mathematics, one-fourth philosophy, and one-seventh observe silence. There are three females besides." How many pupils were there?

28. I go out for an afternoon with a certain sum in my pocket. I spend $\frac{1}{3}$ of it on a railway ticket, $\frac{1}{4}$ of what remains in a rowing boat, $\frac{1}{6}$ of what now remains in tea, $\frac{4}{5}$ of what now remains on a ticket home; and find myself left with 1*s*. What was the price of the tea, and how much has my afternoon's holiday cost?

29. An army loses 1/10 in the first battle, and 1/5 of the remainder in the second. What fraction has it lost altogether? What fraction remains? If 18,000 remain, what was the original number, and the loss in each battle?

30. A tourist loses his purse, containing $\frac{3}{4}$ of his money, and decides to return home at once. His ticket costs $\frac{1}{3}$ of what remains. After buying the ticket he finds that he has only £1 left. How much did he lose?

31. A man spends $\frac{1}{2}$ of his money in one shop, $\frac{1}{3}$ of the remainder in another shop, and $\frac{3}{4}$ of what is left he gives away. If he still has 36 shillings, how much had he at first?

32. I spend $\frac{5}{9}$ of my money: what fraction of the remainder must I spend so as to have left $\frac{1}{6}$ of the whole?

33. A man bequeathed $\frac{5}{12}$ of his money to one son, $\frac{1}{2}$ of the remainder to another, and the surplus to his widow. The difference of his sons' legacies was £1568; how much did the widow receive?

34. Each month a man loses 1/10 of his weight at the beginning of the month. What fraction has he lost after 3 months? What fraction remains?

35. An elastic ball was found to rebound to a height which is 2/9 of that which it fell; on the third rebound it rises to a height of 4/11 feet; from what height did it first fall?

36. 1000 metres are 5/8 mile. Express $2\frac{1}{2}$ miles in metres.

37. A ball of string contains 100 feet. How many pieces $5\frac{1}{2}$ feet long can be cut from it, and what will be left?

38. From an iron rod 5 ft. 11 in. long, how many bolts $3\frac{1}{2}$ in. long could be cut; and what will be left?

39. To a gallon of mixture consisting of 5 parts of milk and 1 part of water is added another quart of water. What fraction of the resulting mixture is milk?

40. Find a fraction half-way between 3/4 and 5/6, having denominator 24. Prove that it is half the sum of the two fractions.

41. Find the difference between:

(*a*) $\frac{2}{3}+\frac{4}{5}$ and $\frac{2+4}{3+5}$. (*b*) $\frac{4}{7}-\frac{1}{3}$ and $\frac{4-1}{7-3}$.

(*c*) $\frac{3}{4}\times\frac{7}{8}$ and $\frac{3\times 7}{4\times 8}$. (*d*) $\frac{9}{25}\div\frac{3}{5}$ and $\frac{9\div 3}{25\div 5}$.

(*e*) $\frac{3}{5}+\frac{3}{10}$ and $\frac{3+3}{5+10}$. (*f*) $\frac{5}{12}-\frac{4}{11}$ and $\frac{5-4}{12-11}$.

42. Find, by trying three cases, whether adding 1 to the numerator and denominator of a proper fraction makes the fraction greater or less.

43. Repeat Ex. 42 for an improper fraction.

44. Choose any fraction less than 1, and add to numerator and denominator any integer (except 1). Is the fraction increased or diminished?

45. Repeat Ex. 44 with a fraction greater than 1.

46. Repeat Ex. 44 with a fraction equal to 1.

47. Find whether or no 2/3 lies half-way between 1/2 and 3/4.

48. Find the least fraction which, added to 47/13, will make the answer an integer.

*Harder Fractions

§77. In dealing with complicated fractions, it is convenient to use the following method of dividing by a fraction.

Consider $\frac{3}{5}\div\frac{7}{11}$. This is equal to $\frac{3\times 11}{55}\div\frac{7\times 5}{55}=\frac{3\times 11}{7\times 5}$. But this is $\frac{3}{5}\times\frac{11}{7}$. Therefore, to divide by $\frac{7}{11}$ is the same as to multiply by $\frac{11}{7}$.

The rule thus suggested is as follows:

To divide by a fraction, multiply by the fraction formed by interchanging numerator and denominator of the divisor (i.e. "invert and multiply").

* This may be postponed.

The same rule applies to the simplification of a complex fraction, which of course is equivalent to the division of one fraction by another. Thus

$$\frac{3}{5}\bigg/\frac{7}{11}=\frac{3}{5}\times\frac{11}{7}=\frac{33}{35}.$$

Example. Simplify $\dfrac{\dfrac{21}{44}\div\dfrac{35}{121}}{\dfrac{12}{25}}$.

$$\text{Expression}=\frac{\overset{3}{\cancel{21}}}{\underset{4}{\cancel{44}}}\times\frac{\overset{11}{\cancel{121}}}{\underset{5}{\cancel{35}}}\times\frac{\overset{5}{\cancel{25}}}{\underset{4}{\cancel{12}}}=\frac{55}{16}=\underline{3\frac{7}{16}}.$$

Example. Simplify

$$\frac{91}{122}\left(4\frac{1}{8}\times6\frac{2}{7}+\frac{3}{14}\right)\div\left\{4\frac{1}{8}\text{ of }\left(6\frac{2}{7}+\frac{3}{14}\right)\right\}.$$

Note that $6\frac{2}{7}+\frac{3}{14}$ in the last bracket can be simplified to $6\frac{1}{2}$ at once, for $\frac{2}{7}=\frac{4}{14}$.

$$\begin{aligned}\text{Expression}&=\frac{91}{122}\left(\frac{33}{8}\times\frac{44}{7}+\frac{3}{14}\right)\div\left(4\frac{1}{8}\times6\frac{1}{2}\right)\\&=\frac{91}{122}\times\frac{1452+12}{8\times7}\div\left(\frac{33}{8}\times\frac{13}{2}\right)\\&=\frac{91}{122}\times\frac{1464}{7\times8}\div\frac{33\times13}{8\times2}\\&=\frac{\cancel{91}\times\overset{\overset{8}{\cancel{488}}}{\cancel{1464}}\times\cancel{8}\times\cancel{2}}{\underset{\cancel{61}}{\cancel{122}}\times\cancel{7}\times\cancel{8}\times\underset{11}{\cancel{33}}\times\cancel{13}}=\underline{\frac{8}{11}}.\end{aligned}$$

*EXERCISE VI aa

Simplify:

1. $\dfrac{49}{72}\div\dfrac{35}{48}$. **2.** $\dfrac{15}{34}\div\dfrac{45}{119}$. **3.** $4\dfrac{5}{12}\div3\dfrac{25}{27}$.

4. $3\dfrac{6}{25}\div\dfrac{27}{35}$. **5.** $10\dfrac{9}{16}\Big/1\dfrac{11}{80}$. **6.** $2\dfrac{34}{55}\Big/\dfrac{48}{77}$.

* This may be postponed.

7. $3\frac{6}{70}\Big/1\frac{32}{49}$.

8. $2\frac{10}{27}\Big/\frac{40}{99}$.

9. $\dfrac{\frac{2}{3}\times\frac{4}{5}}{\frac{12}{25}}$.

10. $\dfrac{1\frac{1}{2}\times\frac{3}{8}}{\frac{15}{16}}$.

11. $\dfrac{\frac{3}{7}\div\frac{4}{5}}{\frac{45}{49}}$.

12. $\dfrac{4\frac{1}{2}\div\frac{4}{9}}{10\frac{1}{8}}$.

13. $\left(2\frac{2}{9}+4\frac{5}{6}\right)\times\frac{2}{5}+6\frac{1}{2}$.

14. $5+\dfrac{6}{5+\frac{2}{5}}$.

15. $\dfrac{1\frac{3}{4}-\frac{17}{16}}{3\frac{1}{4}\times 2\frac{1}{2}}$.

16. $\dfrac{1\frac{4}{7}-\frac{2}{3}}{\frac{4}{49}\times 7\frac{7}{12}}$.

17. $\dfrac{1\frac{1}{4}-2\frac{1}{8}+1\frac{2}{3}}{5\frac{1}{5}+7\frac{1}{12}}$.

18. $\dfrac{4\frac{2}{9}-2\frac{1}{4}}{3\frac{3}{4}-3\frac{1}{3}+\frac{11}{144}}$.

19. $\dfrac{2\frac{2}{15}-1\frac{1}{6}}{3\frac{1}{4}\times 2\frac{1}{5}}$.

20. $\dfrac{1\frac{1}{6}-\frac{7}{18}+\frac{1}{7}}{\frac{1}{2}+\frac{3}{7}-\frac{5}{12}}$.

21. $\frac{10}{9}\times\dfrac{\frac{1}{5}-\frac{1}{8}}{\frac{1}{2}-\frac{15}{32}}$.

22. $\dfrac{2\frac{5}{12}}{\frac{5}{6}+\frac{3}{8}}\times\dfrac{\frac{5}{6}-\frac{3}{8}}{4\frac{3}{8}}$.

23. $\dfrac{5\frac{1}{2}\times 3\frac{1}{3}\times 2\frac{1}{7}-6\frac{1}{4}\times 3\frac{1}{5}\times 1\frac{1}{6}}{7\frac{4}{9}\times 3\frac{4}{7}}$.

24. $\dfrac{\left(3\frac{1}{2}\times 3\frac{3}{5}\right)\div 2\frac{3}{4}}{2\frac{3}{5}+3\frac{4}{7}}$.

25. $\frac{107}{119}\times\dfrac{15\frac{5}{8}+11\frac{1}{4}}{11\frac{1}{3}-1\frac{1}{7}}$.

26. $\left(\frac{1}{3}+\frac{4}{7}\right)\dfrac{5\frac{1}{16}}{3\frac{6}{7}+2\frac{1}{4}}$.

27. $1\frac{1}{2}-\dfrac{2\frac{1}{2}}{3+\frac{1}{4}\times\frac{6}{5}}$.

28. $\dfrac{\frac{3}{4}+\frac{2}{5}}{\frac{3}{4}-\frac{2}{5}}\times\dfrac{1\frac{1}{3}}{\frac{2}{3}+\frac{3}{7}}$.

29. $\dfrac{\frac{1}{3}+\frac{3}{4}\times\frac{2}{9}}{\frac{7}{3}+\frac{11}{12}}\times\dfrac{4\frac{1}{3}}{\frac{1}{4}+\frac{5}{12}}$.

30. $\frac{3}{7}+\dfrac{3\frac{7}{8}-3\frac{1}{4}}{2\frac{1}{3}\times 9}$.

31. $\dfrac{\left(5\frac{3}{8} - 3\frac{5}{8}\right) \times \frac{29}{30}}{5\frac{3}{8} - 3\frac{5}{8} \times \frac{29}{30}}$.

32. $1\frac{1}{5} + \dfrac{1\frac{2}{5}}{1\frac{3}{5} + \frac{2}{11}}$.

33. $\dfrac{1\frac{6}{11} \times 2\frac{3}{4}}{\frac{17}{36}} + \dfrac{1\frac{3}{4} - \frac{7}{10}}{1\frac{3}{4} + \frac{7}{10}}$.

34. $\dfrac{2\frac{1}{14} - 1\frac{1}{8}}{3\frac{1}{4} + 1\frac{3}{14}} \div \dfrac{1\frac{1}{2} - 1\frac{2}{7}}{2\frac{1}{7} + 1\frac{1}{2}}$.

35. $\dfrac{\frac{1}{3}\left(\frac{1}{2} + \frac{5}{6}\right) + \frac{4}{7} \times 2\frac{1}{10}}{4\frac{3}{4} - \frac{2}{3} \times 5\frac{2}{5}}$.

36. $\dfrac{\frac{1}{2} + \frac{1}{3} \times 5\frac{1}{4}}{4\frac{3}{5} - 2\frac{5}{7}} \div \left(1 + \dfrac{1}{1 + \frac{1}{10}}\right) + 2$.

37. $\dfrac{31\frac{1}{3} - 22\frac{1}{5}}{11\frac{1}{5} - 1\frac{1}{7}} \div 1\frac{49}{88} + 2\frac{5}{12}$.

38. $\dfrac{3\frac{2}{9}}{3\frac{1}{2} + \frac{7}{9} \times \frac{1}{3}} + \dfrac{2\frac{1}{12} - \frac{11}{15}}{4\frac{2}{3} - 1\frac{4}{5}}$.

39. $3 + \dfrac{1\frac{3}{8} + \frac{11}{12}}{2\frac{11}{15} - 1\frac{9}{20}}$ of 14.

40. $\dfrac{\left(\frac{2}{3} - \frac{2}{11} + \frac{16}{21}\right) \div \left(2 + \frac{2}{5}\right)}{\left(\frac{3}{77} + \frac{2}{33} - \frac{1}{21}\right) \times \left(3 - \frac{3}{8}\right)}$.

41. $\dfrac{\frac{5}{14} - \frac{3}{7} \times \frac{1}{2}}{\frac{5}{16} + \frac{7}{12} \times 3\frac{1}{4} - \left(\frac{7}{8} \times \frac{37}{21} - \frac{1}{3}\right)} \div \dfrac{\frac{1}{3} \times \frac{1}{2} + \frac{3}{2} \times 5}{9\frac{1}{3} - 1\frac{2}{3}}$.

42. $\dfrac{5\frac{3}{4} - \frac{3}{7} \times 15\frac{3}{4} + 2\frac{2}{35} \div 1\frac{11}{25}}{\frac{3}{4} \times 7\frac{3}{7} - 5\frac{3}{5} \div 3\frac{4}{15}}$.

CHAPTER VII

UNITARY METHOD

§78. This method was illustrated in Chap. III by easy examples in which fractions were avoided. There follow some examples involving fractions. After a little practice with this method, it should be discarded in favour of the method of § 81.

DIRECT VARIATION

§79. *Example* 1. If an aeroplane travels 324 miles in 10 hours, how far will it travel in 3 hours at the same rate?

The question is—how many miles? Begin by rearranging the statement with "miles" at the end, thus:

In 10 hours the aeroplane travels 324 miles.

In 1 hour " " " $\frac{324}{10}$ miles.

In 3 hours " " " $\frac{\overset{162}{\cancel{324}}\times 3}{\underset{5}{\cancel{10}}}$ miles.

This is $\frac{486}{5}$ miles $= 97\frac{1}{5}$ miles.

Example 2. If 20 lbs. of wedding cake cost £2/16/6, how much can be bought for £6?

Here the most suitable unit of money is 6*d*. £2/16/6 = 56/6 = 113 sixpences; £6 = 120/- = 240 sixpences.

113 sixpences buy 20 lbs. of cake.

$\therefore$ 1 sixpence buys $\frac{20}{113}$ lbs.

$\therefore$ 240 sixpences buy $\frac{20\times 240}{113}$ lbs.

This is $\frac{4800}{113}$ lbs. $= 42\frac{54}{113}$ lbs.

§80. Example 2 illustrates two weaknesses of the method:

(1) The middle step may be absurd: it is probably impossible to buy a sixpennyworth of wedding cake.

(2) The stores list from which the example is taken gives £6/2/6 as the price of a 40 lb. cake. Therefore for £6 one would get less than 40 lbs., whereas the answer obtained was $42\frac{54}{113}$ lbs. In fact, the price per lb. depends on the size of the cake: two 20 lb. cakes cost £5/13/-, but one 40 lb. cake costs £6/2/6. We made the simple assumption that the total cost is **proportional** to the weight.

Such assumptions are made in order to get an approximate answer, but they may not be in accordance with facts.

§81. The middle step (reduction to the unit) should presently be omitted, and the answer written down at once. Thus in Example 1, the distance traversed in 3 hours is $\mathbf{\frac{3}{10}}$ of that traversed in 10 hours; or $324 \times \mathbf{\frac{3}{10}}$ miles. Again, in Example 2, the amount of cake bought for 240 sixpences is $\mathbf{\frac{240}{113}}$ of that bought for 113 sixpences; or $20 \times \mathbf{\frac{240}{113}}$ lbs.

In each case, then, we have to multiply by a certain fraction. Remember that the process will mean an increase or a decrease according as this multiplying fraction is greater or less than 1: as common sense will show whether to look for an increase or a decrease, there should be no danger of having the fraction "the wrong way up."

Inverse Variation

§82. *Example* 3. If 200 men coal a ship in 10 hours, how long will it take 250 men to coal the same ship?

200 men coal the ship in 10 hours.

∴ 1 man coals " " 10×200 hours.

∴ 250 men coal " " $\frac{10 \times 200}{250}$ hours

$= 8$ hours.

In Example 3 again the middle statement is absurd, and it is preferable to write down the answer immediately.

Thus: **more** men, therefore **shorter** time, therefore multiply by a fraction **less** than 1.

As 200 men coal ship in 10 hours.

$\therefore$ 250 men „ „ $10 \times \frac{200}{250}$ hours.

EXERCISE VII a

In each case the rate is supposed to remain the same.

1. If 10 men can mow a field in 1 day, how long will it take (1) 1 man, (2) 15 men, (3) 25 men?

2. If 10 litres of wine cost 35 francs, what will 36 cost?

3. If 5 sheep cost £16. 5*s*., how much will 12 cost? How many can be bought for £22. 15*s*.?

4. A party of 12 men have provisions for a week. How long would the provisions feed 14 men?

5. If a car goes 45 yds. in 3 seconds, how far does it go in $4\frac{4}{5}$ secs.? How long does it take going 120 yds.?

6. If 20 men take 12 days to build a certain wall, how many days would 15 men take? How many men would finish it in 10 days?

7. If a man earns £2. 16*s*. 8*d*. in 17 days, how much can he earn in 21 days? In how many days can he earn £2. 10*s*.?

8. If 14 lbs. of tea cost 18*s*. 8*d*., how much will 20 lbs. cost? How much can be bought for 14*s*. 8*d*.?

9. If 35 men have rations for 20 days, how long will the food last if a reinforcement of 15 men arrives?

10. How many eggs at 14 for a shilling can be bought for £1. 2*s*. 6*d*.? How much will 175 eggs cost?

11. If 28 cows cost £308, what will 11 cost? How many can be bought for £209?

12. A steamer crosses from Milford to Cork in 10 hours, steaming at 12 knots. How long would she take at (1) 6 knots, (2) 10 knots, (3) 8 knots?

13. A garrison of 1200 men has provisions for 60 days; how long will they last if the garrison (1) is reduced to 600 men; (2) reinforced by 800 men?

14. If 8 lbs. of tea cost 13*s.* 4*d.*, how much will 10 lbs. cost? How many lbs. cost 28*s.* 4*d.*?

15. If a man earns £4. 19*s.* 8*d.* in 23 days, how much will he earn in 10 days? In how many days will he earn £2. 16*s.* 4*d.*?

16. How long will 27 men take to do the same job that 30 men do in 18 days?

17. If a motor travels 15 miles in 25 minutes, how far does it go in $1\frac{1}{2}$ hours? How long will it take to go 24 miles?

18. How many eggs at 10*d.* a dozen can be bought for 17*s.* 1*d.*? How much will 138 cost?

19. If a man walks from Winchester to Salisbury in 6 hours, walking at 4 miles an hour, how long would he take if he (1) walked at 3 miles an hour? (2) bicycled at 10 miles an hour? (3) bicycled at 15 miles an hour?

20. If 13 yds. of silk cost £3. 2*s.* 10*d.*, how much will 31 yds. cost? How much can be bought for £3. 12*s.* 6*d.*?

21. If 14 francs are worth 11*s.* 1*d.*, what is the value of 1200 fr.?

22. If 10 gallons of ink just supply a school of 240 for 75 days, how long will it last if the school is increased to 250?

23. If 7 lbs. of tea cost 16*s.* 11*d.*, how much will 10 lbs. cost? How many lbs. can be bought for 10*s.* $10\frac{1}{2}d.$?

24. If a workman earns £1. 17*s.* 7*d.* in 11 days, how much can he earn in 14 days? In how many days can he earn £1. 10*s.* 9*d.*?

25. A garrison has supplies for 200 days. How long will they last if the men are given 1/2 rations? 3/4 rations?

26. If a car goes $3\frac{1}{2}$ miles in 14 minutes, how far does it go in 25 minutes? How long will it take to go $19\frac{1}{4}$ miles?

27. If a train takes $3\frac{1}{2}$ hours to complete a certain journey at 45 miles per hour, what time will it take at 42 miles per hour? At what rate must it travel to complete the journey in $4\frac{1}{2}$ hrs.?

28. How many things at $10\frac{1}{2}$*d.* a dozen can be bought for 2*s.* 11*d.*? How much do 300 cost?

29. If the duty on 1 cwt. of tobacco is £20. 15*s.* 4*d.*, what is the duty on 74 lbs.?

30. A garrison of 600 men has supplies for 40 days. How long will the supplies last if the garrison (1) is reduced to 500 men, (2) increased to 800 men?

31. If 120 dollars are worth £25, what is the value of 27 dollars?

32. How many books $1\frac{3}{4}$ inches thick can be placed in a shelf which holds 44 books each $2\frac{5}{8}$ inches thick?

33. If the scale of a map is 25 miles to 4 inches, what distance on the map is represented by $2\frac{1}{2}$ inches? What length represents a distance of 65 miles?

34. A rope is wound 48 times round a drum whose circumference is 6 ft. 3 in. How many turns will there be if it is wound on a drum whose circumference is 4 ft. 2 in.?

§83. *Example* 4. If 100 miners working 9 hours a day win 500 tons of coal in 5 days, how many miners working 8 hours a day are needed to win 960 tons of coal in 6 days?

Fewer hours a day mean more men, ∴ multiply by $\frac{9}{8}$.

More coal means more men, ∴ multiply by $\frac{960}{500}$.

More days mean fewer men, ∴ multiply by $\frac{5}{6}$,

$$\therefore \text{ number of miners required} = 100 \times \frac{9}{8} \times \frac{960}{500} \times \frac{5}{6} = 180.$$

EXERCISE VII b

1. If 3 men can earn £8 in 12 days,

(1) In how many days will 9 men earn £4?

(2) How long will it take 6 men to earn £40?

(3) How much will 12 men earn in 2 days?

(4) How many men will earn £24 in 36 days?

2. 8 men working 8 hours a day earn £8 in 8 days.

(1) What do 8 men earn in 4 days of 12 hours each?

(2) How many hours a day must they work to earn £6 in 12 days?

(3) In how many days would they earn £16 working 10 hours a day?

3. A besieged town has provisions for 100 days. After 20 days 1/3 of the inhabitants escape and the remainder are put upon half-rations. How long will the provisions last?

4. If 6 cats eat 6 rats in 6 days, how many rats will 12 cats eat in 18 days? In how many days will 4 cats eat 12 rats?

5. A garrison of 10,000 men in a besieged town has provisions to last 28 weeks at the rate of 18 oz. per man per day. Just before the siege starts, they are reinforced by 4000 men, and the rations are cut down to 15 oz. per man per day. How long will their provisions last now?

6. If, by travelling $7\frac{1}{2}$ hours a day, I pass over 375 miles in 10 days, how many hours a day must I travel in order to accomplish 210 miles in 9 days?

7. If the freight on 5 tons 13 cwts. 84 lbs. of goods amounts to £6. 12*s.* $8\frac{1}{2}$*d.* for 350 miles, what is the rate per ton for 100 miles?

8. If 160 horses consume a stack of hay 20 ft. long, 11 ft. 3 in. broad, and 31 ft. 6 in. high, in 9 days, for how many days will a stack 15 yards long, 5 ft. broad, and 14 ft. high, supply 80 horses?

9. If 500 lbs. of mutton cost £17. 3*s.* 9*d.*, what will 600 lbs. cost when the price of mutton has risen $1\frac{1}{4}$*d.* a lb.?

10. A man experimenting with fuel found that when he was using only coal 5 tons lasted 7 weeks, but when he was using only coke 4 tons lasted 6 weeks. If coal costs 28*s.* per ton and coke 26*s.* per ton, how much money would he save during a year by using the cheaper of these two kinds of fuel instead of the other?

11. *A* can run 100 yards in $10\frac{4}{5}$ secs., and *B* in 11 secs.; which will win if *B* has 2 yards start?

12. A ship's chronometer is known to lose 5 seconds every 15 hours. The ship sails at noon and arrives at her destination in 7 days 15 hours. What time will the chronometer show immediately on her arrival, if it is right when she starts?

13. A clock is set right at 9 a.m. every morning. At 10.45 a.m. (correct time) the clock is observed to read 10.33 a.m.; what will it read at 9 a.m. next morning?

14. At noon on December 15th a clock showed exactly right time. At 6 a.m. on December 18th it showed 6.5 a.m. What was the time to the nearest second shown by the clock at 6 p.m. December 16th, correct time?

15. A chronometer which gains 17 seconds per hour was 12 minutes 45 seconds slow at 8 a.m. on July 30th; on what date and at what time will it indicate the right time?

16. A watch which gains regularly is set right at midday on Monday. On the following Saturday at noon it is found to have gained 5 minutes. What will be the time by it at 4 p.m. on the following Monday?

17. I put my clock 5 minutes fast at 10 p.m. on Sunday, and at 10 p.m. on the following Sunday I find it 3 minutes slow. When was it right?

18. A watch which loses regularly was found to be 3 minutes fast at 10 a.m. on Ap. 4th. At 3 p.m. on Ap. 9th it was 2 minutes slow. When did it show the correct time, and what was the time by it at 9 p.m. on Ap. 7th?

19. A chronometer was 4 mins. fast at 11 a.m. on March 2nd and one minute slow at noon on March 18th. Find when it was showing the correct time and also what time it showed on March 28th at noon.

EXERCISE VII c

Can the following examples be solved by the unitary method? If not, give your reasons.

1. A stone dropped into a well 16 ft. deep reaches the bottom in 1 second. Would it take 2 secs. if the well were 32 ft. deep?

2. 15 sheep can be grazed in a field 100 yds. square. How many sheep can be grazed in a field 200 yds. square?

3. The diameter of a halfpenny is 1″. What is the diameter of one penny?

4. A camel can carry a load of a cwt. 20 miles in a day. How far can it carry a load of 15 tons in a day?

5. The "Hope" diamond, weighing 48 carats, has been valued at £24,000. If it were cut into 10 equal diamonds, what would each one be worth?

6. A boy, when ten, weighed 5 stone; when twenty he weighed 10 stone. What will he weigh when he is sixty?

CHAPTER VIII

DECIMAL FRACTIONS

Notation

§ 84. Consider the meaning of 653.

The **3** means 3 units. The **5** does not mean 5 units, but 5 tens; the **6** does not mean 6 units, but 6 hundreds. The meaning of each figure depends not only on the figure, but also on the **position** of the figure. The meaning of 653 may be shown thus:

Hundreds	Tens	Units
6	5	3

In the series—hundreds, tens, units, each denomination is $\frac{1}{10}$ of its neighbour on the left. We could continue the series as follows:

Hundreds, tens, units, tenths, hundredths, etc.,

for a tenth is $\frac{1}{10}$ of a unit, a hundredth is $\frac{1}{10}$ of a tenth, etc.

Suppose that we want to express

6 hundreds 5 tens 3 units 2 tenths 7 hundredths 8 thousandths we could write this in a tabular form:

Hundreds	Tens	Units	**Tenths**	**Hundredths**	**Thousandths**
6	5	3	2	7	8

This is written more simply as follows:

6 5 3 · 2 7 8

Notice the dot between the 3 and the 2. Immediately to the left of the dot are the units; immediately to the right are the tenths. This dot is called the **point**; we read the above notation thus:

Six five three **point** two seven eight.

It means $653 + \frac{2}{10} + \frac{7}{100} + \frac{8}{1000}$.

It is called a **decimal fraction,** or a **decimal** (from Latin *decimus* = tenth).

If we have to express 6 hundreds and 3 tens (the units missing) we write 63**0**; the **0** "keeps the place" for the missing units. Similarly, if we want to write 7 tenths 5 thousandths we write ·7**0**5; the **0** keeps the place for the missing hundredths.

Again, $\frac{9}{100} + \frac{3}{1000}$ would be written ·**0**93; the **0** represents the missing tenths. If we omitted this **0** and wrote ·93, this would mean $\frac{9}{10} + \frac{3}{100}$.

On the other hand, there is no reason to retain the **0** in ·4**0**; it does not affect the meaning of the 4. It is no more useful than would be the **0** in **0**12 (twelve).

Note, however, that when there are no figures to the left of the point, it is very usual to insert a **0**: thus **0**·17 instead of ·17.

EXERCISE VIII a (oral)

Read off as decimals:

	Ten-thousands	Thousands	Hundreds	Tens	Units	Tenths	Hundredths	Thousandths	Ten-thousandths
1.						9			
2.						6			
3.						5			
4.					2	3			
5.					1	7			
6.					4	5			
7.				2	4	6	2		
8.			1	2	7	4	9		
9.				1	9	4	7	2	
10.			1		9	4	7	2	
11.			1		9	4		2	
12.			1		9			2	
13.			1					2	
14.				2			7		1
15.	1		3		2	5		6	
16.		3		6		4	9		
17.		3	0	6		4	9		
18.			9		2	6		1	

EXERCISE VIII b (oral)

Read off as decimals:

1. Three tenths. **2.** Five hundredths.

3. Seven hundredths. **4.** One thousandth.

5. Six ten-thousandths. **6.** $\frac{7}{10}$.

7. 9/10. **8.** 1/10. **9.** 1/100.

10. 3/100. **11.** 7/1000. **12.** 5/1000.

13. $\frac{9}{10000}$. **14.** $\frac{1}{10}+\frac{3}{100}$. **15.** $\frac{3}{10}+\frac{2}{100}$.

16. $\frac{7}{10}+\frac{9}{100}$. **17.** $\frac{3}{10}+\frac{4}{100}+\frac{9}{1000}$.

18. $30+9+\frac{4}{10}+\frac{6}{100}$. **19.** $300+9+\frac{4}{10}+\frac{6}{100}$.

20. $7+\frac{6}{100}$. **21.** $9+\frac{3}{100}$.

22. $\frac{5}{10}+\frac{7}{1000}$. **23.** $\frac{6}{10}+\frac{9}{1000}$.

24. $\frac{1}{10}+\frac{4}{1000}+\frac{5}{10000}$. **25.** $60+\frac{1}{100}+\frac{5}{1000}$.

26. $4+\frac{3}{100}+\frac{5}{10000}$. **27.** $50+\frac{4}{100}+\frac{7}{10000}$.

Express each of the following decimals as the sum of fractions with denominators 10, 100, 1000, etc. Thus

$$2{\cdot}467 = 2+\frac{4}{10}+\frac{6}{100}+\frac{7}{1000}.$$

28. ·3. **29.** 2·3. **30.** 2·34.

31. ·04. **32.** 3·04. **33.** 14·007.

34. 100·605. **35.** ·003. **36.** ·0001.

37. 4·1237. **38.** 400·70. **39.** 400·07.

40. ·340507. **41.** 41·20406. **42.** 5·7204.

43. 14·02701. **44.** 0·7035. **45.** 0·0604.

MULTIPLICATION AND DIVISION BY POWERS OF 10

§ 85. The chief advantage attaching to the use of decimal notation is the ease with which a decimal can be multiplied or divided by 10, 100, 1000, ..., i.e. by 10, 10^2, 10^3,..., the **powers** of 10.

Let us multiply 278·56 by 10.
6 hundredths × 10 = 6 tenths;
5 tenths × 10 = 5 units;
8 units × 10 = 8 tens;
7 tens × 10 = 7 hundreds;
2 hundreds × 10 = 2 thousands.

		Ten-thousands	Thousands	Hundreds	Tens	Units	Tenths	Hundredths
Thus	278·56 =			2	7	8	5	6
	278·56 × 10 =		2	7	8	5	6	
Similarly	278·56 × 100 =	2	7	8	5	6		

To multiply by 10, therefore, is the same as to shift the number multiplied one place to the left in the table; to multiply by 100 shifts the multiplicand two places to the left.

Thus 278·56 × 10 = 2785·6,

278·56 × 100 = 27856·

etc.

We may say that to shift the figures one place to the *left* on the table is equivalent to shifting the decimal point one place to the *right*; and so on. When the point arrives at the right-hand end of the number, as in 27856·, we have a whole number, and may omit the point: 27856.

To multiply by 10 once more would again move the point one place to the right; but now we must put in a **0** to keep the units' place:

278·56 × 1000 = 27856**0**· = 278560.

§ 86. Now consider 278·56 ÷ 10, or $\frac{278{\cdot}56}{10}$.

2 hundreds ÷ 10 = 2 tens;
7 tens ÷ 10 = 7 units;
8 units ÷ 10 = 8 tenths;
5 tenths ÷ 10 = 5 hundredths;
6 hundredths ÷ 10
= 6 thousandths.

	Hundreds	Tens	Units	Tenths	Hundredths	Thousandths	Ten-thousandths
Thus 278·56 =	2	7	8	5	6		
278·56 ÷ 10, or $\frac{278{\cdot}56}{10}$ =		2	7	8	5	6	
And 278·56 ÷ 100, or $\frac{278{\cdot}56}{100}$ =			2	7	8	5	6

In decimal notation:

$$278{\cdot}56 \div 10, \text{ or } \frac{278{\cdot}56}{10} = 27{\cdot}856,$$

$$278{\cdot}56 \div 100, \text{ or } \frac{278{\cdot}56}{100} = 2{\cdot}7856,$$

etc.

We see that to *divide* by 10, 100, 1000, etc. necessitates shifting the point one, two, three, etc. places to the *left*.

Continuing: $\frac{278{\cdot}56}{1000} = {\cdot}27856,$

$$\frac{278{\cdot}56}{10000} = {\cdot}\mathbf{0}27856.$$

The **0** has been put in to keep the tenths' place vacant.

$$\frac{278{\cdot}56}{100000} = {\cdot}\mathbf{00}27856,$$

etc.

A whole number is divided by a power of 10 in exactly the same way. Thus

$$\frac{4720}{100} = 47{\cdot}20 \text{ or } 47{\cdot}2.$$

In dividing by a power of 10, *never cancel.*

EXERCISE VIII c (oral)

A. Multiply each of the following by 10, 100, 1000:

1. 2·9. **2.** 1·65. **3.** 3·431. **4.** 7·996, etc.

B. Divide each of the above by 10, 100, 1000.

C. Give the value of:

1. $3·7 \times 100$. **2.** $5·3 \div 10$. **3.** $76 \div 100$.
4. $9·15 \times 10$. **5.** $3·01 \times 10$. **6.** $1·05 \div 100$.
7. $15·6 \div 1000$. **8.** $0·0721 \times 1000$. **9.** $6·073 \div 100$.
10. $401·2 \times 100$. **11.** $11·92 \div 1000$. **12.** $180·5 \div 100$.
13. $·0821 \times 1000$. **14.** $0·1385 \div 10$. **15.** $88·01 \times 100$.
16. $7·011 \div 1000$. **17.** $60·03 \div 10$. **18.** $476·9 \div 1000$.
19. $0·1549 \times 1000$. **20.** $3143 \div 1000$. **21.** $7·447 \times 100$.
22. $0·02518 \times 10000$. **23.** $·00969 \div 100$.
24. $3·601 \times 10000$. **25.** $0·00735 \div 10$. **26.** $6543·2 \div 100$.
27. $12·345 \times 1000$. **28.** $21·007 \div 1000$. **29.** $7·6304 \div 1000$.
30. $3000·3 \times 100$. **31.** $0·4989 \div 1000$. **32.** $169·05 \times 100$.
33. $30·403 \div 100$. **34.** $0·08901 \div 1000$.
35. $0·70001 \times 10000$. **36.** $50·401 \div 1000$.
37. $73569·1 \div 10000$. **38.** $4·3124 \times 10000$.
39. $0·972561 \div 1000$. **40.** $114·002 \times 10000$.
41. $\frac{213·57}{100000}$. **42.** $\frac{·09}{10000}$, etc.

ADDITION OF DECIMALS

§ **87.** The process is the same as in the addition of whole numbers. Write down the numbers to be added, keeping *point under point*; add up, and carry if necessary in the ordinary way.

Thus, to find the sum of 274·3, 0·078, 107·42, 2·965.

Say 5, **13**; write down 3 (thousandths); there remain 10 thousandths = 1 hundredth; therefore carry 1. Continuing: 7, 9, **16**; write down 6 and carry 1; 10, 14, **17**; write down 7 and carry 1, etc.

$$\begin{array}{r} 274·3 \\ 0·078 \\ 107·42 \\ 2·965 \\ \hline 384·763 \end{array}$$

EXERCISE VIII d

Find the sum of each of the following:

1.	**2.**	**3.**	**4.**
23·1	30·157	1·677	0·0508
0·415	5·786	0·5095	2·77
7·02	0·04	28·35	11·908
304·297	19·2	7·9	70·8903

5. 0·568, 19·38, 1·163, 0·2218.
6. 5·607, 38·72, 605·5, 0·7496.
7. 46·09, 3·307, 0·1013, 1·169.
8. 0·2076, 1·986, 0·0431, 0·2917, 5·54.
9. 8·41, 47·24, 2·2375, 6·71, 0·9572.
10. 87·5, 8·75, 0·875, 0·0875, 0·00875.
11. 7·57, 0·757, 14·93, 1·493, 0·1493.
12. 2·0814, 2·0847, 2·0913, 2·0787, 2·0869.
13. 18·98, 1·694, 0·224, 246·4, 17·935.
14. 23·066, 2·507, 2·4409, 202·26, 21·627.
15. 0·0734, 0·9094, 0·0987, 0·00765, 0·4987.

SUBTRACTION OF DECIMALS

§ 88. Write the number to be subtracted under the other in the usual way, *point under point*; subtract. If there is no figure in the upper line to subtract from, supply a zero.

Thus, to find 912·7 – 403·642.
(The added zeros need not be written down, after a little practice.)

912·7**00**
403·642
509·058

EXERCISE VIII e

Simplify:

1. 4·302 – 3·479.
2. 2·145 – 0·786.
3. 9·5 – 7·867.
4. 3·09 – 2·8654.
5. 40·4 – 4·047.
6. 5·87 – 3·892.
7. 11·215 – 6·9431.
8. 98·65 – 9·752.
9. 0·464 – 0·07546.
10. 1·000037 – 0·7001.
11. 1·89098 – 0·9006.
12. 7·509 – 5·7455.
13. 0·8645 – 0·00786.
14. 0·12122 – 0·112774.
15. 0·098765 – 0·0856789.
16. 10·00605 – 8·0679.

Subtract 3·762584 from each of the following:

17.	5·2969.	**18.**	7·000935.	**19.**	10.
20.	11·08431.	**21.**	4·762473.	**22.**	10·111111.
23.	9·9999.	**24.**	12·050511.	**25.**	17·1428.

EXERCISE VIII f (oral)

Simplify:

1.	1·4 + 2·03.	**2.**	0·6 + 3·0.	**3.**	3·02 + ·007.
4.	60 + 102·5.	**5.**	4·21 + 0·3	**6.**	2·4 − 0·2.
7.	4·3 − 1·2.	**8.**	2 − 1·6.	**9.**	3 − 2·4.
10.	4 − 0·1.	**11.**	14 − 7·5.	**12.**	3·2 − 0·01.
13.	1 − ·001.	**14.**	6 − ·75.	**15.**	8 − 1·5.
16.	10 − 0·2.	**17.**	4 − 3·7.	**18.**	1 − ·05.
19.	2 − 1·06.	**20.**	6 − 2·3.	**21.**	12 − 6·01.
22.	100 − ·02.	**23.**	7 − 6·001.	**24.**	17 − 10·50.
25.	2 − 0·06.	**26.**	1·0 − 0·1.	**27.**	·02 + 7·2.
28.	13 − 4·3.	**29.**	7·1 − 6·9.	**30.**	6·2 − 4·1.
31.	7·1 − 2·3.	**32.**	13 − 6·4.	**33.**	·25 + 7·2.
34.	·10 + 12·4.	**35.**	21 − 6·01.	**36.**	7·0 − 1·6.
37.	9·1 − 1·6.	**38.**	4·2 + 5·9.	**39.**	5·1 + 6·92.
40.	14·6 − 10·8.	**41.**	5·01 + ·6.	**42.**	7·1 − 1·04.
43.	12·1 − 5·7.	**44.**	11·06 − 2·9.	**45.**	10·1 − 1·2.
46.	11·7 − 5·8.	**47.**	12·6 − 7·8.	**48.**	10·3 − 3·5.
49.	·7 + 14·8.	**50.**	10·2 + 10·9.	**51.**	16·5 − 8·7.
52.	11·4 − 4·1.	**53.**	1·4 + 6.	**54.**	6 − 4·5.
55.	15·5 − 7.	**56.**	7·2 − 1·02.	**57.**	17 − 10·2.
58.	4·3 + 2 − 5·2.	**59.**	2·7 + 3 − 4·1.	**60.**	400 − 200·1.
61.	21·4 − 3·5.	**62.**	3·1 − ·09.	**63.**	0·3 + 4·09.
64.	3·09 + 4·5.				

EXERCISE VIII g

Find the value of:

1.	37·21 − 5·041 + 0·5730.	**2.**	32·116 − 26·83 + 313·6.
3.	0·5476 + 4·681 − 3·752.	**4.**	1·04961 − 0·053 + 1·75.
5.	4·4371 − 3·654 + 2·073.	**6.**	73·46 − 9·4315 + 2·051.
7.	4·3215 + 7·778 − 9·63.	**8.**	1·01007 − 0·1007 + 3·007.
9.	6·3051 + 0·077 − 5·146.	**10.**	135·69 − 72·4681 + 1·87.

MULTIPLICATION OF DECIMALS

§ 89. **Multiplier an integer not greater than 12.** To find the product of 0·342 and 4.

2 thousandths × 4 = 8 thousandths, ∴ write 8 under 2; and so on, keeping *point under point*.

$$\begin{array}{r} 0{\cdot}342 \\ 4 \\ \hline 1{\cdot}368 \end{array}$$

EXERCISE VIII h (oral)

Simplify:

1. ·2 × 4.	**2.** ·07 × 9.	**3.** ·12 × 2.
4. ·003 × 5.	**5.** ·09 × 7.	**6.** ·5 × 8.
7. ·03 × 5.	**8.** ·008 × 11.	**9.** ·13 × 3.
10. ·04 × 6.	**11.** ·9 × 12.	**12.** ·6 × 9.
13. ·05 × 6.	**14.** ·009 × 12.	**15.** ·2 × 4.
16. 1·3 × 8.	**17.** 1·1 × 7.	**18.** ·7 × 11.
19. ·05 × 7.	**20.** ·0011 × 12.	**21.** ·003 × 12.
22. ·07 × 9.	**23.** 1·2 × 5.	**24.** ·8 × 11.
25. ·06 × 8.	**26.** 1·4 × 10.	**27.** 1·2 × 12.

EXERCISE VIII i

Simplify:

1. 2·55 × 2.	**2.** 80·6 × 5.	**3.** ·825 × 7.
4. ·0738 × 8.	**5.** ·142 × 3.	**6.** 25·9 × 6.
7. ·682 × 11.	**8.** ·0162 × 9.	**9.** 6·19 × 4.
10. ·0922 × 7.	**11.** 2·93 × 12.	**12.** ·0492 × 11.

§ 90. **Multiplication. General case.** As a preliminary, notice that if one of two factors is multiplied by any number, say 10, and the other factor is divided by that same number, the product is unaltered. Thus

$$60 \times 20 = 6 \times 200 = 600 \times 2 = 1200.$$

Similarly, $\cdot 57 \times 40 = 5{\cdot}7 \times 4$,

for the multiplier 40 has been divided by 10, while the multiplicand ·57 has been multiplied by 10.

Again, $9{\cdot}2 \times \cdot 076 = \cdot 092 \times 7{\cdot}6$,

for the ·076 has been multiplied by 100, while the 9·2 has been divided by 100.

EXERCISE VIII j (oral)

Fill in the missing factors below. First discover the number by which the one factor has been multiplied (or divided) and then alter the other factor accordingly.

1. $\cdot 23 \times \cdot 52 = ? \times 5\cdot 2$. **2.** $\cdot 37 \times \cdot 093 = ? \times 9\cdot 3$.

3. $43 \times \cdot 54 = ? \times 5\cdot 4$. **4.** $\cdot 54 \times 65 = ? \times 6\cdot 5$.

5. $\cdot 068 \times \cdot 026 = ? \times 2\cdot 6$. **6.** $\cdot 0074 \times 6700 = ? \times 6\cdot 7$.

7. $85 \times \cdot 078 = ? \times 7\cdot 8$. **8.** $99 \times \cdot 39 = ? \times 3\cdot 9$.

9. $\cdot 25 \times 720 = ? \times 7\cdot 2$. **10.** $\cdot 36 \times 830 = ? \times 8\cdot 3$.

11. $\cdot 0042 \times 44 = ? \times 4\cdot 4$. **12.** $\cdot 056 \times \cdot 85 = ? \times 8\cdot 5$.

§ 91. It will be noticed that in each of the sums of the last exercise, the multiplier has been changed to a number of a certain form, namely to a form in which the point follows the left-hand figure. Such a number is said to be **in standard form**; thus 4·7, 3·26, 7·192 are all in standard form.

In multiplication of decimals, the first step is to change the multiplier to standard form. This is done by moving the point so many places to left or right as the case may be. But, in order to keep the product unchanged, we must now change the multiplicand by moving its point the same number of places *in the opposite direction.*

Example. Find the value of 26·53 × ·00872.

$$\overleftarrow{26}\cdot 53 \times \cdot 00\overrightarrow{872} = \cdot 02653 \times 8\cdot 72$$

I.

```
 ·02653
8·72
 ·21224
 ·018571
 ·0005306
 ·2313416
```

First multiply by the 8, as in § 89, writing the figures of the product under the corresponding figures of the multiplicand and *point under point.* This first line fixes the position of the point. In each of the successive multiplications, the right-hand figure is written one place to the right, as in multiplication of whole numbers. In all partial products after the first, the points and added zeros may be left out, as in II.

II.

```
 ·02653
8·72
 ·21224
    18571
     5306
 ·2313416
```

It may be convenient to interchange multiplier and multiplicand: thus, in finding the value of ·3312 × 98·76, multiply by ·3312, as 3 occurs twice.

EXERCISE VIII k (oral)

Simplify:

1. $200 \times \cdot 3$. **2.** $30 \times \cdot 07$. **3.** $\cdot 4 \times \cdot 3$.

4. $\cdot 05 \times 70$. **5.** $6 \times \cdot 04$. **6.** $\cdot 007 \times 80$.

7. $8 \times \cdot 4$. **8.** $90 \times \cdot 08$. **9.** $\cdot 2 \times 50$.

10. $\cdot 3 \times \cdot 9$. **11.** $\cdot 04 \times \cdot 5$. **12.** $\cdot 5 \times \cdot 9$.

13. $\cdot 06 \times \cdot 06$. **14.** $7 \times \cdot 02$. **15.** $80 \times \cdot 06$.

16. $9 \times \cdot 07$. **17.** $(\cdot 1)^2$. **18.** $(\cdot 2)^2$.

19. $(\cdot 3)^2$. **20.** $(\cdot 4)^2$. **21.** $(\cdot 6)^2$.

22. $(\cdot 8)^2$. **23.** $(1\cdot 1)^2$. **24.** $(1\cdot 2)^2$.

25. $(\cdot 01)^2$. **26.** $(\cdot 03)^2$. **27.** $(\cdot 05)^2$.

28. $(\cdot 09)^2$. **29.** $(\cdot 11)^2$. **30.** $(\cdot 12)^2$.

31. $(\cdot 1)^3$. **32.** $(\cdot 2)^3$. **33.** $(\cdot 3)^3$.

§ 92. **Rough Check.** Remember that in any decimal calculation the worst mistake you can make is to misplace the point; to place it wrongly even by one place makes the answer 10 times too large or too small, and quite worthless.

It is therefore well to begin by forming a rough estimate of the answer, taking one figure from each factor.

Example. $\cdot 225 \times \cdot 374$.

For a rough estimate, we will take $\cdot 2 \times \cdot 4$. Notice that $\cdot 4$ is nearer to $\cdot 374$ than is $\cdot 3$. The whole sum may be shown thus:

Product roughly $= \cdot 2 \times \cdot 4 = \cdot 08$

$$\overleftarrow{\cdot 3}74 \times \cdot\overrightarrow{22}5 = \cdot 0374 \times 2\cdot 25$$
$$= \cdot 08415$$

$$\begin{array}{r} \cdot 0374 \\ 2\cdot 25 \\ \hline \cdot 0748 \\ 748 \\ 1870 \\ \hline \cdot 084150 \end{array}$$

Why may the final zero of the product be dropped?

EXERCISE VIII l

Find the value of each of the products in Ex. VIII j.

Simplify:

1. $9{\cdot}4 \times 20$. **2.** $6{\cdot}5 \times 500$.

3. $470 \times 0{\cdot}6$. **4.** $3{\cdot}6 \times 0{\cdot}08$.

5. $0{\cdot}88 \times 0{\cdot}3$. **6.** $0{\cdot}052 \times 70$.

7. $0{\cdot}29 \times 700$. **8.** $0{\cdot}013 \times 0{\cdot}9$.

9. $0{\cdot}0071 \times 0{\cdot}04$. **10.** $510 \times 0{\cdot}071$.

11. $24 \times 0{\cdot}0201$. **12.** $4{\cdot}6 \times 43{\cdot}02$.

13. $0{\cdot}87 \times 480$. **14.** $0{\cdot}062 \times 0{\cdot}205$.

15. $0{\cdot}0033 \times 650{\cdot}7$. **16.** $0{\cdot}099 \times 10{\cdot}97$.

17. $9{\cdot}8 \times 0{\cdot}0803$. **18.** $730 \times 0{\cdot}00609$.

19. $4{\cdot}296 \times 0{\cdot}0843$. **20.** $\cdot 7261 \times 75{\cdot}4$.

21. $89{\cdot}27 \times 0{\cdot}0469$. **22.** $40{\cdot}21 \times 30{\cdot}08$.

Division of Decimals

§ 93. **Divisor an integer not greater than 12.** To divide 62·68 by 4.

$$4\overline{)62{\cdot}68}$$
$$15{\cdot}67$$

4 into 6, **1**; 4 into 22, **5**; 4 into 26 *tenths*, **6** tenths, 4 into 28 hundredths, **7** hundredths.

In fact, divide as in short division of integers, and bring down the point when you come to it. The result is

$$62{\cdot}68 \div 4, \text{ or } \frac{62{\cdot}68}{4} = 15{\cdot}67.$$

It may happen that the last partial division is not exact; e.g. ·054 ÷ 8; 8 into 54, **6**; with 6 remaining. This means 6 thousandths; but we may add zeros at the tail of the dividend and call the remainder 60 ten-thousandths; 8 into 60, **7**; 8 into 40, **5**. These zeros need not be written down, except perhaps by beginners.

$$8\overline{)\cdot 05400}$$
$$\cdot 00675$$

Take the following case:

$$42 \div 4, \text{ or } \frac{42}{4} = 10{\cdot}5.$$

$$4\overline{)42}$$
$$10{\cdot}5$$

It is worth while to consider what this result means. In earlier work we were accustomed to say $42 \div 4 = 10$ with remainder 2. Suppose that we are dividing 42 pence among 4 children. Each may receive 10 pence apiece, and now there are 2 pence left over. But these 2 pence may be divided among the 4, and there will be $\frac{1}{2}d.$ more for each child. Thus $42d. \div 4 = 10\frac{1}{2}d.$, with no remainder.

In decimal notation, $42d. \div 4 = 10{\cdot}5d.$; for $\cdot 5d.$ means $\frac{5}{10}d. = \frac{1}{2}d.$

EXERCISE VIII m (oral)

Express as a single decimal:

1. $4{\cdot}8 \div 4$. **2.** $\frac{3{\cdot}6}{3}$. **3.** $\frac{3{\cdot}6}{2}$. **4.** $\frac{3{\cdot}6}{9}$.

5. $3{\cdot}6 \div 12$. **6.** $\frac{\cdot 48}{2}$. **7.** $\cdot 48 \div 3$. **8.** $\frac{\cdot 48}{6}$.

9. $\cdot 48 \div 8$. **10.** $\frac{\cdot 48}{12}$. **11.** $4{\cdot}2 \div 2$. **12.** $\frac{4{\cdot}2}{3}$.

13. $4{\cdot}2 \div 6$. **14.** $\frac{4{\cdot}2}{7}$. **15.** $4{\cdot}2 \div 4$. **16.** $\frac{4{\cdot}2}{5}$.

17. $\frac{\cdot 1}{2}$. **18.** $\frac{\cdot 01}{2}$. **19.** $\frac{1}{2}$. **20.** $\frac{\cdot 1}{4}$.

21. $\frac{1}{4}$. **22.** $\frac{\cdot 03}{4}$. **23.** $\frac{3}{4}$. **24.** $\frac{30}{4}$.

25. 1/8. **26.** 3/8. **27.** 5/8. **28.** 7/8.

29. 1/5. **30.** 3/5. **31.** 12/5. **32.** 34/5.

Divide:

33. 20·5 by 5. **34.** 21·0 by 5. **35.** 31 by 5.

36. 62·4 by 4. **37.** 62 by 4. **38.** 1·25 by 8.

39. 71·3 by 2, 4, 5. **40.** 2·7 by 2, 3, 5, 9.

41. 0·21 by 2, 3, 5, 7. **42.** 6·3 by 2, 3, 5, 7, 9.

43. ·84 by 2, 3, 4, 7, 12. **44.** ·0035 by 2, 5, 7.

EXERCISE VIII n

Simplify:

1. $\frac{967}{4}$. **2.** $\frac{51{\cdot}6}{8}$. **3.** $\frac{{\cdot}204}{3}$. **4.** $\frac{{\cdot}0924}{6}$.

5. $\frac{6{\cdot}75}{5}$. **6.** $\frac{{\cdot}203}{8}$. **7.** $\frac{6{\cdot}43}{8}$. **8.** $\frac{307}{5}$.

9. $\frac{4{\cdot}83}{3}$. **10.** $\frac{{\cdot}732}{6}$. **11.** $\frac{706}{4}$. **12.** $\frac{3{\cdot}48}{6}$.

13. $\frac{{\cdot}0391}{8}$. **14.** $\frac{{\cdot}405}{5}$. **15.** $\frac{{\cdot}984}{12}$. **16.** $\frac{{\cdot}0561}{11}$.

§ 94. **Recurring decimals.** Consider the quotient 13·7 ÷ 3.

$$\begin{array}{l} 3\underline{)13{\cdot}7} \\ 4{\cdot}5666\ldots \end{array}$$

We find that however many zeros we supply after the 7, the process never ends; we have an unending succession of 6's. If we stop at any stage, e.g. 4·5666, we have a decimal which is nearly equal to 13·7 ÷ 3, but not quite. If we go further, we have a decimal more nearly equal to 13·7 ÷ 3, and so on; but we never obtain a decimal exactly equal to 13·7 ÷ 3. This fact is expressed by writing

$$13{\cdot}7 \div 3 = 4{\cdot}5\dot{6},$$

which is read "four point five six **recurring.**"

Take another example, namely, $\frac{23}{7}$. Here the figures that "recur" are 285714; accordingly we put a dot over the first and last of the recurring figures; thus

$$\begin{array}{l} 7\underline{)23} \\ 3{\cdot}285714\ 285714\ldots \end{array}$$

$$\frac{23}{7} = 3{\cdot}\dot{2}8571\dot{4}.$$

The result of a division sum must either terminate, or recur.

§ 95. **Results correct to a given number of places.** For practical purposes, we never need recurring decimals; what we generally need is a result correct to a specified number of places. For instance, if the final result is a sum of money expressed in pounds and decimals of a pound, anything less than $\frac{1}{1000}$ of a £ is unimportant, for $\frac{1}{1000}$ of a £ is almost a farthing $\left(\frac{1}{4}d. = \frac{1}{960}£\right)$, and nothing less than a farthing need be taken into account in the final answer. Now every thousandth of a £ is represented by 1 in the third place of decimals; thus £·103 means $\frac{1}{10}£ + \frac{3}{1000}£$; or roughly 2 shillings and 3 farthings, or $2/0\frac{3}{4}$. Accordingly, the final result of a money sum needs to be correct only to three places.

(Note however that this does not necessarily hold for a money result that occurs *before* the end of the sum. Suppose that we have obtained a "semi-final" result that has to be multiplied by 1000 to give the final result; the figure that finds its way into the 3rd place of decimals in the final result was in the 6th place in the semi-final, and 6 places would have to be retained in the semi-final.)

If it is required to find $23 \div 7$ correct to one place of decimals, it might be supposed that we are at liberty to stop as soon as we have obtained one place; thus $23 \div 7 = 3{\cdot}2\ldots$. But this is not sufficient. The next figure in the division is 8, for $23 \div 7 = 3{\cdot}28\ldots$. Now 28 is nearer to 30 than to 20; 3·30... is a more nearly accurate result than 3·20...; and we must say that the value of $23 \div 7$ correct to one place is 3·3. *To obtain a result correct to a given number of decimal places, always carry the calculation to one place further.*

If we have to write down the number with 2 places of decimals that is the nearest approach to 6·235, then 6·23 and 6·24 are equally close and there is no preference between them. In such a case, it is best to leave the 5 and refuse to curtail the result.

¶**Ex.** Remembering that $23 \div 7 = 3{\cdot}\dot{2}8571\dot{4}$, what is the value of $23 \div 7$ correct to 2, to 4, to 6, to 7 places of decimals?

EXERCISE VIII o

Express each of the following as a recurring decimal:

1. $\frac{1}{3}$.	**2.** $\frac{1}{6}$.	**3.** $\frac{1}{7}$.	**4.** $\frac{1}{9}$.	**5.** $\frac{1}{11}$.
6. 2/3.	**7.** 2/7.	**8.** 2/9.	**9.** 2/11.	
10. 3/7.	**11.** 3/11.	**12.** 4/7.	**13.** 4/9.	
14. 4/11.	**15.** 5/6.	**16.** 5/7.	**17.** 5/9.	
18. 5/11.	**19.** 6/7.	**20.** 6/11.	**21.** 7/9.	
22. 7/11.	**23.** 8/9.	**24.** 8/11.	**25.** 9/11.	

Express each of the following in decimals correct to 3 places:

26. $\frac{13}{3}$.	**27.** $\frac{13}{4}$.	**28.** $\frac{13}{7}$.	**29.** $\frac{13}{8}$.	**30.** $\frac{11}{9}$.
31. $\frac{29}{3}$.	**32.** $\frac{23}{9}$.	**33.** $\frac{23}{11}$.	**34.** $\frac{41}{5}$.	**35.** $\frac{52}{7}$.
36. $\frac{24}{11}$.	**37.** $\frac{71}{8}$.	**38.** $\frac{47}{12}$.	**39.** $\frac{29}{9}$.	**40.** $\frac{51}{8}$.
41. $\frac{126}{11}$.	**42.** $\frac{245}{6}$.	**43.** $\frac{75}{4}$.	**44.** $\frac{145}{12}$.	**45.** $\frac{37}{7}$.
46. $\frac{297}{8}$.	**47.** $\frac{162}{11}$.	**48.** $\frac{47}{7}$.	**49.** $\frac{259}{12}$.	**50.** $\frac{439}{11}$.
51. $\frac{90}{7}$.	**52.** $\frac{199}{9}$.	**53.** $\frac{359}{12}$.	**54.** $\frac{237}{11}$.	**55.** $\frac{350}{9}$.

§ 96. **Divisor a decimal in standard form.** It is required to divide 0·8775 by 3·25.

```
        0·27
3·25)0·8775
      ·650
      2275
      2275
```

The divisor 3·25 is only a little greater than the whole number 3. Just as in long division of integers we make a trial division: ·8 divided by 3. This gives the first figure in the quotient as **·2**.

We will write the quotient *over* the dividend, with *point over point*; thus we shall write the **2** over the 8. This gives us a start, and the rest of the division is performed in the ordinary way, as in division of integers.

$$0{\cdot}8775 \div 3{\cdot}25 = 0{\cdot}27.$$

Now let us find the quotient 2·08 ÷ 3·25. 2 divided by 3, "won't go"; write **0** over 2. 20 tenths divided by 3 gives **6** tenths; write point over point and **6** over 0. Multiply 3·25 by 6 (taking no notice of the point); there will be *four* figures in this product, so that we must imagine a zero after the 8. On subtracting we have 130, and then bring down the next figure of the dividend, which is a 0. 13 divided by 3 gives **4**, which turns out to be the final figure,

```
         0·64
3·25)2·080
       1·950
        1300
        1300
```

$$2·08 \div 3·25 = 0·64.$$

The advantage of writing the quotient over the dividend is that it recalls short division. Compare the sum worked above with the division 2·08 ÷ 3. In the one case the quotient is above and in the other it is below; otherwise, the process of obtaining the figures of the quotient and of placing the point is the same in both cases.

```
3)2·08
  0·69̇3̇
```

EXERCISE VIII p

To divide by a decimal of standard form

(Before working the following exercises, it may be well to go through them orally and make a rough estimate of the quotient in each case: thus

$$52·5 \div 2·5 = 52 \div 2 \text{ roughly} = 26.)$$

Simplify:

1. 52·5 ÷ 2·5. **2.** 15·21 ÷ 1·17.

3. 72·36 ÷ 1·44. **4.** 672·88 ÷ 6·47.

5. 7·3807 ÷ 2·3. **6.** 0·4472 ÷ 4·3.

7. 0·0288 ÷ 2·4. **8.** 0·0062 ÷ 2·5.

9. 0·651021 ÷ 3·207. **10.** 13·014 ÷ 2·41.

11. 264·708 ÷ 3·24. **12.** 1326 ÷ 4·25.

13. 748 ÷ 8·5. **14.** 28·782 ÷ 3·69.

15. 0·023328 ÷ 3·6. **16.** 0·032 ÷ 6·4.

17. 0·7704256 ÷ 9·28. **18.** 5·06016 ÷ 7·53.

19. $2{\cdot}76766 \div 3{\cdot}71$. **20.** $1{\cdot}024 \div 2{\cdot}56$.

21. $2 \div 1{\cdot}25$. **22.** $16 \div 2{\cdot}56$.

23. $0{\cdot}11 \div 1{\cdot}28$. **24.** $0{\cdot}12 \div 3{\cdot}75$.

25. $0{\cdot}02 \div 6{\cdot}25$. **26.** $1 \div 5{\cdot}12$.

§ 97. **Division. General case.** The method adopted is to modify the problem in such a way that the divisor may be of standard form.

Example. Evaluate $0{\cdot}00724 \div 0{\cdot}0892$, correct to 3 places.

In effect, we are dealing with a complex fraction $\frac{0{\cdot}00724}{0{\cdot}0892}$. We do not alter the value if we multiply numerator and denominator by 100; thus $\frac{0{\cdot}00724}{0{\cdot}0892} = \frac{0{\cdot}724}{8{\cdot}92}$: the denominator is now in standard form, and we proceed as in the preceding paragraph. Arrange the work as follows:

$$\frac{0{\cdot}00724}{0{\cdot}0892} = \frac{0{\cdot}724}{8{\cdot}92}.$$

Quotient roughly $= \frac{0{\cdot}72}{9} = 0{\cdot}08$.

Quotient $= \underline{0{\cdot}081}\ldots$ to 3 places.

```
          0·0811...
   8·92)0·7240
          7136
          ----
           1040
            892
           ----
           1480
            892
```

EXERCISE VIII q

Simplify:

1. $1{\cdot}9248 \div 0{\cdot}008$. **2.** $700727 \div 0{\cdot}029$.

3. $1024 \div 25{\cdot}6$. **4.** $0{\cdot}1292 \div 32{\cdot}3$.

5. $906{\cdot}5 \div 0{\cdot}185$. **6.** $0{\cdot}4496 \div 11{\cdot}24$.

7. $240{\cdot}204 \div 444$. **8.** $11{\cdot}9385 \div 0{\cdot}525$.

9. $197{\cdot}896 \div 232$. **10.** $1769{\cdot}08 \div 0{\cdot}47$.

11. $55{\cdot}5676 \div 17{\cdot}3$. **12.** $367{\cdot}848 \div 46{\cdot}8$.

13. $232{\cdot}379 \div 0{\cdot}0373$. **14.** $6601{\cdot}68 \div 76320$.

15. $516571 \div 75500$. **16.** $210{\cdot}15984 \div 466{\cdot}4$.

17. $0{\cdot}119385 \div 22{\cdot}74$. **18.** $91{\cdot}008 \div 379{\cdot}2$.

19. $0{\cdot}06828467 \div 0{\cdot}6781$. **20.** $542{\cdot}913 \div 0{\cdot}05271$.

Express as a decimal, correct to the number of places stated:

21. $\frac{132\cdot6248}{3\cdot08}$ (to 1 pl.). **22.** $\frac{0\cdot51}{6\cdot25}$ (to 3 pl.).

23. $\frac{10\cdot7358}{0\cdot174}$ (nearest integer). **24.** $\frac{101}{814}$ (to 2 pl.).

25. $\frac{20\cdot677}{0\cdot0713}$ (nearest 10). **26.** $\frac{87}{426\cdot8}$ (to 2 pl.).

27. $\frac{5\cdot42913}{527\cdot1}$ (to 3 pl.). **28.** $\frac{1}{846}$ (to 4 pl.).

CONVERSION OF DECIMAL TO VULGAR FRACTIONS, AND VICE VERSA

§ 98. **To convert a non-recurring decimal fraction into a vulgar fraction in its lowest terms.**

Example. 0·125. This is equal to

$$\frac{1}{10} + \frac{2}{100} + \frac{5}{1000} = \frac{100+20+5}{1000} = \frac{125}{1000}.$$

But you ought to write down this result at once without intermediate steps, by fixing your attention first of all on the last figure, 5, and noticing that this means $\frac{5}{1000}$.

$$0\cdot125 = \frac{\begin{array}{c}1\\ \cancel{5}\\ \cancel{25}\\ \cancel{125}\end{array}}{\begin{array}{c}\cancel{1000}\\ \cancel{200}\\ \cancel{40}\\ 8\end{array}} = \frac{1}{8}.$$

The conversion of recurring decimals into fractions cannot be explained properly till you have studied higher algebra.

EXERCISE VIII r (oral)

Express as vulgar fractions in their lowest terms:

1. 0·7. **2.** 0·79. **3.** 0·249. **4.** 0·053.

5. 0·907. **6.** 9·63. **7.** 10·7. **8.** 0·00907.

etc., etc.

EXERCISE VIII s

Express as vulgar fractions in their lowest terms:

1. 0·5.	**2.** 0·25.	**3.** 0·75.	**4.** 0·0625.
5. 0·625.	**6.** 0·35.	**7.** 0·175.	**8.** 0·15.
9. 0·075.	**10.** 5·65.	**11.** 9·64.	**12.** 1·28.
13. 0·0256.	**14.** 3·504.	**15.** 10·375.	**16.** 12·408.

§ 99. **To convert a vulgar fraction into a decimal fraction.** To express a vulgar fraction, e.g. $\frac{3}{14}$, as a decimal is the same problem as to find the quotient $3 \div 14$ as a decimal. The problem has been dealt with in § 93.

Remember that $\frac{1}{2} = \cdot 5$, $\frac{1}{4} = \cdot 25$, $\frac{3}{4} = \cdot 75$,

$\frac{1}{5} = \cdot 2$, $\frac{1}{3} = \cdot\dot{3}$, $\frac{2}{3} = \cdot\dot{6}$.

DECIMALIZATION OF MONEY, ETC.

EXERCISE VIII t (oral)

1. What fraction of £1 is 2*s*.?
2. Express 2*s*. as a decimal of £1.
3. Express 1*s*. as a decimal of £1.

Express as decimals of £1:

4. 4*s*.	**5.** 6*s*.	**6.** 8*s*.	**7.** 10*s*.
8. 9*s*.	**9.** 5*s*.	**10.** 15*s*.	**11.** 17*s*.
12. 19*s*.	**13.** 6*d*.	**14.** 2*s*. 6*d*.	**15.** 3*s*. 6*d*.
16. 7*s*. 6*d*.	**17.** 12*s*. 6*d*.	**18.** 15*s*. 6*d*.	**19.** 17*s*. 6*d*.
20. 19*s*. 6*d*.	**21.** 1*s*. 6*d*.	**22.** £2/4/-.	**23.** £5. 5*s*. 6*d*.

§ 100. *Example.* Express 8*s*. $6\frac{1}{2}d$. as the decimal of £1.

Write $6\frac{1}{2}d. = 6{\cdot}5d.$

To reduce pence to shillings, divide by 12.

$6\frac{1}{2}d. = 0{\cdot}54166\ldots$ shillings.

$\therefore$ 8*s*. $6\frac{1}{2}d. = 8{\cdot}54166\ldots$ shillings.

12 | 6·5*d*.
20 | 8·54166...*s*.
·427083...£

To reduce shillings to pounds, divide by 20.

$\therefore$ 8*s*. $6\frac{1}{2}d. = £0{\cdot}427\ldots$

EXERCISE VIII u

Express as a decimal of £1, correct to 3 places of decimals:

1.	5*s.* 3*d.*	**2.**	1*s.* 7*d.*	**3.**	2*s.* 4*d.*
4.	4*s.* 11*d.*	**5.**	11*s.* 7*d.*	**6.**	12*s.* 5*d.*
7.	14*s.* 3*d.*	**8.**	15*s.* 7*d.*	**9.**	17*s.* 8*d.*
10.	19*s.* 2*d.*	**11.**	18*s.* 4*d.*	**12.**	18*s.* $4\frac{3}{4}$*d.*
13.	£2. 17*s.* 2*d.*	**14.**	£1. 17*s.* $2\frac{1}{2}$*d.*	**15.**	£5. 8*s.* 3*d.*
16.	£4. 8*s.* $3\frac{1}{2}$*d.*	**17.**	£6. 8*s.* $3\frac{3}{4}$*d.*	**18.**	£7. 8*s.* 4*d.*
19.	$1\frac{1}{2}$*d.*	**20.**	4*s.* $1\frac{1}{2}$*d.*	**21.**	4*s.* 9*d.*
22.	7*s.* 3*d.*	**23.**	15*s.* $1\frac{1}{2}$*d.*	**24.**	6*s.* $7\frac{1}{2}$*d.*
25.	6*s.* 9*d.*	**26.**	15*s.* $10\frac{1}{2}$*d.*	**27.**	3*s.* $1\frac{1}{2}$*d.*

§ 101. *Example.* To reduce £17·694 to £ *s.* *d.*, to the nearest penny.

To reduce £0·694 to shillings, multiply by 20.

£0·694 = 13·88 shillings.

To reduce 0·88 shillings to pence, multiply by 12.

0·88 shillings = 10·56*d.* = 11*d.* to the nearest penny.

$$\begin{array}{r} £17,\cdot694 \\ 20 \\ \hline 13,\cdot88s. \\ 12 \\ \hline 10\cdot56d. \end{array}$$

∴ £17·694 = £17. 13*s.* 11*d.*

Remember that

(i) £0·1 = 1 florin, (ii) £0·001 = £$\frac{1}{1000}$ = about 1 farthing.

EXERCISE VIII v

Find the value (to the nearest penny) of:

1.	£0·5.	**2.**	£0·25.	**3.**	£0·6.
4.	£0·65.	**5.**	£0·8.	**6.**	£0·95.
7.	0·5*s.*	**8.**	£0·125.	**9.**	£0·142.
10.	£0·725.	**11.**	£0·178.	**12.**	£0·925.
13.	£0·882.	**14.**	£2·286.	**15.**	£3·142.
16.	£17·441.	**17.**	£18·405.	**18.**	£24·003.
19.	£17·643.	**20.**	£17·644.	**21.**	£17·645.
22.	£17·646.	**23.**	£17·647.	**24.**	£8·296.

§ 102. In decimalizing other quantities the same methods should be adopted.

Example. To reduce 13 mins. 37 secs. to a decimal of an hour (correct to 3 places):

13 mins. 37 secs. = 0·227 hrs.
(correct to 3 places).

```
60 | 3,7         secs.
60 | 1,3·6166...mins.
     ·2269...   hrs.
```

Example. Express 7·654 hrs. in hrs., mins., secs. to the nearest second.

7·654 hrs. = 7 hrs. 39 mins. 14 secs.

```
 7,·654 hrs.
     60
39,·24 mins.
     60
  14·4 secs.
```

Example. Express 12 stone 9 lbs. as a decimal of 1 ton (correct to 3 places).

12 stone 9 lbs. = 0·079 ton
(correct to 3 places).

```
14 {2 |  9·0         lbs.
   {7 |  4·5
    8 | 12·6428...stone
   20 |  1·5803...cwts.
         ·07901...ton
```

Notice that in dividing by 14 we have divided in succession by 2 and 7, in spite of the advice given in § 8. Where there is no question of finding "remainders," there is no objection to using the simple factors of the divisor.

Example. Express 0·616 ton in cwts., qrs., lbs. (to the nearest lb.).

0·616 ton = 12 cwts. 1 qr. 8 lbs.

```
 0,·616 ton
     20
12,·32 cwts.
      4
 1,·28 qrs.
     28
    56
   224
  7·84 lbs.
```

§ **103. Practice sums** can sometimes be shortened by the use of decimals, as follows:

Example. Find the cost of 243 things at £2/4/3½ each.

	£243	= cost at	£1	each
	£486	= cost at	£2	each
4/- = $\frac{1}{10}$ of £2	48·6	= ,,	4*s.*	,,
1/- = $\frac{1}{4}$ of 4/-	~~12·15~~	= ,,	~~1*s.*~~	,,
3*d.* = $\frac{1}{4}$ of 1/-	3·0375	= ,,	3*d.*	,,
½*d.* = $\frac{1}{6}$ of 3*d.*	0·5062	= ,,	½*d.*	,,
	£538,·144	= ,,	£2/4/3½	,,

```
£538,·144
       20
   ------
    2,·88
       12
   ------
   10,·56
```

∴ total cost = £538/2/11.

Exercises on Practice will be found in Chap. II.

EXERCISE VIII w

Express as a decimal of an hour (correct to 3 places):

1. 15 minutes.
2. 45 minutes.
3. 10 mins.
4. 12 mins.
5. 2 hrs. 55 mins.
6. 50 mins.
7. 15 mins. 30 secs.
8. 40 mins. 15 secs.
9. 50 mins. 45 secs.
10. 52 mins. 14 secs.
11. 29 mins. 19 secs.
12. 16 mins. 11 secs.

Express in hours, minutes, seconds (to the nearest second):

13. 2·5 hrs.
14. 2·56 hrs.
15. 13·575 hrs.
16. 6·235 hrs.
17. 12·002 hrs.
18. 9·147 hrs.

Express as a decimal of a foot (correct to 3 places):

19. $8\frac{1}{2}$ in. **20.** $4\frac{1}{4}$ in. **21.** $5\frac{1}{8}$ in.

22. $6\frac{3}{8}$ in. **23.** $14\frac{1}{16}$ in. **24.** $15\frac{3}{16}$ in.

Express in yds., ft., inches (to nearest inch):

25. 3·375 yds. **26.** 2·6 feet. **27.** 2·675 feet.

28. 0·333 mile. **29.** 0·1 mile. **30.** 0·01 mile.

Express as a decimal of 1 ton (correct to 3 places):

31. 3 cwts. **32.** 8 cwts. **33.** 10 stone.

34. 15 lbs. **35.** 100 lbs. **36.** 15 st. 10 lbs.

37. 3 cwts. 5 stone 12 lbs. **38.** 11 cwts. 3 qrs. 17 lbs.

Express in cwts., lbs. (to nearest pound):

39. 0·1 ton. **40.** 2·32 tons. **41.** 0·01 ton.

42. 0·001 ton. **43.** 0·057 ton. **44.** 0·034 ton.

EXERCISE VIII x

Miscellaneous Exercises on Decimalization

1. What decimal fraction is 3*s*. $10\frac{1}{2}$*d*. of £1?

2. Express £13·614 in £ *s*. *d*. to the nearest penny.

3. Express 6/2 as a decimal of £1 to 3 places.

4. Reduce £4·356 to £ *s*. *d*. (correct to nearest $\frac{1}{2}$*d*.).

5. Express 1 ton 2 cwts. 11 lbs. as a decimal of a ton correct to 3 places.

6. Express £5·608 in £ *s*. *d*. correct to the nearest penny.

7. Express 3 cwts. 2 qrs. 12 lbs. as a decimal of a ton correct to 3 places.

8. Find, to the nearest penny, the value of £0·67.

9. What decimal of a stone is 1 lb. 5 oz.? (5 places.)

10. Express 0·3567 ton in cwts., qrs., lbs. (nearest lb.).

11. Reduce £17. 4*s*. 6¼*d*. to a decimal of £1 (to 3 places); and £17·226 to £ *s*. *d*. (to the nearest farthing).

12. Find the value of 7·8125 of 8 cwts., in lbs.

13. Find the value of 0·16 of 5 guineas, to the nearest 1*d*.

14. Calculate the value of 5·21 tons to the nearest penny if a ton is worth £6. 12*s*. 0*d*.

15. Find the cost of 5·24 tons of coal at 18*s*. 6*d*. per ton.

16. In 1902 there were 462,593 children attending London Board Schools. The amount levied in school rates was £2,339,540. How much per child does this amount to?

17. Express 4 cwts. 3 qrs. 21 lbs. as a decimal of a ton, and find its value at 4 guineas a ton, correct to a penny.

18. £512 is to be divided among 29 people. How much will each person get (correct to the nearest penny)?

19. Express the area of a courtyard measuring 22 yds. by 37 yds. as a decimal of an acre, correct to 2 places.

20. On a rubber estate in the East, during the year ending June 30, 1912, there were produced 303,410 lbs. of rubber. The cost of production was £12,009. 18*s*., and the rubber was sold at an average price of 4*s*. 3·6*d*. per lb. Find the amount received for the sale of the rubber, and the profit.

21. The dining-room of a certain house is lighted with a 32 candle-power lamp for an average of 1·5 hours a day. The consumption of light for the rest of the house is as follows:

	Average hours per day	Candle-power
Hall	4	16
Drawing-room	4	32
Kitchen	5	20
Two sitting-rooms, each	4	16
Passages	4	24
Bedrooms, etc.	1	112

If the cost of the light works out at 0·006*d*. per candle-power per hour, find the annual cost of lighting the house on the basis of the above estimate. Take a year as 365 days.

22. Find, to a tenth of a penny, the rate per £ represented by the following income tax payments for the year 1914–15: (i) £200 pays £2, (ii) £350 pays £9/10/-, (iii) £501 pays £19/1/-, (iv) £1001 pays £58/7/10, (v) £2501 pays £208/8/4, (vi) £20,000 pays £3052/15/6, (vii) £100,000 pays £16,830/11/1.

23. What will be the income tax on the following incomes at the rates per £ quoted: (i) £475 at 8·2*d.*, (ii) £660/10/- at 10·7*d.*, (iii) £2435 at 18·7*d.*, (iv) £62,435 at 39·8*d.*?

EXERCISE VIII y

Problems on Decimals

1. The points A, B, C, D, and E lie along a road. The distances from A to B, from B to C, from C to D, and from D to E are 2·15, 0·74, 2·4, and 0·96 miles. What is the distance from A to E in miles and decimals of a mile?

2. The rainfall in October was 7·2 inches and during the whole year 36 inches. What fraction of the year's rain fell in October?

3. A gallon of water weighs 10·26 lbs. How many gallons (to the nearest gallon) are there in a ton of water?

4. A cubic foot of water weighs 62·43 lbs. Find the weight of 38·25 cubic feet of water (i) correct to the nearest lb., (ii) correct to the nearest hundredth of a lb.

5. The moon's mean distance from the earth is 238,800 miles. How many earth's radii is this, to 1 place of decimals? (Earth's radius = 3960 miles.)

6. The following prices are quoted for soda: 7 lbs., $3\frac{1}{2}$*d.*; 14 lbs., $6\frac{1}{2}$*d.*; 28 lbs., 1/1; $\frac{1}{2}$ cwt., 2/-; 1 cwt., 3/10. Find the price per lb. in each case, as a decimal of a penny correct to 2 places.

7. On a particular day the sun was above the horizon for 10·55 hours, and was visible for 2·4 hours. How many hours was it hidden?

8. Find (to the nearest mile) the average rate per hour in each of the following non-stopping railway journeys:

	Railway	Distance in miles	Time in hours
London to Bournemouth ..	L. S. W. R.	107·75	2·1
,, Bristol	G. W. R.	118·5	2
,, Cardiff	G. W. R.	145·25	2·92
,, Exeter	G. W. R.	194	3·5
,, Plymouth ..	G. W. R.	247	4·32
,, Doncaster ..	G. N. R.	156	2·82
,, Crewe	L. N. W. R.	158	2·92
,, Sheffield	G. C. R.	164·75	3
Paris to Calais	Nord	185	$3\frac{1}{4}$
,, Orleans	Orleans R.	$75\frac{1}{4}$	$1\frac{1}{2}$
,, Marseilles (stopping)	P. L. M.	536	$10\frac{1}{2}$

9. A motor is found to pass two points on the Brighton Road, 23 miles apart, at 11.58 a.m. and 1.3 p.m. What was its average speed, in miles per hr. to 1 place?

10. In 1895 the journey from London to Aberdeen (540 miles) was run in 8 hrs. 40 mins. The Great Northern in October 1914 ran:

London to Edinburgh (393 miles) in 7 hrs. 50 mins.
,, Aberdeen (540 miles) in 11 hrs. 40 mins.

Find the average speed in each of these journeys. Compare with the rate in the following journey: New York to Buffalo, 437 miles, in 440 minutes. (Answers to 1 place.)

11. The length of a gun-barrel is 47·6 feet. If the average velocity of the shell while in the barrel is 1423 feet per second, find how long the shell is in the barrel during one shot. Assuming that the gun is rendered useless in 150 shots, find the total firing life of the gun.

12. For engines working at 40,000 horse-power there is a coal consumption of 36 tons per hour. What is this consumption equivalent to in lbs. per horse-power?

13. In England, 0·77 of the whole population lived in towns, in 1901; in France 0·41; in Germany 0·543; in U.S.A. 0·3311. Find, in each case, what decimal fraction lived in the country.

14. Of the total imports into the United Kingdom in 1900, 0·1025 came from France, 0·1196 from Germany and Holland, 0·2653 from the United States. What decimal fraction came from all other countries together?

15. Of the total exports in 1900 from the United Kingdom 0·0697 went to France, 0·1305 to Germany and Holland, 0·7 to the United States. What decimal fraction went to all other countries combined?

16. A map is drawn to a scale of 1 in. to a mile. How far apart are two towns which are distant on the map 1·738 in.? Give your answer to the nearest yard.

17. Mount Everest is 29,000 ft. high; Mont Blanc 15,780 ft.; Ben Nevis 4,406 ft.; Snowdon 3,560 ft. If the height of Ben Nevis is taken as unit, express the other heights in terms of this unit, to 2 places.

18. A cubic foot of water weighs 62·3 lbs. Express this in ounces to the nearest ounce.

19. The years of the planets Jupiter, Saturn, Uranus and Neptune are respectively 11·9, 29·5, 84, 164·6 of our years. Express the years of each of the last three named in terms of Jupiter's year (correct to 1 decimal place).

20. Three lbs. of water at 4° C. occupy ·048 cubic foot, how much do 29 lbs. occupy?

21. Find the cost of 25 lbs. of treacle, at $11\frac{1}{2}$*d.* per 4 lb. tin, to the nearest penny.

22. Find the cost of 11 lbs. of dog food at 16/6 per cwt.

23. 40 gallons of lamp oil cost 28/4; what is the cost of 9·4 gallons, to the nearest penny?

24. 5 turns of a screw are found to drive it in 0·3 inch; how many turns will drive it in 1 inch? $1\frac{1}{2}$ inches?

25. 13·5 lbs. of honey can be bought for £1·25. What would be the cost of 62·75 lbs.? (Answer in £ *s. d.*)

CHAPTER IX

THE METRIC SYSTEM

§104. This system of measuring quantities is in use in most civilized countries; the chief exceptions are the British Empire and the United States. Its great advantage is that Compound Rules are entirely dispensed with.

The units of area, volume (capacity) and weight are all derived from the unit of length which is called **a Metre** (written 1 m.). This is about one forty-millionth of the circumference of the earth; it is rather more than a yard, is nearly 40 inches, and is almost exactly 39·37 inches.

The beginner should acquire clear ideas of the units employed and with this in view suggestions for oral work have been inserted at each stage.

A metre rule subdivided into centimetres and millimetres, and a cubical wooden block with an edge 10 cm. long divided as in the figure and fitting closely into an open tin box, will be found very useful in the following exercises.

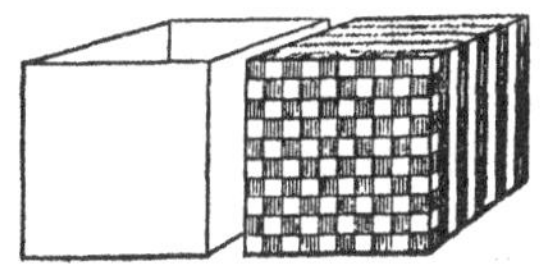

EXERCISE IX a (oral)

1. Hold your hands about 1 metre apart.

2. Taking 1 m. = 40 inches approximately, how many metres are there in 20 feet? in 120 yards?

3. Express each of the following approximately as (1) a fraction of 1 m., (2) a decimal of 1 m.: 10 inches, 1 inch, 8 inches, 4 inches.

§105. The Latin prefixes *deci-*, *centi-*, *milli-*, are used throughout the Metric System to denote one-tenth, one-hundredth, one-thousandth; and the Greek prefixes *deca-*, *hekto-*, *kilo-*, are used to denote ten times, a hundred times, a thousand times.

Thus a decimetre (1 dm.) is 1/10 of a metre,
a centimetre (1 cm.) is 1/100 of a metre,
a millimetre (1 mm.) is 1/1000 of a metre,
a decametre (1 Dm.) is 10 metres,
a hektometre (1 Hm.) is 100 metres,
a kilometre (1 Km.) is 1000 metres.

For another way of expressing these facts see the Tables at the beginning of the book.

The **millimetre, centimetre, metre** and **kilometre** are the units most commonly used.

EXERCISE IX b (oral)

1. Hold up two fingers (1) about 1 cm. apart, (2) about 10 cm. apart.

2. About how many inches are there in 10 cm.?

3. About how many (1) inches, (2) cm. are there in the length of a page of this book?

4. What fraction of 1 mile is a kilometre? (8 Km. = about 5 miles.)

5. Express a mile as a kilometre and a fraction of a kilometre.

6. About how many Km. are there in 20 miles, 35 miles?

7. About how many miles are there in 24 Km., 80 Km.?

§106. In the following exercise a few suggestions are given for practice in the relations between metric units of length. Constant drill is recommended in this work.

EXERCISE IX c (oral)

(i) What fraction, (ii) what decimal is

1. 1 mm. of 1 m.
2. 1 m. of 1 Km.
3. 1 cm. of 1 m.
4. 1 cm. of 1 Km.
5. 1 Dm. of 1 Km.
6. 3 cm. of 1 dm.
7. 4 m. of 1 Km.
8. 40 cm. of 1 Km.
9. 41 cm. of 1 Km.
10. 70 m. of 1 Km.
11. 75 m. of 1 Km.
12. 114 m. of 1 Km.?

Express in (i) metres and decimals of a metre,
(ii) cm. and decimals of a cm.:

13.	1 dm.	**14.**	4 dm.	**15.**	7 dm.
16.	1 cm.	**17.**	2 cm.	**18.**	8 cm.
19.	4 mm.	**20.**	6 mm.	**21.**	1 dm. 2 cm.
22.	3 dm. 5 cm.	**23.**	74 cm.	**24.**	4 dm. 3 cm.
25.	43 cm.	**26.**	3 cm. 5 mm.	**27.**	4 cm. 1 mm.
28.	62 mm.	**29.**	9 cm. 1 mm.	**30.**	67 mm.
31.	2 dm. 46 mm.	**32.**	24 cm. 6 mm.	**33.**	246 mm.
34.	1 m. 4 dm.	**35.**	1 m. 46 cm.	**36.**	1 m. 469 mm.

Express as kilometres and decimals of a Km.:

37.	5 Km. 4 m.	**38.**	5 Km. 4 Dm.	**39.**	5 Km. 40 m.
40.	5 Km. 41 m.	**41.**	7 Km. 56 m.	**42.**	1 Km. 156 m.
43.	3 Km. 300 m	**44.**	3 Km. 347 m.	**45.**	462 m.
46.	56 m.	**47.**	75 m.	**48.**	750 m.

Express as metres and decimals of a m.:

49.	4 cm. 357 mm.	**50.**	246 m. 472 mm.
51.	40 Km. 562 m.	**52.**	1 Km. 43 m.
53.	53 m. 72 mm.	**54.**	70 m. 34 mm.

§ 107. The units of **area** are the **square metre**, the square centimetre, the square kilometre and so on, written 1 sq. m., 1 sq. cm., 1 sq. Km., etc.

EXERCISE IX d (oral)

1. Point out some area which is about 1 square metre.

2. Guess the number of sq. m. in the ceiling of the room.

3. Show some surface whose area is about 1 sq. cm.

4. If a square metre is divided into strips 1 cm. wide, how many such strips will there be?

5. Into how many sq. cm. can each strip be divided?

6. How many sq. cm. are there in 1 sq. m.?

7. Guess the number of sq. cm. in the page of this book.

8. Point out some surface whose area is about 1 sq. dm.

9. How many sq. cm. are there in 1 sq. dm.?

10. How many sq. dm. are there in 1 sq. m.?

11. How many sq. m. are there in 1 sq. Dm.?

12. How many sq. m. are there in 1 sq. Km.?

§108. The units of **volume** are the **cubic metre**, cubic centimetre and so forth, written 1 cub. m., 1 cub. cm. (or 1 c.c.).

EXERCISE IX e (oral)

1. If a cubic metre were standing on the floor in the corner of the room show the approximate position of its top corners.

2. Is there anything in the room which contains about 1 cub. m. though it is not exactly cubical?

3. Point out something about the size of 1 c.c.

4. Indicate with your hands the size of a cubic dm.

5. If there were a cubical box containing exactly 1 cub. m., how many cub. dm. could be put in a row along one of the bottom edges? How many cub. dm. could be packed in a layer one deep on the bottom? How many such layers could be packed in the box? How many cub. dm. are there then in 1 cub. m.?

6. Show in a similar way how to find the number of (1) c.c. in 1 cub. m., and (2) c.c. in 1 cub. dm.

7. Guess the number of cub. m. of air in this room.

§109. The unit of **capacity** is the **litre** (written 1 l.) which is the same as a cub. dm. (See the tin box containing the wooden cub. dm.)

EXERCISE IX f (oral)

1. What kind of bottle contains about the same amount as the tin box (i.e. 1 litre)?

2. How much water (pints) does an ordinary tumbler hold? How many tumblerfuls do you think go to 1 litre? What fraction of a litre is a tumblerful?

§ 110. The unit of **weight** is the **gram** (written 1 gm.). *A gram is the weight of a cubic centimetre of pure water* under certain special conditions of temperature and pressure.

EXERCISE IX g (oral)

1. Use fingers to indicate 1 cm. and 1 c.c.

2. If a 1 gm.-weight be made of brass (which is about 8 times as heavy as water), will it be larger or smaller than the same weight of water? How big will it be?

3. How many gm. does 1 c.c. of brass weigh?

4. How many c.c. are there in 1 litre?

5. How many gm. does 1 litre of water weigh?

6. Mercury is about 13 times as heavy as water. What does a litre of mercury weigh approximately?

A thousand grams is called a **Kilogram** (written 1 Kgm. and often abbreviated to "Kilo"). Thus a litre of pure water weighs a Kilogram (= about 2 lbs.).

7. What is the weight of 5 litres of water?

8. On certain foreign railways 56 lbs. of luggage is allowed free. About how many Kilos is this?

9. If you wanted to buy about 10 lbs. of sugar in a French shop, what weight would you ask for?

10. About how many Kilos does a boy weigh if he is known to weigh 7 stone?

EXERCISE IX h

Suggestions for Oral Revision

1. How many dm. in 3 m.? cm. in 7 m.? sq. dm. in 3 sq. m.? sq. cm. in 7 sq. m.? cub. dm. in 3 cub. m.? c.c. in 7 cub. m.?

2. What is the weight of 1 cub. dm. of wood ($\frac{1}{2}$ as heavy as water)? 10 c.c. of cork (0·2 times as heavy as water)? 4 c.c. of lead (11 times as heavy as water)?

3. What volume of wood weighs 40 gm.? of cork weighs 35 gm.? of lead weighs 22 gm.?

4. What is the weight of 2·63 l. of water? 90 l. of oil (0·9 times as heavy as water)? 13 l. of mercury (13 times as heavy as water)?

To vary the quantities above, the following facts will be found useful. Gold is 19 times as heavy as water; the **specific gravity** of Gold is 19. Other specific gravities are: Silver 10, copper 9, brass 8, steel $7\frac{1}{2}$, aluminium $2\frac{1}{2}$, wood from 0·5 to 1·5, stone from 2 to 3, mercury 13·6 more exactly.

Decimal Coinage

§ 111. Many foreign countries have a decimal coinage.

In France a **franc** is divided into 100 **centimes.**

In America a **dollar** is divided into 100 **cents.**

In Germany a **mark** is divided into 100 **pfennig.**

EXERCISE IX i (oral)

1. Express in centimes:

16 fr. 50 c.	4·75 fr.	0·25 fr.
3 fr. 5 c.	7·26 fr.	4·5 fr.

2. Express in francs:

2 fr. 35 c.	65 c.	20 c.
1 centime.	20 fr. 5 c.	201 c.

3. Express in cents:

16 dollars 50 cents.	4·75 dollars.	0·25 dollar.
3 dollars 5 cents.	7·26 dollars.	4·5 dollars.

4. Express in dollars:

2 dollars 35 cents.	65 cents.	20 cents.
1 cent.	20 dollars 5 cents.	201 cents.

5. Express in pfennig:

M 1,30.	M 20,45.	M 1,00.
M 13,25.	M 123,05.	M 5,04.

6. Express in marks:

17 pf.	2005 pf.	194 pf.
2 pf.	202 pf.	2020 pf.

Addition and Subtraction

EXERCISE IX k

Perform the following additions:

	Km.	Hm.	Dm.	m.	dm.	cm.	mm.
1.				2	4	5	
			1	3	7	9	
2.			1	7	9		
			2	4	6		
3.		2	0	4	6		
		1	7	2			
		9	0	7			
4.	7	2	4				
	9	6	3				
	1	4	6				

5. 7 m. 3 dm. 4 cm. + 9 dm. 3 cm. (answer in m.).

6. 6 m. 4 dm. + 9 dm. 3 cm. (answer in metres).

7. 16 m. 4 dm. + 5 m. 4 dm. (answer in metres).

8. 9 Km. 24 m. + 6 Km. 135 m. (answer in Km.).

9. 125 m. 624 mm. + 71 m. 937 mm. (answer in mm.).

10. 42 Dm. + 724 mm. (answer in metres).

11. 67 m. + 1435 mm. (answer in metres).

12. 4 Dm. + 47 m. + 6421 mm. (answer in metres).

13. 4·568 m. + 7·246 m. (answer in mm.).

14. 14·42 m. + 8·735 m. (answer in Km.).

15. 8·723 Kgm. + 479·2 gm. (answer in Kgm.).

16. 8·2 Km. + 81·9 m. + 642·3 mm. (answer in metres).

17. 9·24 Hm. + 16·2 m. + 42000 mm. (answer in Km.).

18. 9 Km. + 4 Hm. + 3 Dm. + 7 m. (answer in Km.).

19. 8 Dm. + 4 m. + 3 cm. (answer in Km.).

20. 6 fr. 43 c. + 50 c. + 2 fr. 15 c. (answer in francs).

21. 7·25 fr. + 18·5 fr. + 0·5 fr. (answer in francs).

Subtract (expressing the answer in metres):

22. 4 m. 3 dm. 6 cm. from 6 m. 9 dm. 8 cm.

23. 6 m. 9 dm. 4 cm. from 7 m. 1 dm. 2 cm.

24. 9 m. 423 mm. from 12 m. 678 mm.

25. 6 mm. from 2 m.

26. 9 cm. from 1 m.

27. 8 cm. from 2 dm.

28. 4 cm. 2 mm. from 85 dm.

29. 91 mm. from 64 cm.

30. 53 m. from 4 Km.

Subtract (expressing the answer in kilometres):

31. 64 m. from 1 Km.

32. 743 cm. from 2 Km.

33. 486 m. from 1724 m.

34. 7840 m. from 8 Km.

35. 9 Km. 432 m. from 10 Km.

36. 45 Dm. from 6 Hm.

Subtract:

37. 5 centimes from 5 francs.

38. 5 pf. from M 9,00.

39. 0·25 francs from 5·5 fr.

40. 7 c. from 6·2 fr.

41. A man walks in succession through streets of length 472 m., 671 m., 124 m., 96 m., 521 m.; find how many kilometres he has walked.

42. The distance from a tee to the hole on a certain golf course is half-a-kilometre; a player drives, in succession, 173 m., 16 m., 184 m. and 97 m. How far is he from the hole he is approaching, supposing him to have driven straight?

43. From a town *A* to a town *B* is 14·75 Km.; of this distance 5 Km. 650 m. is downhill; 4 Km. 89 m. is level. How much of the road is uphill?

44. A man is found to be 1·895 m. high with his boots on and 1·87 m. high with his boots off. Find, in centimetres, the thickness of the heel of his boot.

45. A tunnel is being bored from both ends simultaneously; when completed its length will be 3 Km. 466 m. The workmen at the two ends are found to have penetrated 1831 m. and 1624 m. How much rock remains to be bored?

46. A detective finds that the external depth of an open box is 1 metre, and the internal depth 910 mm. Reckoning the wood to be 1 cm. thick, he decides that there must be a false bottom. On his calculation, what should be the depth of the hollow space between the two thicknesses of wood at the bottom?

47. A packing case measuring 2·49 m. by 1·45 m. by 0·93 m. arrives at the door of a house. The door is 2 metres high and 81 cm. wide. Can the case be got through the door?

48. The thickness of a book including the covers is found to be ·038 metre; and the thickness measured without the covers is ·029 metre. Find the thickness of each cover (in millimetres).

49. The external diameter of the barrel of an air-gun is found to be 1·62 cm. and the thickness of the metal is 7 mm. Find (in centimetres) the internal diameter.

MULTIPLICATION

EXERCISE IX 1

1. Find the height of a pile of 45 books, if each book is 0·023 metre thick.

2. Find the height of a pile of 35 bricks, if each brick is 0·058 metre thick.

3. Find the cost of 10·05 gr. of gold at 2·625 fr. per gram.

4. Find the cost of 720 kilos of tea at 4·5 fr. per kilo.

5. Find the cost of 943 mm. of platinum wire at 8 fr. 45 c. per metre.

6. Find the cost of 1 Kgm. 247 gm. of mercury when a bottle of 10 Kgm. costs 67 francs.

7. If 100 litres of mercury cost 6530 francs, what is the cost of 472 decilitres?

8. If 20 litres of mercury cost 1317 francs, what is the cost of 14·72 decilitres?

What is the weight of:

9. 20 c.c. of aluminium (sp. gr. 2·6. See p. 188)?

10. 42·7 c.c. of glycerine (sp. gr. 1·26)?

11. 12·46 litres of olive oil (sp. gr. 0·915)?

12. 0·52 c.c. of lead (sp. gr. 11·3)?

13. 1·24 c.c. of gold (sp. gr. 19·3)?

14. 52·7 c.c. of platinum (sp. gr. 21·5)?

15. 1460 cubic decimetres of ice (sp. gr. ·917)?

16. 728·5 c.c. of cork (sp. gr. 0·2)?

17. A policeman finds that a motor car traverses a measured 100 metres in 6 secs. At how many kilometres per hour is it moving?

18. Light travels with a velocity of 3×10^{10} cm. per sec. Find, in kilometres, the distance of the pole-star, given that light takes 50 years to traverse the distance.

19. If the thickness of a penny is 2 mm., find in metres the height of a pile of £1 worth of pennies.

20. Find the cost of 25 tickets at 35 fr. 40 c. apiece.

21. My luggage weighs 97 kilos and I am allowed 26 kilos free. Excess luggage is charged at the rate of 0·06 c. per kilo per Km. What shall I have to pay for luggage between Brussels and Cologne (225 Km.)?

22. Find the total cost of 10 kilos of tea, 17 of coffee, and 42 of sugar at 5·5 fr., 4·75 fr. and 75 c. per kilo respectively.

23. I want a second-class ticket from Brussels to Antwerp (44 Km.), the fare being 6·25 c. per Km. What change shall I receive from a 20-franc piece?

24. At Charing Cross Station I am told that to-day's rate of exchange is £1 = 25·35 fr. If I change £42. 10*s.* for French money, what shall I get? How many francs should I waste if I kept my English money till I reached France, and then changed it at the rate of 25 francs for £1?

Division

EXERCISE IX m

1. If 21·6 metres of twine are required to wrap round the handle of a cricket bat 200 times, what is the average circumference of the handle (in centimetres)?

2. The 2nd class fare from Paris to Brussels ($192\frac{1}{2}$ miles) is 23 fr. 30 c. How much is this per mile (to the nearest centime)?

3. Divide 47 fr. 60 c. among 36 people.

4. If 75 lbs. of mercury cost 209 fr. 14 c., find the cost of 1000 lbs., to the nearest franc.

5. A pile of paper consisting of 350 sheets is found to be 1·75 cm. thick. Find the thickness of a sheet.

6. A cyclist rides 10 Km. in 53 minutes. At this rate how long will he take to ride 23 Km.?

7. How many pieces of 5 cm. can be cut off from 36 cm. of string; and what is left?

8. How many pieces of 6·7 cm. can be cut off from 400 cm. of string; and what remains over?

9. How many pieces of 4·5 cm. can be cut off from a rod of length 3·46 metres, and how many mm. remain?

10. How many railway tickets costing 1 fr. 75 c. apiece can be bought with a 20-franc piece; what change is left?

11. A lift can carry any load up to 100 kilos. How many books, each weighing ·53 kilo, will it bear?

EXERCISE IX n

Miscellaneous Exercises on Metric System

1. Express 8 m. 705 mm. as the decimal of a kilometre.

2. Express 2 m. 29 cm. in mm.

3. The points A, B, C, D and E lie along a road. The distances from A to B, from B to C, from C to D, and from D to E are 2150, 740, 2400 and 960 metres. What is the distance from A to E in kilometres?

4. The contents of 8 equal bottles just fill a 10-litre vessel. How many litres and centilitres does each bottle hold?

5. A motor car travels for the first 20 minutes at the rate of 6 kilometres 126 metres per hour, then for 2 hrs. 20 mins. at the rate of 30 kilometres per hour, and afterwards for 25 minutes at the rate of 7 kilometres 152 metres per hour; find the total distance traversed by the car.

6. Subtract 93 milligrams from 64 grams and give your answer in kilograms.

7. Find the cost of 5 m. 65 cm. of wire at 2 fr. per m.

8. Express in grams the difference between 82 gm. 527 mgm. and 79 gm. 529 mgm.

9. A motor car is running at the rate of 65 kilometres per hour. How far does it go in 1 sec. (to the nearest m.)?

10. Find the price of 3 metres 465 mm. of platinum wire at 8*s*. per metre (to the nearest penny).

11. Find in francs and centimes the cost of 12·4 metres of silk at 9·8 francs the metre.

12. A man flew a distance of 24 kilometres in 31·6 minutes. Find, to the nearest tenth of a minute, how long that is for each kilometre.

13. White paint costs 2/9 a kilogram. What will 3 kilograms 500 grams cost?

14. Find the weight in Kgm. of 2 Km. 250 m. of wire, a metre of which weighs 13·5 gm.

15. A railway is being built at the rate of 1350 metres in 6 days. How many days will it take to build 50 kilometres? You may neglect fractions of a day.

16. If 100 kilograms of tea cost 453 francs, find the cost of 633 kilograms.

17. The fare for a distance of 223 kilometres is 23 fr. 30 c. Find to the nearest centime the rate per kilometre.

18. If a kilogram of a certain kind of tobacco costs 12*s*. 6*d*., what would be the cost of 5 kilograms 300 grams?

19. If 47 things cost 145 fr. 70 c., what do 12 cost?

20. Find the cost of 5·4 litres of wine at 2/6 per litre.

21. If 17 machines produce 1·7 Km. of cloth in 5 days, in how many days will 16 machines produce 640 m.?

22. Find the cost in dollars and cents of 7 dozen articles costing 9 dollars 37 cents each.

23. If three-eighths of an article weighs 18·6 grams, find the weight of four-fifths of the article.

24. Find the cost of 163 litres at £3/6/8 per 100 litres.

25. The area of a rectangular sheet of drawing paper is 35 square dm. If the breadth of the paper is 450 mm., find its length to the nearest mm.

26. What is the weight to the nearest ounce of 5·24 litres of a liquid which weighs 2 lbs. 3 oz. the litre?

27. How many times can a piece 3 cm. 4 mm. long be cut off from a rod 9 dm. long, and what remains?

28. The following relations are approximately true:

1 metre = 39·4 inches. 1 kilometre = 5/8 mile.
1 Kgm. = $2\frac{1}{5}$ lbs. 1 litre = $1\frac{3}{4}$ pints. 25 francs = £1.

Express:

(i) 1 mile in Km. (ii) $2\frac{1}{2}$ miles in Km.
(iii) 3·4 Km. in miles. (iv) 1/4 mile in metres.
(v) 100 yards in metres. (vi) 1 cm. in inches.
(vii) 1000 Kgm. in lbs. (viii) 1 lb. in Kgm.
(ix) 1 ton in Kgm. (x) 1 ounce in grams.
(xi) 4·5 litres in pints. (xii) 1 pint in litres.
(xiii) $2\frac{1}{2}$ quarts in litres. (xiv) 1 franc in pence.
(xv) 1*s.* in francs and centimes.
(xvi) 1*d.* in centimes. (xvii) 20 fr. in £ *s.* *d.*
(xviii) Which is the greater—1 ton or 1000 Kgm.?
(xix) Which is the greater—5 cm. or 2 in.?
(xx) How many miles from the North Pole to the Equator, a distance of 10,000,000 m.?
(xxi) English railways allow 60 lbs. free luggage, French railways 30 Kgm.; which are the more generous?

29. Taking an inch as 2·54 cm., find to the nearest 1/10 inch the inside diameter of the French 75 mm. field gun and the German 42 cm. howitzer.

30. If a metre is equal to 39·37 inches find (correct to two places of decimals) the number of inches in 0·76 metre.

31. An ounce weight is found to weigh 28·4 grams. How many pounds (to 2 places) are there in 1 Kgm.?

32. A man weighed 12 stone at Dover and 76 kilogrammes at Calais; which is the greater?

33. The second class fare from Paris to Marseilles (536 miles) is 65·25 francs; from London to Winchester (66 miles) is 7*s*.; which gives the greater cost per mile?

34. A kilolitre contains 35·32 cubic feet, and a gallon contains about 277 cubic inches; how many gallons (to the nearest gallon) are there in a kilolitre?

35. Find to the nearest yard the difference between 81 kilometres and 50 miles, taking 1 m. = 39·37 in.

36. Calculate whether posting at a shilling per mile is less or more expensive than posting at 65 centimes per Km., given that 1 metre = 39·37 in. and 25 fr. = £1.

37. The third class railway fare in France is 5 centimes per kilometre and in England 1*d*. per mile. Given that 1 yard = ·9144 metre and £1 = 25·17 francs, find (in English money) the difference of the fares for a journey of 100 miles in the two countries, correct within a farthing.

38. A man when starting for a holiday in France changes £15 into French money at the rate of 25·2 francs for £1. He is in France for 21 days, and pays on the average 7·5 francs a day for board and lodging: he also spends 62·7 francs for railway fares and 87 francs on other expenses. On returning he exchanges what French money he still has back into English money at the rate of 25 francs for £1. Find to the nearest penny how much he then has.

CHAPTER X

RATIO AND PROPORTION

§112. The length of H.M.S. *Lion* is 660 ft.; and a model is made 31 ft. in length. The model's length is therefore $\frac{31}{660}$ of the ship's length; and the lengths of all the other parts are **in proportion.**

In making the model, all the dimensions of the ship are diminished in the ratio of 31 to 660.

The ratio of 31 to 660 is written 31 : 660.

Consider the "beam" (breadth) of ship and model.

The ratio of model's beam : ship's beam = 31 : 660..(i).

In other words, model's beam $= \frac{31}{660}$ of ship's beam,

$$\text{or } * \quad \frac{\text{model's beam}}{\text{ship's beam}} = \frac{31}{660} \quad \text{.............(ii).}$$

(If this step seems difficult, take a simple case such as: half-a-crown = $2\frac{1}{2}$ of a shilling. And $\frac{\text{half-a-crown}}{\text{a shilling}}$ means "the number of times a shilling is contained in half-a-crown," i.e. $2\frac{1}{2}$.)

On comparing (i) and (ii) above, it appears that for statements about ratios we may substitute statements about fractions; and as fractions are more familiar to us than ratios this is a convenience.

It is usual, in fact, to regard $a : b$ and $\frac{a}{b}$ as meaning one and the same thing, provided that a and b are numbers, or quantities of the same kind........................(iii).

§113. We cannot speak of the ratio of two quantities *not* of the same kind: e.g. 15 feet : 10 seconds has no meaning.

* See § 56.

For no amount of magnifying or diminishing will change 10 seconds into 15 feet. Nor has $\frac{15\text{ feet}}{10\text{ seconds}}$ any meaning for you at present.

On the other hand, though 15 feet : 10 has no meaning, since 15 feet is a length and 10 an abstract number, yet $\frac{15\text{ feet}}{10}$ is intelligible, and equal to 1·5 feet.

These two illustrations explain the proviso in (iii) above.

For if a and b are *not* of the same kind, then $a : b$ is unmeaning, while $\frac{a}{b}$ may in certain cases have a meaning. We cannot therefore regard $a : b$ and $\frac{a}{b}$ as equivalent in *all* cases.

§ 114. **To express a ratio in its simplest form,** reduce the corresponding fraction to its lowest terms.

EXERCISE X a (oral)

Express each of the following ratios in its simplest form:

1. A length of 3 cm. to a length of 10 cm.

2. A weight of 4 lbs. to a weight of 20 lbs.

3. An area of 5 sq. cm. to an area of 20 sq. cm.

4. A volume of 2 litres to a volume of 100 litres.

5. A population of 10,000 to a population of 20,000.

6. The number 5 to the number 10. **7.** 6 to 30.

8. 2 : 4. **9.** 50 : 25. **10.** 25 : 100.

11. 250 : 1000. **12.** 90 : 270. **13.** $\frac{1}{4} : 2$.

14. $2 : \frac{1}{4}$. **15.** $\frac{1}{2} : \frac{1}{8}$. **16.** $\frac{1}{8} : \frac{1}{2}$.

17. $\frac{1}{6} : \frac{1}{3}$. **18.** ·2 : ·5. **19.** ·9 : 1.

20. ·75 : 1. **21.** ·125 : 10. **22.** 7·4 : 10.

23. 6 yds. to 4 yds. **24.** 18 m. to 90 m.

25. 15 m. to 7·5 m. **26.** 15 m. to 1 Km.

27. 14 cm. to 1 m. **28.** 25 cm. to 50 mm.

29. 3 yds. to 2 ft. **30.** 3 sq. ft. to 1 sq. yd.

31. 1 lb. : 24 oz. **32.** 6*s*. : 6*d*.

33. 2*s*. 6*d*. : 10*s*. **34.** £1/10/0 : 2/6.

Comparison of Ratios

§115. At one date, the number of capital ships in the navies of Great Britain and of Germany was 65 and 40 respectively. It was proposed to limit shipbuilding in future so as to preserve this ratio, which is roughly the same as 1·6 : 1. But it is not quite the same, for

$$65 : 40 = \frac{65}{40} = \frac{1{\cdot}625}{1} = 1{\cdot}625 : 1.$$

Ratios are most easily compared by expressing them in this form, with 1 as the second term; in fact, by expressing them in the form n : 1.

EXERCISE X b

Which is the greater ratio of the following pairs?—

1. 5 : 1 or 4 : 1.

2. 1 : 5 or 1 : 4.

3. 3 : 2 or 4 : 3.

4. 7 : 10 or 8 : 11.

5. 3 : 5 or 0·66 : 1.

6. 41 : 83 or 371 : 743.

7. 22 : 7 or 355 : 113.

8. $3^2 : 2^2$ or 3 : 2.

9. $12^2 : 7^2$ or 12 : 7.

10. $5^2 : 7^2$ or 5 : 7.

11. $2^3 : 3^3$ or $2^2 : 3^2$.

12. 1 : 0·5 or 1 : $(0{\cdot}5)^2$.

Express in the form n : 1 the following ratios:

13. 1 guinea : £1.

14. £$5\frac{1}{4}$: £$2\frac{3}{16}$.

15. £1 : 20 francs.

16. 107 : 100.

17. 96 : 100.

18. $97\frac{3}{4}$: 100.

19. 1 mile : 1 kilometre.

20. 1 metre : 1 yard.

21. 1 litre : 1 quart.

22. 1° F. : 1° C.

23. 2/6 : 2/-.

24. 3 tons 5 cwts. : 2 tons.

25. £1/10/- : £30.

26. $17^2 : 10^2$.

27. $25^2 : 15^2$.

28. $14^2 : 35^2$.

29. $135^2 : 45^2$.

30. 1 year : 1 fortnight.

31. £1 : 13/4.

32. £9/16/8 : £7/7/6.

33. In a certain battle 35 officers and 735 men are killed or wounded. What is the ratio of (1) officers to men, (2) officers to the total casualties?

34. 40 men are killed, 480 wounded. What is the ratio of (1) killed to wounded, (2) killed to total casualties?

35. What is the ratio of A's income to B's income if A earns £10 per month and B £200 per annum?

36. What is the ratio of the average rates of two trains, one of which travels 420 miles in 8 hours, and the other 450 miles in 12 hours?

37. Find the ratio of (i) the area of a square of side 2 in. to the area of a square of side 4 in.; (ii) the area of a square of side 3 in. to the area of a square of side 6 in.; (iii) the area of a square of side 2 in. to the area of a square of side 6 in.

38. Find the ratio of the volumes of:

(i) A cube of side 1 in. to a cube of side 2 in.

(ii) ,, ,, 3 in. ,, ,, 6 in.

39. Mars rotates on its axis in a period 41 minutes greater than our day. Find the ratio of the Martian day to our day, correct to 2 places, in the form $n : 1$.

40. Venus revolves round the sun in 225 of our days. Find, correct to 2 places, the ratio of our year to Venus' year, in the form $n : 1$.

41. The longest radius of the earth is 20926000 ft. and the shortest is 20855000 ft. Find (in the form $1/n$) the ellipticity, i.e. the ratio of the difference of these radii to the larger.

42. On the shortest day of the year the sun rises at Greenwich at 8.6 a.m. and sets at 3.52 p.m.; on the longest day at 3.44 a.m. and 8.18 p.m. Find the ratio of the periods in which the sun may be seen on the longest and the shortest day (in form $n : 1$).

43. Hungarian flour costs 1/3 per 7 lbs., and £2/7/6 per sack of 280 lbs. Find the ratio of the former to the latter price per lb.

44. Soft soap costs 5*d.* per lb., but a firkin of 64 lbs. costs 14/3. Find the ratio of the latter to the former price per lb.

45. In France 50 wax matches cost 1*d.* In England a dozen boxes each containing 500 cost 2/6. Find the ratio of the French to the English price.

46. A cubic foot of water weighs 62·3 lbs., and 41 cubic feet of teak weigh a ton. Find the ratio of the weights of equal volumes of water and teak in the form $n : 1$.

47. The ratio of the weights of equal volumes of ice and water is 0·917 : 1; 1 cubic foot of water weighs 62·3 lbs. Find the difference in the weights of a cubic yard of ice and of water (to the nearest pound).

48. At the beginning of a war, the numbers of capital ships possessed by two powers are in the ratio of 1·6 : 1, the weaker power having 40. In a general engagement, each power loses the same number of ships, but the ratio is changed to 2 : 1. How many ships does each lose?

49. In Milan I bought 100 grams of English tobacco for 3 lire; in England the price of this tobacco is 5*d.* an ounce. Find, to one place of decimals, the ratio of the Italian to the English price (1 lira = $9\frac{1}{2}$*d.*; 1 kilogram = 2·2 lbs.).

50. A man walks over a mountain, going up hill at the rate of 2 miles an hour and down (the same distance) at 5 miles an hour; he returns along the flat round the mountain at $3\frac{3}{4}$ miles an hour, in the same time as he took to go; compare the distances over and round the mountain.

51. A sovereign contains 113 grains of gold, and a shilling 87 grains of silver. If 1 oz. of gold be worth 30 oz. of silver, and the gold in a sovereign be worth £1, what is the value of the silver in a shilling (to the nearest $\frac{1}{4}$*d.*)?

PROPORTION

§ 116. Reverting to the case of H.M.S. *Lion* and its model (§ 112), we saw that

model's beam : ship's beam = model's length : ship's length.

Two pairs of quantities related in this way are said to be **in proportion.**

If three of the quantities are known, the fourth can be calculated. Suppose that we know that ship's beam = 88 ft., model's length = 31 ft., ship's length = 660 ft.; then we can find the model's beam. For

$$\frac{\text{model's beam}}{88 \text{ ft.}} = \frac{31 \text{ ft.}}{660 \text{ ft.}} = \frac{31}{660},$$

$$\therefore \text{ model's beam} = \frac{31}{660} \text{ of } 88 \text{ ft.}$$

$$= 4\tfrac{2}{15} \text{ ft.}$$

§ 117. It is advisable to write the unknown member of the proportion first. Thus, suppose that we wish to ascertain, by measuring the model, the height of the ship's mast. We find the model's mast to be 91 inches high. Then

$$\frac{\text{ship's mast}}{91 \text{ in.}} = \frac{660}{31},$$

$$\therefore \text{ ship's mast} = \frac{660 \times 91}{31} \text{ in.}$$

$$= \frac{55 \times 91}{31} \text{ ft.}$$

$$= 161 \text{ ft. to the nearest foot.}$$

§ 118. The type of exercise that is usually worked by unitary method is really a case of proportion. Thus: If a man earns 18/6 in 5 days, how long will it take him to earn £2? It is clear that, if the number of days is increased in a certain ratio, the sum earned is increased in the same ratio.

$$\therefore \frac{\text{number of days required}}{5} = \frac{£2}{18/6} = \frac{40}{18\frac{1}{2}},$$

$$\therefore \text{ number of days required} = \frac{200}{18\frac{1}{2}} = \frac{400}{37} = 10{\cdot}8 \text{ approx.}$$

This method, in fact, is the same as that explained in § 81.

EXERCISE X c

Ratio and Proportion

1. A picture measures 10′ by 7′. The picture is photographed, and the short side of the photograph is 6″; what is the long side?

2. The length and breadth of a room are 21′ 6″ and 17′ 9″ respectively. On a plan the length appears as 12″; what is the breadth on the plan?

3. A park is 2·5 miles long and 0·75 mile broad. On a map the length is 6 cm.; what is the breadth?

4. A rectangular wall-map 3′ 6″ long and 5′ broad is photographed. If the reproduction is 14 cms. long, how broad will it be? and how far apart on the photograph will two towns be which on the map were 17·5 inches apart?

5. The scale of a map is 1 : 1250000. Find in kilometres the real distance represented by a centimetre on the map; and as the decimal of an inch the distance on the map which represents a mile.

6. In drawing to a scale of 1 inch to 20 feet, how long must a line be drawn which is to represent a length of 78 ft. 6 in., and what is the height (to the nearest inch) of a tower represented in the drawing by a line 3·77 inches in length?

7. On a model the scale of which is 6 inches to a mile, what represents the height of Scawfell Pike, which is 3212 feet high? Also, what is the distance (in miles and yards) between two places 23·13 inches apart on the model?

8. In the ordnance map of 1 in. to a mile a nobleman's park appears as a rectangle 1·25 in. long and 0·5 in. broad; how many acres does it contain?

9. In a map of the Solent on the scale of 4 statute miles to the inch, the distance from Portsmouth to Ryde is 1·06″; give this distance in miles. The Osborne woods cover a fifth of a square inch on the same map. What is their acreage?

10. The area of Great Britain is 90,000 sq. miles; that of Ireland is 30,000 sq. miles; and that of the Isle of Wight is 150 sq. miles.

(i) Express in its simplest form the ratio between these three.

(ii) How large would these be on a map of scale 10 miles to the inch?

11. A map of a certain district is drawn to scale, one inch representing 60 yards. (1) A viaduct in the district being 134 feet long, how long will its representation on the map be? (2) The district contains a square field, each of whose sides is 132 yds. long. Find the area in sq. in. of the square on the map which represents this field.

12. A map is drawn to a scale of 5 inches to a mile. How many acres are there in a field which is represented on the map by a square whose side is 1·2 inches?

13. A map is drawn to a scale of 6″ to 1 mile. What area on the map will represent an estate of 5000 acres?

14. A map is drawn to a scale of 4 inches to the mile. Find the area on the map of a district of 950 acres.

15. The map of a country is drawn on the scale of 1/10″ to the mile. What area on the map will represent a lake 4000 acres in extent?

16. In a map, scale 2 in. to 1 mile, find the distance represented by a line 8·54 inches long, and the area of a district 2·82 inches long and 2·5 inches wide on the map.

17. On a map drawn to a scale of 1 : 2500 the sides of an oblong field are 4·7 and 3·2 cm. Find the area of the field in sq. metres.

18. A map drawn to a scale of 10 Km. to 1 cm. measures 11·4 cm. by 7·6 cm. Find in square kilometres the area that the map represents.

19. The distance between two towns, which are known to be 45 miles apart, measures 3·75 inches on a map. Find the scale of the map (in miles to an inch).

On the same map a piece of country measures 2·67 square inches; find the actual area of the piece of country.

20. A map is drawn on a scale of 25 inches to the mile. Find in acres the area of a rectangular field whose sides on the map are 15 and 7·5 inches.

21. A farmer splits up his farm into arable and pasture land in the ratio 3 : 8. If the area of his arable land is 50 acres, what is the total area of his farm?

22. A German map drawn to a scale of 1 cm. to 10 Km. measures 24 cm. by 18 cm. If the areas of land and water represented are in the ratio of 7 : 2, find in square Km. the land area represented.

23. The new Ordnance Survey plans of towns are on a scale of $\frac{1}{500}$. How many inches to the mile is this? Find the area on this scale of the map of a football ground 150 yards by 80 yards (in square inches).

24. A reduction of the Sugar Duty from 4/2 to 1/10 per cwt. entailed a loss of revenue amounting to £3,650,000. What was the revenue received after the reduction, and how many cwts. of sugar were imported?

25. Bedford, Wellingborough and Kettering are on the Midland Railway main line, 50, 65 and 72 miles respectively from the London terminus. A train which leaves Bedford at 9.15 a.m. arrives at Kettering at 9.42 a.m. At what time (to the nearest min.) will it pass through Wellingborough, assuming that its speed is uniform?

26. In a school there are 218 boys and 256 girls. If the proportion of boys to girls is as nearly as possible the same at another school at which there are 183 girls, how many boys are there at it?

27. An incline rises at a constant gradient for a distance of 2·25 miles. At a point 252 yds. from the foot of the incline the rise is found to be 2 ft. 4·5 in. Find what the total rise must be at the top.

28. During last year the cars of a certain tramway company were running on 312 days, and the total receipts from fares amounted to £3757. Each passenger paid either 1*d.* or 2*d.*, and the number of penny fares was to the number of twopenny fares as 7 to 5. Find the average number of penny and twopenny passengers carried each day.

29. The number of electric lights (of given strength) needed for the illumination of a class-room is proportional to the area of the room. If a room 16′ by 20′ needs 8 lights, how many are needed for a room 21′ by 28′?

30. Do the following two pairs of numbers form a proportion: 17, 13; 13, 10? If not, what number must take the place of 17 in order that the first may be in proportion? And what number would serve instead of 10?

Work the above question substituting the following pairs of numbers: 5, 8; 8, 13.

Proportional Division

§ 119. This matter is best illustrated by an example.

Example. Three partners have to divide a profit of £9752 in the proportion 3 : 4 : 5. How much does each get?

Divide the profit into (3 + 4 + 5) shares, i.e. 12 shares. They will then take respectively $\frac{3}{12}$, $\frac{4}{12}$, $\frac{5}{12}$; i.e. $\frac{1}{4}$, $\frac{1}{3}$, $\frac{5}{12}$; i.e. £2438, £3250/13/4, £4063/6/8.

EXERCISE X d

1. In a fishing smack the catch is divided between the captain, mate and crew in the proportion of 5 : 2 : 1; after a haul of 520 fish, how many will the mate get?

2. Some prize-money was divided between four men in the proportion of 10 : 6 : 3 : 1. If the total amount to be divided was £891, how much did each man get?

3. 2400 tons of coal are to be distributed among three ships in the ratios 1 : 3 : 4. How much coal should be supplied to each ship?

4. In a sweepstake 432 half-crown tickets are sold, and the prizes are in the ratio of 8 : 3 : 1. Find the value of the second prize.

5. Divide £10,000 amongst *A*, *B*, and *C* in the proportion of the numbers 2, 3, 5. What (1) fraction, (2) decimal, of the whole does each get?

6. Three men work for 15, 10 and 8 days respectively; divide £33. 2*s*. 9*d*. fairly between them.

7. Divide £5. 12*s*. 6*d*. between A, B, and C, so that their shares are in the ratio 7 : 5 : 3.

8. Divide £143. 11*s*. 2*d*. amongst three persons in the proportions of 2, 7, 19.

9. A line AD is 30 in. long and is divided into 3 parts AB, BC, and CD so that $AB : BC : CD = 3 : 4 : 8$. It is also divided at X so that the ratio $AX : XD = 2 : 3$. Find the lengths of AB, BC, and BX.

10. Two men rent a field for ten months at a rental of £47. The first grazes 15 horses for 3 months, and then 12 horses for 5 months; while the other grazes 13 horses for the whole time. How much ought each to pay?

11. A clock loses at the rate of 8·5″ an hour when the fire is alight, and gains at the rate of 5·2″ an hour when the fire is not burning, but on the whole it neither gains nor loses. How long in the 24 hours is the fire burning?

12. A man's salary is paid by the quarter, the quarters ending on 24th March, 24th June, 24th September, and 24th December, the differences in length of the quarters being ignored. Up to 10th July, 1910, his salary was at the rate of £770 per annum, and from 11th July, 1910, at the rate of £775. Calculate his salary for each quarter of the year ending 24th December, 1910.

CHAPTER XI

AREA. VOLUME

Area of Rectangle

§120. In Chapter IV it was shown that in order to find the number of units of area in a rectangle, we must multiply together the numbers of units in the length and breadth.

This was proved only for the case when all the numbers involved are whole numbers. Does the rule hold always, e.g. for a rectangle whose sides are 1·2 cm. and 1·5 cm.?

Since 1·2 cm. = 12 mm. and 1·5 cm. = 15 mm., we can express the length and breadth as whole numbers of millimetres. The area is therefore 12 × 15 sq. mm. But 1 sq. cm. = 100 sq. mm.

$$\therefore \text{ the area} = \frac{12 \times 15}{100} \text{ sq. cm.}$$

$$= \frac{12}{10} \times \frac{15}{10} \text{ sq. cm.}$$

$$= 1{\cdot}2 \times 1{\cdot}5 \text{ sq. cm.}$$

In this particular case therefore the rule holds good; and as a matter of fact it always holds good, though the general proof is too difficult at this stage.

Example. A tin plate measuring 17″ by $12\frac{1}{2}$″ costs $3\frac{1}{2}d.$; what is the cost per square yard?

$$17 \times 12\tfrac{1}{2} \text{ sq. in. cost } 3\tfrac{1}{2}d.,$$

$$\therefore \text{ 1 sq. yd., or } 144 \times 9 \text{ sq. in cost } \frac{7}{2} \times \frac{144 \times 9}{17 \times 12\frac{1}{2}} \text{ pence.}$$

This is $\frac{7\times 144\times 9}{17\times 25} = \frac{28\times 144\times 9}{17\times 100}$ pence

$= \underline{1/9 \text{ approx.}}$

```
         144
          28
        ----
         288
        1152
        ----
        4032
           9
       -----
1700)36288(21·3
     34
     --
      22
      17
      --
       58
       51
```

Example. Find the cost of covering a floor 16′ 4″ by 15′ 9″ with Brussels carpet 27″ wide at 1/10 per yard.

This carpet is sold in a long strip 27″ wide, and 1 yard *length* of this costs 1/10. Sometimes the price of carpet is quoted per *square* yard, but more usually per yard of length.

The area of carpet needed is $16\frac{1}{3} \times 15\frac{3}{4}$ sq. ft.

A strip of carpet of this area and width 27″ or $\frac{9}{4}$ ft. will be of length

$$\frac{16\frac{1}{3}\times 15\frac{3}{4}}{\frac{9}{4}} \text{ ft.} = \frac{16\frac{1}{3}\times 15\frac{3}{4}}{\frac{9}{4}\times 3} \text{ yds.}$$

This costs $\frac{16\frac{1}{3}\times 15\frac{3}{4}}{\frac{9}{4}\times 3} \times 22$ pence

$$= \frac{\frac{49}{3}\times\frac{63}{4}\times 22}{\frac{27}{4}} d. = \frac{49\times \overset{7}{\cancel{63}}\times 22}{\underset{3}{\cancel{27}}\times 3} d.$$

$= 838d.$ to the nearest penny

$= \underline{£3/9/10.}$

```
    49
     7
   ---
   343
    22
   ---
   686
   686
  ----
 9)7546
12)838..d.
   69s. 10d.
```

EXERCISE XI a

1. Find the area of the rectangles: (i) $2\frac{1}{2}''$ by $3\frac{1}{2}''$; (ii) $1\frac{3}{4}''$ by $2\frac{1}{4}''$; (iii) $1\frac{3}{8}''$ by $2\frac{1}{8}''$; (iv) $2\frac{1}{16}''$ by $1\frac{5}{8}''$; (v) $2'\ 3''$ by $4'\ 6''$, in sq. ft.; (vi) $3'\ 9''$ by $1'\ 8''$, in sq. ft.; (vii) 1 yd. 2 ft. by 3 yds. 1 ft., in sq. yds.; (viii) $5\frac{1}{2}$ yds. square.

2. Find the area of the rectangles: (i) 12·5 cm. by 14·3 cm., in sq. cm.; (ii) 1·25 dm. by 1·43 dm., in sq. dm.; (iii) 0·125 m. by 0·143 m., in sq. m.

3. A postage stamp measures 0·95 in. by 0·8 in. Find to the nearest square inch the area of 10*s*. worth of (1) penny, (2) halfpenny stamps.

4. Find to the nearest sq. cm. the area of (i) a postcard 11·3 cm. by 8·82 cm.; (ii) a letter card 14·6 cm. by 9·12 cm.; (iii) a sheet of note-paper 17·2 cm. by 11·3 cm.

5. Find the number of sq. cm. in a sq. in., given that 1 in. = 2·54 cm. (answer correct to 2 places).

6. A rectangular field is 18·3 chains long and 14·5 chains wide. Give the area in acres to the nearest acre.

7. A rectangular estate measures 1260 yds. by 600 yds., the longer sides running E. and W. It is to be laid out for building, and is intersected by 3 straight roads, each 18 yds. wide—one running E. and W., the two others intersecting it and running N. and S. Find, to the nearest hundredth of an acre, what area would be available for building purposes; and, if each residence is to be allowed 450 sq. yards of ground, how many may be built.

8. Find the weight of a lead sheet $12'\ 7''$ long by $5'\ 3''$ broad, weighing 2·4 lbs. per sq. ft.

9. The area of a rectangle is 511 sq. cm. The length is 28·1 cm. Find the breadth to the nearest mm.

10. A book contains 360 pages which measure 8 in. by $5\frac{1}{2}$ in. Every page has a blank margin 1 in. deep along each short side and 3/4 in. deep each long side. Find the total area available for printing.

11. The ceiling of a room is in the form of a rectangle, 27 ft. 6 in. long and 13 ft. 8 in. wide, with a smaller rectangle (over the fireplace) cut out. If this smaller portion is 7 ft. 4 in. long and 1 ft. 3 in. wide, calculate the area of the ceiling in sq. ft.

12. How many stones each 1 ft. by 7·5 in. would be required to pave a courtyard of 100 ft. by 90 ft.?

13. Find the total area of the four walls of a room, given the length, breadth and height of the room to be (i) 10 m., 7·2 m., 1·8 m.; (ii) 20 ft. 6 in., 18 ft., 12 ft. 6 in.; (iii) 27 ft., 16 ft., 12 ft. 9 in.

14. How much cardboard is needed to make boxes whose length, breadth and height are (i) 8 cm., 5 cm., 4·2 cm. (no lid); (ii) 8·2 cm., 3·5 cm., 3·5 cm. (no lid); (iii) 8 cm., 4·25 cm., 4·25 cm. (with lid); (iv) 4 cm., 2·25 cm., 2·5 cm. (with lid)?

15. Repeat Ex. 14, doubling all the dimensions.

16. The external dimensions of an open tank are: length 5 ft. 4 in., breadth 4 ft. 6 in., depth 3 ft. Find the cost of painting the outside at 3*d*. per square yard.

17. Find the cost of carpeting the following floors:

Length	Breadth	Price of carpet
24 ft.	20 ft.	6*s*. per sq. yd.
18 ft.	12 ft.	2*s*. 6*d*. ,, ,,
25 ft.	22 ft.	3*s*. ,, ,,
15 m.	12 m.	8 f. per sq. m.
12·5 m.	10 m.	8·25 f. per sq. m.

18. A carpet 20 ft. by 12 ft. costs £10. What is the cost per sq. yd.?

19. Find the cost of the following carpets: (i) 24 ft. 6 in. by 18 ft. 9 in. at 3*s*. per sq. yd.; (ii) 19 ft. 8 in. by 12 ft. 10 in. at 4*s*. 6*d*. per sq. yd. (to the nearest penny); (iii) 10·45 m. by 8·55 m. at 5·275 f. per sq. m. (to the nearest centime).

20. Wall paper is usually sold in strips 21 inches wide. What is the price

(1) Per square foot if a length of 1 yard costs $1\frac{1}{2}d.$?
(2) Per sq. ft. if a piece 12 yds. long costs 2/6?
(3) Per yard if one sq. foot costs 1*d*.?

21. Oak planks 16 cm. wide are placed side by side to form a paling. How many will be required for a paling 1 kilometre long? If the paling is 1 m. 5 dm. high what length of planking will be required?

22. The floor of a room is 26 ft. long by 18 ft. broad. What will it cost to carpet it at 6*s*. 6*d*. a square yard, so as to leave uncovered a width of one yard all round the walls?

Find the extra cost of covering this space with matting at 1*s*. 4*d*. per square yard.

23. What is the cost of papering the walls of a room 30′ long, 19′ broad and 15′ high, with paper 21″ wide, at 2*s*. per 'piece' of 12 yards?

24. A room is 12′ 2″ long, 9′ 10″ wide, and 11′ 2″ high: find the cost of painting the four walls at $2\frac{1}{4}d$. per sq. ft.

25. A room is 20 ft. 6 in. long, 17 ft. 4 in. broad and 12 ft. high, and the windows, door and fireplace together measure 162 square feet. How much will it cost to paint the walls and ceiling at 2*s*. per square yard?

26. Calculate the cost of paper for a room 18 feet long, 17 feet broad and 12 feet high, allowing 40 square feet for doors and windows, the price of the paper being 2*s*. 4*d*. per piece 12 yards long and 21 inches broad.

27. What will it cost to paper the walls of a recess 4 feet deep, 6 feet long, and 15 feet high with paper sold in rolls, 12 yards long and 21 inches wide, at 1*s*. 10*d*. a roll?

28. Find the cost of a carpet 2 ft. 3 in. wide at 5*s*. a yard, for a room 20 ft. 3 in. by 13 ft. 4 in., and of paper 2 ft. 7 in. wide at $4\frac{1}{2}d$. a yard, the height of the room being 10 ft. 6 in.

29. A room is 15 ft. 6 in. long and 12 ft. 6 in. wide. Find the cost of carpeting it with carpet 30 in. wide at 2*s*. 11*d*. per yard.

30. A room 15 feet long by 12 ft. 6 in. wide has a carpet in the centre which leaves a margin of 9 inches all round the room. This margin is covered with linoleum, 6 ft. wide, at 3*s*. a yard. Find the cost of the linoleum.

31. Find the area of carpet required for the floor of a room 15′ by 13′, leaving a margin 1′ 6″ wide all round. Find the cost of carpeting the floor and staining the border, the carpet being 30″ wide at 1*s*. 3*d*. per yard, and the staining 1*d*. per square foot.

32. A cricket field is square, measuring 300 yards each way. Four pieces of it, each 26 yds. by 6 yds., are laid with turf at a cost of 2*d*. per sq. ft. The rest is sowed with grass at a cost of 2*s*. per 100 square feet. Find, correct to the nearest £1, the total cost of preparing the field.

33. If a piece of cardboard measuring 16 cm. by 20 cm. weighs 115·2 gm., find the area of a piece which weighs 194·4 gm.

34. If the weight of a piece of cardboard 10 inches long and 6 inches wide is 26·25 gm., find the area of a piece which weighs 16·8 gm.

Volume of Cuboid

§ **121.** In Chapter IV we found a rule for the volume of a cuboid, which was abbreviated to

$$\text{Volume of cuboid} = \text{length} \times \text{breadth} \times \text{height}.$$

This was proved only for the case when all the dimensions are expressed by whole numbers. Let us now consider a case in which the dimensions are not expressed by whole numbers, e.g. a cuboid whose dimensions are 3·45 metres, 4·2 metres, 1·26 metres. It will be noticed that if we express all three dimensions in centimetres, the numbers involved become whole numbers; the dimensions are 345 cm., 420 cm., 126 cm. In accordance with the rule already proved, the volume is

$$345 \times 420 \times 126 \text{ cubic centimetres.}$$

Now 1 m. = 100 cm. $\quad \therefore$ 1 cu. m. = 100^3 cu. cm.

$$\therefore \text{ the volume is } \frac{345 \times 420 \times 126}{100^3} \text{ cubic metres}$$

$$= \frac{345}{100} \times \frac{420}{100} \times \frac{126}{100} \text{ cu. m.}$$

$$= 3{\cdot}45 \times 4{\cdot}2 \times 1{\cdot}26 \text{ cu. m.}$$

This *suggests* that the rule holds in all cases, which is true.

Example. 3000 kilograms of water are poured into a rectangular cistern whose base measures 2·4 m. and 1·8 m.; how deep is the water?

3000 kilograms of water occupy

3000 cu. dm. or 3 cu. m.

There is a cuboid of water, whose depth = its volume ÷ its base

$$= \frac{3}{2{\cdot}4 \times 1{\cdot}8} \text{ m.}$$

= 0·69 m., or <u>69 cm.</u>

```
 1·8
 2·4
 ---
 3·6
  72 |0·694
 ----
4·32 |3·000
      2 592
      -----
       4080
       3888
       ----
        1920
        1728
```

EXERCISE XI b

1. How many 1/4″ cubes can be cut from a 2″ cube?

2. Find the volume of each of the following cuboids—the length, breadth and height being:

(1) 1 in., 1 in., 1/2 in. (2) 2 in., 1 in., 1/2 in.
(3) 2 in., 2 in., 1/2 in. (4) 4 in., 3 in., 1/4 in.
(5) 1·2 dm., 1·1 dm., 1 dm. (6) 1·4 dm., 1·1 dm., 1 dm.
(7) 1·8 cm., 1·4 cm., 1·2 cm. (8) 1·45 cm., 1·23 cm., ·56 cm.

3. How many cubic inches of water can be put inside the following boxes, of which the internal dimensions are:

(i) 6·5 in., 5 in., 3 in. (ii) 6·5 in., 4 in., 2·5 in.
(iii) 6·25 in., 8 in., 1·5 in. (iv) 1·5 in., 2·5 in., 10 in.?

4. How many cubic feet of wood are there in a door 8 ft. 3 in. high, 4 ft. wide and 2 in. thick?

5. An ingot of mild steel measuring 2′ 6″ by 1′ 2″ by 2·5″ costs 8/9. What is the cost of a cubic foot?

6. How many bricks are there in a stack 30 ft. long, 16 ft. 6 in. broad and 9 ft. 3 in. high, if each brick measures 9 in. long, 4·5 in. broad and 3 in. deep?

7. A cube, each of whose edges is 0·3 metre, is formed of metal 7 cubic centimetres of which weigh 50 grams. How many kilograms does the cube weigh?

8. A cistern measures 2·75 metres in length, 2·8 in breadth, and 1·7 in depth. How many litres will it hold?

9. A tank 1·2 metres long and 67 cm. broad is filled with water to a depth of 35 cm. Find the weight of the water to the nearest pound. (A kilogram = 2·204 lbs.)

10. What is the area in sq. m. of the bottom of a truck which is filled to a depth of 80 cm. by 10 cu. m. of sand?

A cubic foot is equivalent to 6·25 *gallons.*

11. A cistern is 16′ long, 6′ 4″ wide, and is filled with water to a depth of 3′ 3″. Find how many gallons of water there are.

12. The volume of a certain brick is 63 cubic inches. If its length is 6 in. and its breadth 4·5 in., find its thickness.

13. A tank is 6 ft. 4 in. long, 5 ft. 3 in. wide, and 4 ft. deep. Find the number of gallons of water it can hold.

14. A tank whose internal length and breadth are 3 feet and 2 feet 8 inches respectively contains 4 cubic feet of water. What is the depth of the water?

15. Find the least depth of a tank whose base is 4 ft. square in order that it may hold 250 gallons.

16. Tin costs £210 per ton; aluminium 1*s.* 2*d.* per lb. Attempts have been made to substitute aluminium plating for tin plating in the manufacture of cans. Find the price of each metal used in plating a surface of 400 sq. ft. with a thickness of 0·001 in. of metal, given that 1 cu. ft. of tin weighs 450 lbs. and 1 cu. ft. of aluminium weighs 165 lbs.

17. A level rectangular roof is 20 ft. long by 14 ft. wide. The rain which falls on it is collected in a rectangular cistern 6 ft. long, 2 ft. 6 in. wide, and 8 ft. deep. What depth of water will be in the cistern after a rainfall of an inch?

18. A cistern 9 ft. long by 5 ft. wide is filled to a depth of 7·5 in. with pulp for making paper. How long a roll, 30 in. wide, can be made from this pulp, the thickness of the paper being 0·005 in., supposing half the volume of the pulp to be lost in the process of drying?

19. A schoolroom is used for a class of 60 children and 1 teacher. For proper ventilation an adult requires 3000 cubic feet of air per hour and a child 3/4 of that amount. If air can be admitted at the rate of 5 ft. per sec., what is the least size of opening necessary to allow sufficient air to enter?

20. Find the cost of making a road 110 yards long and 18 feet wide, the soil being first excavated to the depth of 1 foot at a cost of 1*s.* per cubic yard, rubble being next laid down 9 inches deep at 1*s.* per cubic yard, and gravel placed on the top 3 inches thick, at 2*s.* 6*d.* per cubic yard.

21. Steel plate 0·076 in. thick weighs 3·04 lbs. per sq. ft. of surface: what is the weight of a cubic foot of the metal?

22. Find the weight of an oak box with a cover, the outside dimensions being 4 ft. by 3 ft. 3 in. by 3 ft. and the thickness of the wood 1·5 in. (1 cu. ft. of oak weighs 54 lbs.)

23. The external measurements of a tank, with lid, are 2 ft. 9 in. by 3 ft. 2 in. and 1 ft. 10 in. deep and the metal is 1/2 in. thick. Find (i) how many cubic inches of metal there are in the tank; (ii) its capacity in gallons.

24. A rectangular box with lid is found to be 12·5 inches long, 10 inches wide, and 5·25 inches deep measured externally. If it is made of wood 3/8 in. thick, how many cubic inches of air does it contain? If a square foot of wood of this thickness weighs 1 pound, how much does the empty box weigh?

25. How many cu. ft. of water are there in an open rectangular tank partly filled, the length of the tank being 4 ft. 6 in., the breadth 3 ft., and the water 32 in. deep?

How many inches will the water rise if a stone of 1296 cu. in. in volume is put into it?

26. A tank, 3 ft. 6 in. by 3 ft., contains water to a depth of 2 ft. If a cube of metal, of edge 20 in., is introduced into the tank, through what height (to the nearest tenth of an inch) will the water rise?

27. An open box, made of wood 1 centimetre thick, has a square base 12 cm. by 12 cm. (outside measurement). If the box would exactly hold half a litre of fluid, find its external height.

28. How many cubes of edge 3·5″ can be cut from a block of wood 16″ long, 12″ wide and 9″ thick? How many such cubes could be made from a similar block of metal, which could be melted down and recast?

29. Water is poured into a reservoir, whose horizontal cross-section is a rectangle 20 ft. broad and 30 ft. long, at the rate of 400 gallons a minute. Find the rate at which the water rises. (1 gallon = 277 cub. in.)

30. An ordinary match-box consists of an open tray sliding into a cover open at each end; it is made of thin wood whose thickness may be neglected. It measures 2·5″ by 1·5″ by 0·75″. Find the area of wood used in making the box; also the volume of the interior.

31. Find, to the nearest ton, the weight of an inch of rain over an acre of ground. [1 cu. foot of water weighs 62·5 lbs.]

32. During a rainfall of one inch 706·7 tons of water fell on 7 acres of ground. Deduce the weight of a cubic foot of rainwater in ozs.

SOLID OF UNIFORM CROSS-SECTION

§ 122. Fig. 1 represents a steel rail such as is used on railways. If you look at either end of the rail, you see the shape depicted in Fig. 2. If the rail were cut across at any point, the cut or **cross-section** being at right angles to the length, the same figure would be seen. The rail is therefore said to be a **solid of uniform cross-section.**

Fig. 1. Fig. 2.

Contrast this with an object such as a billiard-cue. Here the cross-section is a circle—for most of the length at any rate—but the circle is smaller and smaller as the section is taken nearer and nearer to the tip. A billiard-cue is therefore not a solid of uniform cross-section.

Among manufactured articles there is a great variety of solids of uniform cross-section, e.g. tubes and pipes, gutters, wire, girders, rulers, picture mouldings.

§ **123.** We will now discover a rule for calculating the volume of any solid of this description. Suppose that we want to find the volume of wood in a yard of wooden spouting (Fig. 1), whose cross-section is the figure shown in Fig. 2.

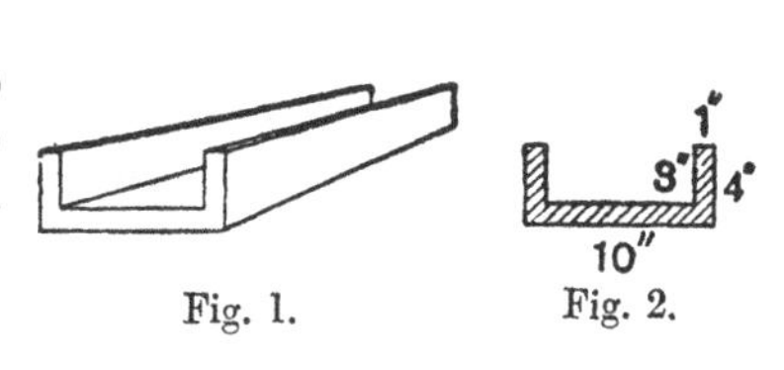

Fig. 1. Fig. 2.

The cross-section is 10 in. wide and 4 in. deep (outside measurements) and the wood is 1 in. thick.

Imagine that a length of 1 inch of the spouting is sawn off.

To find the volume of this one-inch length we may imagine it cut up into cubic inches. The obvious way to cut it up accurately would be to begin by marking off the end, or cross-section, into square inches. As soon as this is done, it is clear that for every square inch of end, we have a cubic inch of wood in the one-inch length. In other words, *the number of cubic inches in the one-inch length is equal to the number of square inches in the cross-section.*

To find the number of cubic inches in a 36-inch length, we multiply by 36. Hence the rule:

The number of units of volume in a solid of uniform cross-section

= number of units of area in the cross-section
$\times$ number of units in the length,

or, briefly,

volume of solid of uniform cross-section
= area of cross-section $\times$ length.

¶**Ex.** See if this rule gives the right result in the case of a cuboid.

EXERCISE XI c

1. Calculate the volume of a solid of uniform cross-section, if (1) cross-section = 1 sq. m. and length = 4 dm.
(2) cross-section = 0·2 sq. m. and length = 1·3 m.

2. What is the length of a solid of uniform cross-section 45 sq. cm. and volume 1 litre?

3. A section of a steel rail has an area of 1 sq. dm. Find the weight of 1 Km. of rail (sp. gr. of steel = 7·8).

4. What is the area of the uniform cross-section of a solid whose length is 35 cm. and volume 45·5 litres?

5. The section of a tunnel is 13·5 sq. metres. Find the number of cu. m. excavated in boring it a distance of 8 m.

6. The area of the section of a boring is 1325 sq. ft., and the excavating machine is driven forward 4 ft. a day. How many cubic yards of earth are excavated in a day?

7. Find the volumes of the solids shown in Figs. 1 and 2. (In Fig. 2 each short edge = 4″.)

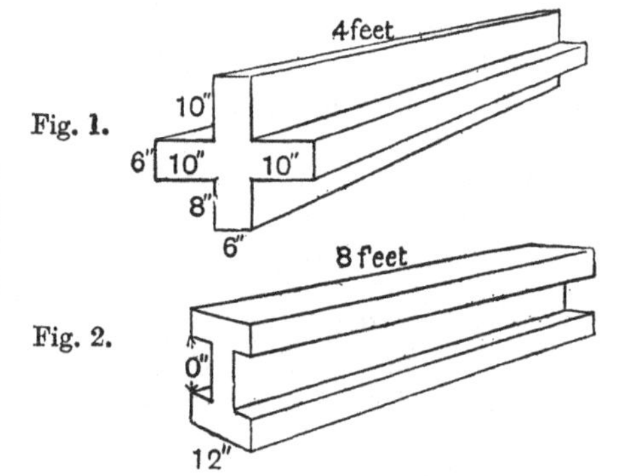

Fig. 1.

Fig. 2.

8. The cross-section of a stream being 210 square feet, find the number of gallons of water that flow under a bridge per hour, the rate of the stream being 1·25 miles per hr.

In the following exercises assume that:

Area of triangle $= \frac{1}{2}$ *base* × *height.*

Area of trapezium $= \frac{1}{2}$ *sum of parallel sides* × *distance between them.*

9. The base of a triangular prism is a triangle of sides 3, 4, 5 in., and the height is 8 in. Calculate the volume.

10. Find the capacity in litres of a vessel whose interior is a prism whose base is a right-angled isosceles triangle of side 10 cm., and whose height is 10 cm.

11. What is the cubic content of a roof in the form of a 3-sided prism with side-edges horizontal, the floor (i.e. the ceiling of the room below) being a rectangle 8·4 m. long and 5·2 m. broad, and the ridge being 3 m. above the floor?

12. Find the number of cubic inches in a sheet of brass 0·15″ thick which is in the shape of a right-angled triangle whose sides are 12·4″, 9·8″, and 7·6″.

13. Water flows in a V-shaped gutter. Find the number of cubic metres discharged in a day if the depth is 20 cm., and the speed a metre per second, the sides of the trough meeting at a right angle.

14. A railway cutting 8 metres deep has to be made with one side vertical and the other inclined at 45° to the vertical; the bottom is to be 9·4 metres broad. How many cubic metres (to the nearest integer) of earth and rock must be removed per kilometre?

15. A railway cutting is 300 feet broad at the top, 100 feet broad at the bottom and 30 feet deep. If the cutting is 150 yards long, how many cubic yards of earth have to be excavated?

16. A trench 22 yds. long has a uniform cross-section 10 ft. wide at the top, 7 ft. wide at the bottom, and 5 ft. deep. Find the weight of earth excavated, reckoning 20 cubic ft. to the ton.

17. A watercourse of uniform cross-section is 10 ft. wide at the top and 2 ft. wide at the bottom; the depth being 3 ft. If the water be running 1 ft. deep, with a mean velocity of 120 ft. per minute, find the number of cubic ft. of water that are passing in a minute.

18. A swimming-bath is 100′ long, 40′ wide, 8′ deep at one end, and 4′ at the other. How many cu. ft. will it hold? How much will the surface of the water fall if 5000 cubic feet of water are run off?

19. Find the number of cubic feet of water in a plunge bath whose vertical section is a trapezium, given the following data: depth at shallow end 4′ 6″; depth at deeper end 8′ 3″; surface of water 9′ 6″ square.

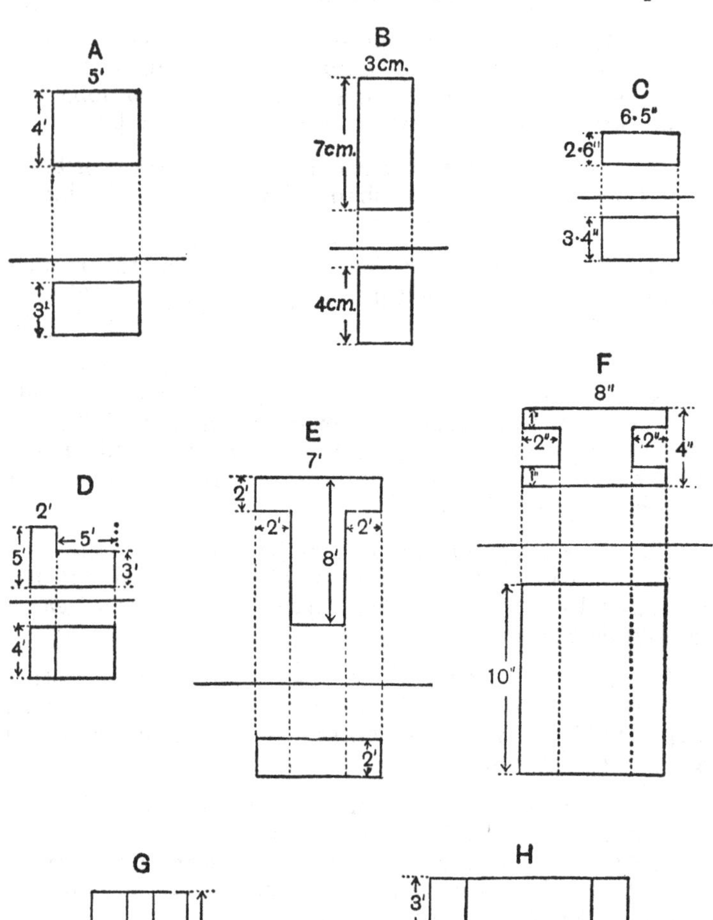
A
5'
4'
3'
B
3cm.
7cm.
4cm.
C
6·5"
2·6"
3·4"
D
2'
5'
5'
3'
4'
E
7'
2'
2'
2'
8'
2'
F
8"
1"
2"
2"
4"
1"
10"

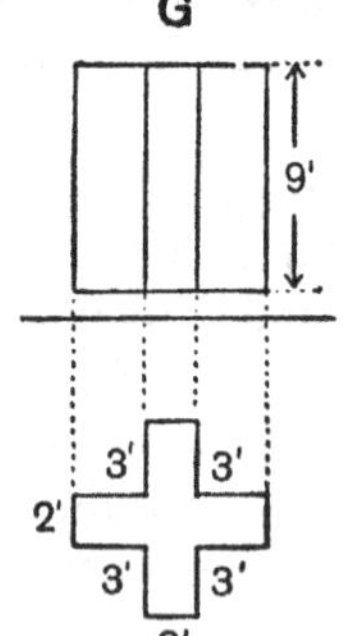
G
9'
3'
3'
2'
3'
3'
2'

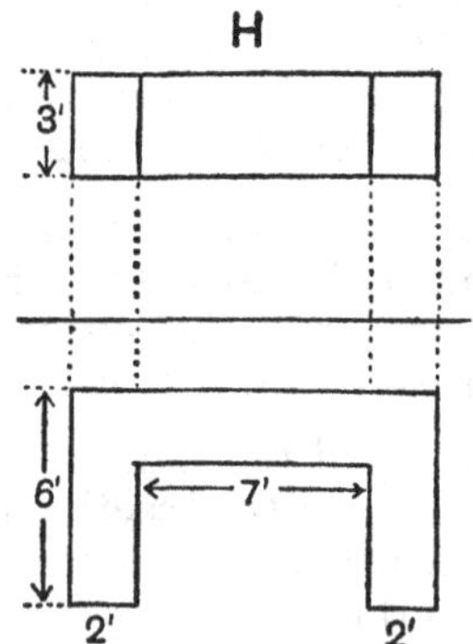
H
3'
6'
7'
2'
2'

20. In the figures A—H each solid is given by its plan and elevation (the elevation being above the plan). Find the volume in each case.

21. A bunker 24 ft. long has a section in the form of a trapezium whose parallel sides 4 ft. 9 in. and 3 ft. are 10 ft. 6 in. apart. Find the number of tons of coal it can hold at 43 cu. ft. to the ton. [To the nearest 1/10 ton.]

22. A canal is 8 feet deep, 18 feet wide at the surface, and 12 feet at the bottom. Find the area of the cross-section (which is trapezium shaped) and the number of gallons of water that pass any spot in an hour, when the water flows at 2 miles per hour. (1 cubic foot = 6·25 gallons.)

23. A haystack is built oblong with vertical ends perpendicular to the length; the length is 30 ft., the width at base 20 ft., the width at the eaves 24 ft., the height to eaves 14 ft., and the height from eaves to ridge 8 ft. Find the weight of hay, reckoning 10 cubic yds. to the ton.

24. A snow-drift 10 ft. wide lies all along a wall 100 ft. long; the snow slopes away from the wall at a uniform angle, and reaches to a height of 2 ft. up the wall. How much snow is there, and how many carts, whose internal dimensions are $2\frac{1}{2}$ yds. by $1\frac{1}{2}$ yds. by 1 yd., will be needed to cart the snow away?

25. A tank 4 ft. by 3 ft. by 2 ft. contains water to a depth of 10 in. By how much will the depth be increased if (1) a 10-inch cube of lead be put in the tank? (2) a 20-inch cube of lead be put in the tank?

CHAPTER XII

SPECIAL TYPES OF PROBLEM

Change of Units of Velocity

§124. *Example.* Express a speed of 60 miles per hour as feet per second.

$$60 \text{ miles per hour} = 60 \times 1760 \times 3 \text{ ft. per hour}$$
$$= \frac{60 \times 1760 \times 3}{60 \times 60} \text{ ft. per sec.}$$
$$= 88 \text{ ft. per sec.}$$

It is useful to remember that 60 miles per hour = 88 feet per second; for this often enables us to do similar exercises mentally; thus 30 miles per hour = 44 ft. per sec.

The phrases "miles per hour," "feet per second," etc., are often abbreviated thus mi./hr., ft./sec., etc.

EXERCISE XII a

1. Express in feet per second (to the nearest foot) the following velocities:

(i) 100 yards in 10 seconds. (ii) 1/4 mile in 48 secs.

(iii) 1 mile in 4 mins. 24 secs. (iv) 25 miles an hour.

(v) 30 nautical miles per hour. [1 n. mile = 6080 feet.]

2. Express, correct to 1 place, (i) 15 cm./sec. as ft./min., (ii) 54 Km./hr. as ft./sec., (iii) 4 mi./hr. as metres per sec.

3. Express in miles per hour (correct to 1 place):

(1) 100 yards in 10·4 secs. (2) 1600 ft. per sec.

(3) 2000 yards in 1·5 minutes. (4) 130 yds. in 4·4 secs.

(5) 1 nautical mile per min. (6) 27 knots (i.e. n. mi. per hr.).

4. At what rate in miles per hour is a train 100 yds. long travelling if it passes a signal in 6·4 seconds?

5. A bullet leaves the rifle with a muzzle velocity of 2200 feet per second. Express this speed in miles per hour.

6. A train was found to travel 1 mile in 58·4 secs. Express its speed (i) in miles per hour, (ii) in ft. per second.

7. In attempting the 5 Km. speed record, an airman flew 5 Km. in 2 mins. 48 secs. Express this speed in miles per hour, given 1 Km. = 0·621 mile.

8. The length of a man's step is 2 ft. 8 in. How many steps does he take in a minute when walking at 4 mi./hr.?

9. From Liverpool to New York is 3250 land miles. If the *Mauretania* keeps up an average speed of 25 knots, how long will she take to cross (days, hours)?

10. Telegraph posts are 55 yards apart. If I pass 14 in a minute, at what rate in mi./hr. do I travel?

11. What is the average speed of a train (in feet per second), if it goes from London to Southampton (69 miles) in an hour and 32 minutes?

12. At what rate in miles per hour is a train 80 yards long travelling if it passes completely over a viaduct 470 yards long in 45 seconds?

13. A train travelling at 60 miles per hour whistles when it is 2330 feet away from a signal box. If the velocity of sound be 1100 feet per second, find what time elapses between the time when the signalman hears the whistle and the time at which the engine reaches the signal box.

14. A ship is firing at a town, at a distance of 6 nautical miles. If sound travels at the rate of 1100′ a second, how far will the ship, which is steaming 12 knots, have travelled before the sound is heard in the town?

AVERAGES

§125. If a boy makes the following scores at cricket: 2, 17, 8, 15, 0, 1, 16, 3, 8, 0, 13, 1; in all he has made 84 runs in 12 innings. His **average** is said to be $84 \div 12 = 7$. Note that if he had in each innings made his average score of 7, his total for the 12 innings would have been 84.

Example. If the average weight of a boat's crew of 8 is 11 stone 2 lbs., what must be the weight of the coxswain to bring down the average to 10 stone 12 lbs.?

The total weight of the 8 is

11 st. 2 lbs. × 8 = 156 × 8 lbs. = 1248 lbs.

The total weight of the 9 is to be

10 st. 12 lbs. × 9 = 152 × 9 lbs. = 1368 lbs.

∴ the coxswain weighs 120 lbs. = 8 st. 8 lbs.

EXERCISE XII b

1. What is the average strength of three regiments, of which the strengths are 620, 655, and 714 respectively?

2. What is a batsman's average when he has made 0, 17, 29, 73, 187, 34, 38, (i) if all the innings are completed, (ii) if he was once "not out"?

3. Find (to the nearest whole number) the averages of the following members of the M.C.C. team in Australia, 1912.

Batting

	Innings	Runs	Times not out
Hobbs	23	1156	2
Douglas	18	481	5
Gunn	17	746	3
Foster	20	683	1

Bowling

	Runs	Wickets
Barnes	1231	59
Douglas	833	42
Foster	1270	66
Hitch	735	37

4. Find the average height of three men who are 6 ft., 6 ft. 1 in., 5 ft. 8 in. high respectively.

5. Find the average weight of a crew of 8 men: bow being 10 stone, No. 2 11 st. 9 lbs., No. 3 11 st. 1 lb., No. 4 12 st. 5 lbs., No. 5 13 st. 1 lb., No. 6 12 st. 9 lbs., No. 7 12 st. 10 lbs., stroke 10 st. 4 lbs. What will be the average if we include the cox, who weighs 7 st. 3 lbs.?

6. In a certain innings the average score of an eleven was 30; and the average of the first six men was 40; what was the average of the last five?

7. A cricketer has averaged 57·3 runs for 34 completed innings. In his next match he makes 1 run his first innings, and 101 runs his second innings. Find his average at the conclusion of each innings.

8. A batsman has made the following scores: 1, 3, 41, 8, 62, 12, 30, 118, 29, 39, 62; he has been four times not out. Find his average, and the score he must make next time (if he gets out) to bring his average up to 60.

9. The average weight of the crew of an eight-oared boat, including the coxswain whose weight is 8 st. 0 lbs. 5 oz., is 6 lbs. 1 oz. less than when he is excluded. What is the average weight of the rowers?

10. On March 22 the sun rises at 6 o'clock and sets at 6.14 p.m.; on April 22 it rises at 4.52 a.m. and sets at 7.6 p.m. What is the average daily lengthening of daylight for this period? Give your answer to the nearest second.

11. The monthly marks and places for a class of 5 boys were as follows:—

A	123 (3)	201 (2)	204 (1)
B	74 (5)	141 (4)	201 (2)
C	112 (4)	204 (1)	186 (4)
D	241 (1)	173 (3)	200 (3)
E	235 (2)	140 (5)	161 (5)

Show that their average places give no indication of their actual places at the end of the term as calculated by adding up their marks.

RELATIVE VELOCITY

§ 126. A cruiser, steaming 30 knots, is chasing a destroyer, steaming 25 knots: the distance apart is 10 (nautical) miles.

A knot is a nautical mile per hour. In 1 hour the cruiser will have gained $30 - 25 = 5$ miles; the "velocity of overtaking" is therefore 5 miles per hour. This is called the **relative velocity**; in the case of two objects moving along the same line *in the same direction* the relative velocity is the *difference* of the velocities.

As the "distance to be overtaken" is 10 miles, the time required is 2 hours.

Next, take a case of two objects moving to meet one another; say two locomotives approaching one another and travelling at 35 and 45 ft. per sec. In one second, the distance apart will have diminished by $35 + 45 = 80$ ft.; the "velocity of approach" is 80 ft. per sec. This again is the relative velocity; in the case of two objects moving along the same line *in opposite directions* the relative velocity is the *sum* of the velocities.

If the two drivers sight one another at 1 mile distance, this distance will be consumed in $\frac{3 \times 1760}{80}$ secs. $=$ 66 secs.

In general, then, we have this rule: time required to cover distance between two objects moving along the same line

$$= \text{distance apart} \div \text{relative velocity}.$$

Example. Two trains of length 150 ft. and 200 ft. are passing one another at speeds of 30 and 40 miles per hour respectively. How long do they take to pass?

$$30 \text{ miles per hour} = \frac{30 \times 1760 \times 3}{60 \times 60} \text{ ft. per sec.} = 44 \text{ ft. per sec.}$$

$$40 \text{ miles per hour} = 44 \times \tfrac{4}{3} \text{ ft. per sec.} = \frac{176}{3} \text{ ft. per sec.}$$

$$\text{The relative velocity is} \left(44 + \frac{176}{3}\right) \text{ ft. per sec.} = \underline{\frac{308}{3} \text{ ft. per sec.}}$$

Fixing our attention on the fronts of the engines, we see that these two points are, to begin with, level; when the trains are just clear of one another, the distance between these points is the sum of the lengths of the trains, namely 350 ft.

The distance between these two points has increased from 0 to 350 ft. at a relative velocity of $\frac{308}{3}$ ft. per sec.

The time of passing is therefore $350 \div \frac{308}{3}$ secs.

$$= \frac{\overset{25}{\cancel{\overset{50}{\cancel{350}}}} \times 3}{\underset{22}{\cancel{\underset{44}{\cancel{308}}}}} = \frac{75}{22} = \underline{3\tfrac{9}{22} \text{ secs.}}$$

Example. At what time between 2 p.m. and 3 p.m. are the hands of a clock in a line?

This can be reduced to a problem of relative velocity if we imagine the hands to be of the same length, so that the tip of each hand describes the same circle. Take $\frac{1}{60}$ of this circle as unit of length: then the speeds of the tip of the large and small hands respectively are 60 units per hour and 5 units per hour. The relative speed is $60 - 5$, or 55 units per hour.

At 2 p.m. the small tip is 10 units in front of the large tip. The large tip will overtake it in a little more than 10 minutes; and when the hands are in a line, the large tip will be 30 units in front of the small. Since 2 p.m. the large tip has gained $10 + 30$ or 40 units; and the time taken is $\frac{40}{55}$ hours $= \frac{8}{11}$ hours $= \frac{480}{11}$ minutes $= 43\frac{7}{11}$ minutes.

The hands will therefore be in a line at $\underline{2.43\frac{7}{11}\text{ p.m.}}$

§ 127. **Train problems** are of some practical interest: if an express is overtaking a slow train, arrangements must be made to put the latter into a siding when the express comes by. Railwaymen work out these time-tables **graphically**, as follows. First, let us take the case of an express travelling at 60 miles per hour which passes Rugby going north at

12.20 p.m. We will draw a graph showing along the horizontal axis "time after 12 p.m.," and along the perpendicular axis "distance north of Rugby."

At 12.20 p.m. we mark a point *on* the horizontal axis; at 12.30 p.m. a point 10 miles "up"; at 12.40 p.m. a point 20 miles up, etc. Clearly all these points lie in a straight line: in fact, *the space-time graph for a train moving with uniform speed is a straight line.* We can therefore draw the graph with a ruler after marking *two* points on it; and generally speaking the obvious points to mark are a point for the beginning, and a point for an hour later.

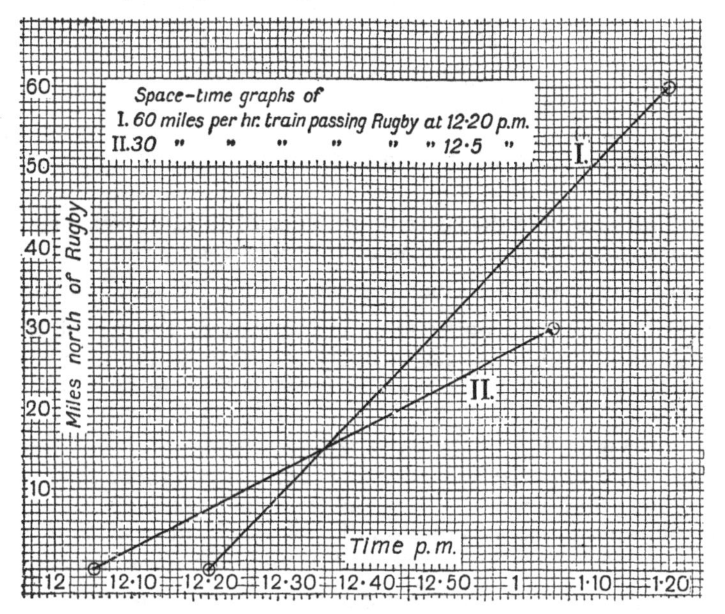

Now suppose that a slow train starts from Rugby at 12.5 p.m. and travels north at 30 miles per hour. We draw its space-time graph by marking the points corresponding to 12.5 p.m. and 1.5 p.m.

Notice that the faster train has the steeper graph.

The intersection of the graphs shows that the express will pass the slow train between 12.34 p.m. and 12.35 p.m., at a distance of about 14½ miles from Rugby.

EXERCISE XII c

1. Two men start walking towards one another at the same time from two towns 12 miles apart. One walks at the rate of $3\frac{1}{2}$ miles an hour, the other at the rate of 4 miles an hour. How long will it be before they meet?

2. Newport is 4 miles from Osborne. At 2 p.m. A starts from Newport for Osborne, walking at 3 miles per hour. At 2.15 p.m. B starts from Osborne at 2 miles per hour to meet A. Where and when do they meet?

3. The down express leaves Paddington at 10.30 and arrives at Plymouth at 2.36. The up express leaves Plymouth at 12.30 and arrives at Paddington at 4.42. Plymouth is 225 miles from London. Assuming that each travels at a uniform rate, find the time at which they meet, and the distance of the meeting-place from Plymouth.

4. A train leaves London for Rugby at 2.45 p.m. and travels at 41 miles an hour, whilst another train leaves Rugby for London at 1.45 p.m. and goes at 25 miles an hour. If the distance between London and Rugby is 80 miles, when and where will these trains meet?

5. A steamer leaves port at the rate of $8\frac{1}{2}$ miles an hour at 3 a.m. on Monday. At midnight a despatch boat starts in pursuit at the rate of 15 miles an hour. When will they be in sight of one another, supposing them to be visible 2 miles off?

6. A tricycle going at the rate of 5 miles an hour passes a milestone, and 14 minutes afterwards a bicycle, going in the same direction at the rate of 12 miles an hour, passes the same milestone; when and where will the bicycle overtake the tricycle?

7. A steamer passes the Lizard lights at 11.30 p.m., steering a course due W., and steaming at 18 knots. At 3.15 a.m. on the same night a destroyer, steaming at 30 knots, passes the same spot in pursuit of her. At what time and at what distance from the Lizard (in nautical miles) will the destroyer overtake her?

8. A and B are two towns 116 miles apart. A non-stop express travelling at 40 miles an hour leaves A for B at 12.10 p.m., and another non-stop express travelling at 50 miles an hour leaves B for A at 12.22 p.m. At how many miles from A will the trains meet?

9. A train of speed 30 miles an hour starts for London from Manchester at 8.30 a.m. A train of speed 45 miles an hour starts for Manchester from London at 9.18 a.m. The distance from London to Manchester being 189 miles, how far from Manchester, and when, do they meet?

10. A man leaves a house and walks along a road at 5 miles per hour. 10 minutes afterwards the owner of the house starts in pursuit on a bicycle, riding at the rate of 10 miles per hour, but he has to dismount after the first six minutes, and wastes 3 minutes blowing up his tyre. He then continues the chase. How long does the chase last?

11. At 8 a.m. a person started from Cowes to walk to Sandown—a distance of 12 miles—at the rate of 4 miles per hour. At 8.30 a.m. two cyclists started, one from Cowes, the other from Sandown, each at a speed of 8 miles per hour. At what time did each pass the pedestrian?

12. Two trains start from a station together. One travels at the rate of 40 miles per hour, the other at 44 feet per second. How far apart will they be in 3/4 of an hour?

13. A thin candle 8 inches long which burns at the rate of 3 inches in 2 hours, and a thick candle $5\frac{1}{2}$ inches long which burns at the rate of 2 inches in 3 hours, are lighted at the same time. When will (1) both be the same length, (2) the thick candle be one inch longer than the thin one?

14. How long will it take a train 160 yards long to pass another train 140 yards long, travelling in the opposite direction, if their speeds are 30 and 20 mi./hr. respectively?

15. A cyclist travelling along a straight level road on a foggy day, when objects were just visible 220 yds. off, saw a milestone in front of him for half a minute. Afterwards he overtook a carriage, the back of which he saw for a minute and a half. What was the speed of the carriage?

16. Two clocks are set right at 2 a.m. on January 1st. One gains 6 minutes 40 seconds, and the other loses 8 minutes 20 seconds in 24 hours. When will they again show the same time if they are kept working regularly?

17. One train is 150 yards long, another 200; they go 25 and 30 miles an hour respectively; how long do they take to pass one another (1) if they are going the same way, (2) if they are meeting one another?

18. A train 88 yards long takes 7 seconds to pass completely over a bridge 66 yards long. How fast is it going in miles per hour?

19. If a man is walking along the side of the track in an opposite direction at 3 miles an hour, how long will the train take to pass him?

20. If the train overtakes a man and passes him in 5 seconds, at what rate is the man running in miles per hour?

21. A train 88 yards long overtakes a man walking along the line at the rate of 4 miles an hour, and passes him in 10 seconds: what is the rate of the train?

22. If 20 minutes later the train overtakes another man, and passes him in 9 seconds, how long after this will it be before the first man overtakes the second?

23. Two stations, A and B, on a railway are $31\frac{1}{4}$ miles apart. Expresses pass both stations without stopping, moving at uniform speed. An express which passes A at 5.33 p.m. reaches B at 6.5 p.m., and an express, going in the opposite direction, passes B at 5.45 p.m. and reaches A at 6.22 p.m. Find the time at which the trains meet, and the distance of the point of meeting from A.

24. A cyclist leaves home at 11 o'clock and, cycling at an average speed of 12 miles an hour, reaches an inn at 1.15. He rests till 2, and continues his journey at 8 miles an hour until he is overtaken by a motor which started from his home at 1.39, and has not stopped on the journey. If the motor averaged 20 miles an hour up to the inn, and then 12 miles an hour till it overtook the cyclist, when did the motor pass the inn, and what was the total length of the journey?

25. Two persons 7 miles apart set out to walk at the same time. If they walk in opposite directions they meet in 1 hour. If they walk in the same direction they are together in 7 hours. Find the rates of walking.

26. At what times between 4 and 5 o'clock will the hands of a watch (i) be in the same straight line, (ii) coincide, (iii) be at right angles?

(*In problems on currents, the speed of the current is to be added to or subtracted from the speed of the vessel in still water.*)

27. In one of the channels of the Straits of Magellan for 4 hours the tide runs due west at 3 knots; there is then slack tide for 2 hours, after which it sets due east at $2\frac{1}{2}$ knots for another period of 4 hours. A steamer enters the Straits, steaming westward, at 2 p.m. on a certain day after the flood tide has been running west for 40 minutes. She steams so that she would make 10 knots in slack water. How far through the Straits will she get by 8.30 p.m.?

28. The distance from Liverpool to New York is 3025 nautical miles. A Cunarder steams at a rate of 22 knots. For the first 300 miles she does not lose or gain by currents, but after steaming 300 miles she meets the Gulf Stream, the average speed of which is 2 knots. How long would the steamer take to do the whole voyage? (Days, hours.)

29. A pinnace takes $12\frac{1}{2}$ minutes to go from a ship to the landing stage at slack tide, the distance being 3200 yds. How long will she take to go there and come back on an ebb tide running at 3 miles (land miles) per hour, supposing she has to wait 10 minutes at the landing stage?

30. A boy can paddle his canoe 3/4 of a mile down the river in 10 minutes, but without the aid of the stream would take 1/4 of an hour; what is the rate of the stream per hour? and how long will it take him to return against it?

31. A steamer does a passage in still water in 5 hours. A log of wood drifts the same distance in a current in 40 hours. How long would the steamer take to do the same passage with the current? (Answer in hours and fractions.)

32. In the Mersey a steamer takes $1\frac{1}{4}$ hours to clear the estuary in still water. A log floating with the tide takes 4 hours. How long will the steamer take with the tide?

MEN SHARING WORK, PIPES FILLING TANKS, ETC.

§ **128.** *Example.* Two men can dig a garden in 4 and 5 days respectively: how long will they take if they work together?

In one day they dig $\frac{1}{4}$ and $\frac{1}{5}$ of the garden respectively.

$\therefore$ if they work together they dig $\frac{1}{4}+\frac{1}{5}$, i.e. $\frac{9}{20}$ in one day.

$\therefore$ they dig 9 such gardens in 20 days,

or 1 garden in $\frac{20}{9}$ days $= 2\frac{2}{9}$ days.

The last step but one *should* be superfluous.

Example. A tap fills a bath in 10 minutes, and the waste empties it in 2 minutes. If the bath is full and both tap and waste are by accident left open, in what time will the bath be emptied?

In **one minute** the tap adds $\frac{1}{10}$ and the waste subtracts $\frac{1}{2}$. Therefore when both are on, in one minute $\frac{1}{2}-\frac{1}{10}$, or $\frac{2}{5}$, will be lost.

$\therefore$ the whole bath will be emptied in $\frac{5}{2}$, or $2\frac{1}{2}$ minutes.

EXERCISE XII d

1. *A* can eat a ham in a week, and *B* can eat it in 5 days. How long will the ham last them?

2. A bath is provided with a tap that will fill it in 5 minutes, and a waste pipe that will empty it in 8 minutes. If both are open, what fraction will be filled in a minute? In what time will the bath be filled if both are open?

3. An inflow pipe fills a cistern in 40 minutes, the waste empties it in 10 minutes. If both are turned on when the cistern is full, how soon will it be empty?

4. Two pipes separately fill a cistern in 9 and 12 minutes respectively; the waste pipe empties it in 6 minutes: in what time will the two pipes together fill the cistern when the waste pipe is open?

5. A cistern has 3 pipes, one of which can fill it in 10 minutes, another in 18 minutes, while a third empties it in 7 minutes: how long will it take to fill the cistern if they are all opened together?

6. Two pipes A and B fill a tank in 30 minutes and 25 minutes respectively. A third pipe C empties it in 20 minutes. If A and B are turned on together for $1\frac{1}{2}$ minutes, and then C is turned on also, find the time taken in filling the bath.

7. A could do a piece of work by himself in 24 days, and B could do it by himself in 36 days. How long would it take them to do it working together? If, after A has worked at it for 14 days, B joins him, and they both work on together, how soon will it be finished?

8. A can do a piece of work in 72 days, B in 96 days, C in 108 days. They all work together for 24 days and then C leaves off. How soon will the work be finished?

9. If 4 men or 10 women can mow a field in 14 days when the men work 10 and the women 8 hours a day, how long will 2 men and 15 women, working 6 hours a day, take to mow the field?

10. Three equally good workmen A, B, C, can together do a job in 5 days working 8 hours a day. How long will they take if A works full time for 2 days and then leaves off, B and C working 6 and 7 hours respectively all through?

11. A boat springs a leak that would fill it up to sinking point in a quarter of an hour. The crew could bail out this quantity of water in 1/2 hour if the leak were stopped. But they cannot stop the leak. How long will the boat float?

12. A water tank is 4′ long, 3′ wide, and 5′ deep. An exhaust pipe when open lowers the depth of water 2″ in one minute; an inflow pipe when open raises the water 1/2″ in one minute. If the tank were full and both pipes were opened simultaneously, how long would be taken in emptying it? Find the number of gallons the tank will hold, and the number which flow through each pipe per minute.

13. *A* can copy 150 pages in 30 hours. *A* and *B* together copy 280 pages in 35 hours. In what time would *B* copy 200 pages?

REVISION PAPERS

Paper 21 (*on Chaps. I–XII*)

1. State which is the greatest and which the least of the fractions 1/13, 7/90, and 3/40.

2. Multiply 789 by 574. Hence find the value of $78{\cdot}9 \times {\cdot}574$ correct to two places of decimals.

3. Express as decimals:

(i) $\frac{1}{10} + \frac{2}{100} + \frac{3}{1000}$. (ii) $\frac{3}{10^2} + \frac{4}{10^5}$.

(iii) $\frac{6}{2^2 \cdot 5^4}$. (iv) $\frac{1}{7}$. (v) $\frac{2}{3}$.

4. Divide:

·24 by ·012, 240 by ·12, and ·024 by 120.

5. Express £5. 6*s*. 9*d*. as a decimal of £1.

6. A motor car travels for 2 hours 10 minutes at the rate of 20 kilometres per hour. Find the distance travelled.

7. If eggs are sold at 16 for 1/-, what do 12 cost?

8. Write down the approximate equivalent of: (*a*) A metre, in inches; (*b*) a kilometre, in miles; (*c*) an inch, in centimetres; (*d*) a kilogram, in pounds; (*e*) a litre, in pints.

9. How many feet per second is equal to 50 miles per hour?

Paper 22 (*on Chaps. I–XII*)

1. Simplify $4\frac{2}{9} \times 6\frac{3}{7} \div \frac{2\frac{1}{2}}{7}$.

2. Express as decimals $\frac{3}{10^5}$ and $\frac{1}{10} + \frac{5}{10^2} + \frac{3}{10^4}$; and add the two together.

3. Divide 0·18496 by 0·0512.

4. Write down the values of:

(i) $\frac{1}{2}$ of $\frac{5}{16}$. (ii) $2\frac{1}{3} + 3\frac{2}{3}$. (iii) $8\frac{2}{5} - 5\frac{1}{5}$. (iv) $\frac{2}{3} \times \frac{3}{2}$.

(v) $(\cdot 03)^2$. (vi) $\frac{100}{2\cdot 5}$. (vii) $\frac{2\cdot 5}{100}$. (viii) $\frac{\cdot 09}{\cdot 03}$.

5. Express 26 cwts. 56 lbs. as tons and decimals of a ton.

6. If a ton of sugar sells for £22. 8*s.*, how many pounds does one get for 2*s.*?

7. What is the amount of space occupied by a litre of water? its weight? the weight of a cubic centimetre of water?

8. £58. 6*s.* 2*d.* is divided among 28 people: find the share of each person to the nearest halfpenny.

9. There is a chest divided up into rows of lockers so that there are 5 lockers in a row. If the width of the whole chest is 5 feet 10 inches and all the wood of which it is made is 3/4 inch thick, what is the inside width of each locker?

Paper 23 (*on Chaps. I–XII*)

1. Arrange in descending order 11/24, 41/84, 49/108.

2. The expenditure on the Navy in a certain year was 30 million pounds. Of this sum $6\frac{1}{4}$ million pounds was for Pay, $17\frac{1}{2}$ million pounds for Shipbuilding and $1\frac{1}{5}$ million pounds for Pensions. Express each of these three items as a fraction (in its lowest terms) of the whole expenditure.

3. Find the value of (i) $163\cdot 04 \times 15\cdot 42$.
(ii) $137\cdot 56 \times 0\cdot 00178$.

4. Divide 2461·906 by ·000703.

5. Express £3. 17*s.* $4\frac{1}{2}$*d.* as a decimal of £1.

6. What is the value of £·81875?

7. If a watch loses 15 secs. in 7 days, what will it lose in 10 days at the same rate?

8. A ream of squared paper (480 sheets) costs 4*s.* Each sheet is 8 in. by 10 in. What is the cost per sq. inch?

9. Express in mi./hr. a velocity of 2640 ft./sec.

Paper 24 (*on Chaps. I–XII*)

1. Simplify: (i) $4 \times (3 - 2) \div (5 - 1)$.
(ii) $4 \times 3 - 2 \div (5 - 1)$. (iii) $4 \times 3 - 2 \div 5 - 1$.

2. Write down as decimals $\frac{1}{2}, \frac{1}{4}, \frac{1}{8}, \frac{1}{16}, \frac{7}{8}, \frac{1}{40}, \frac{1}{50}, \frac{7}{50}, \frac{3d.}{£1}$.

3. Divide ·00409773 by ·0273.

4. Simplify: (*a*) $4 \times {\cdot}03$; (*b*) ${\cdot}5 \times {\cdot}6$; (*c*) ${\cdot}24 \times {\cdot}2$; (*d*) $({\cdot}3)^2$; (*e*) $\sqrt{{\cdot}25}$; (*f*) $({\cdot}1)^3$.

5. Express 7 stones as a decimal of a ton.

6. If $4\frac{2}{3}$ yards of cloth cost 8*s*. 2*d*., find the cost per yard.

7. If the Government pay a contractor £2800 for 42 horses, what will they have to pay for 15 horses?

8. If 150 grammes of water leak from a cask every minute, how many kilograms will be lost in 4 hours?

9. How many 1/2″ cubes can be cut from a cu. ft.?

Paper 25 (*on Chaps. I–XII*)

1. Arrange in order of magnitude:

1/3, 2/7, 8/22, 21/77, 8/21, 5/14.

2. Multiply 9·064385 by 0·37. Hence write down the answer to $906{\cdot}4385 \times {\cdot}037$.

3. Find, to two places, the value of $124 \div 0{\cdot}36$.

4. Express 3 hrs. 52 mins. 21 secs. as a decimal of a day, correct to 4 decimal places.

5. Express as a decimal $2^7 \times 3^3 \div 5^5$.

6. A naval squadron consists of battleships, cruisers, and torpedo-boats. The torpedo-boats are 3/4 of the whole number, and the cruisers are 1/6; what fraction of the whole are battleships? If there are five battleships, how many ships are there of the other two classes?

7. If 17,360 tons of coal can be got out of a mine in 7 days, in how many days may 64,480 tons be got out by the same number of miners working at the same rate?

8. How many times can a piece 1·5 inch long be cut from a bar 32 inches long, and how much will remain over?

9. The internal dimensions of a tank are 2 m. by 2 m. by 1 m., and its weight when empty is 500 kilograms. How many cu. dm. will it hold? What will be the weight when it is full of water?

Paper 26 (*on Chaps. I–XII*)

1. Simplify $1\frac{1}{2} - \frac{1}{3} \times \left(\frac{1}{2} - \frac{1}{3}\right) - \left(\frac{1}{2} - \frac{1}{3}\right) \div \frac{1}{2} - \frac{1}{3}$.

2. Divide 165·973 by 12·34, 1·65973 by 1·234, and 1659·733 by 0·001234.

3. Find values of: (i) $\frac{3}{4} \times \frac{10}{21}$; (ii) $\frac{3}{4} - \frac{10}{21}$; (iii) $\frac{3}{4} \div \frac{10}{21}$; (iv) $\left(\frac{1}{2}\right)^3$; (v) $(1{\cdot}2)^3$.

4. Find the value of £8·3165 to the nearest penny.

5. Reduce 18 miles per hour to feet per sec.

6. Which is the greater, 0·625 or 16/25? Express their difference—(i) as a fraction; (ii) as a decimal.

7. *A* owns 3/5 of a piece of land, *B* owns 2/7 of it, and *C* the remainder. If *C*'s share is worth £34, what are *A*'s and *B*'s shares each worth?

8. Find the value of 29 things such that 7 of them cost £26. 18*s*. $10\frac{1}{4}$*d*.

9. The *Times* contains 16 pages, each measuring 24 in. by 18 in. How many square yards of information are contained in one copy?

Paper 27 (*on Chaps. I–XII*)

1. Find the value of $2\frac{2}{3} + 1\frac{3}{4} - 3\frac{7}{10}$.

2. Multiply 432·52 by ·053.

3. Find the value of:

(i) $\frac{3}{5} + \frac{2}{5} \div \frac{3}{4}$. (ii) $7\frac{9}{13} - 4\frac{51}{65}$. (iii) $6\frac{3}{8} \times 7\frac{1}{17}$.

(iv) $4\frac{9}{25} \div 2\frac{3}{5}$. (v) $\frac{0{\cdot}125 \times 1{\cdot}6 + 0{\cdot}8}{0{\cdot}25 \times 0{\cdot}4}$.

4. Which is the greater: ·7 or ·74?

Find the value of 90·37 × 67·3. Hence *write down* the values of : (i) 9037 × 6·73, (ii) 0·9037 × 0·0673.

5. Express 0·8125 of a ton in cwts. and lbs.

6. After paying 1/15 of his income in rates and taxes, and 1/12 of the remainder in insurance, a man has left £385. What is his income?

7. A man swims 3 miles in $5\frac{1}{2}$ hours. He goes 1 ft. 4 in. at each stroke; how many strokes does he take to the minute on an average?

8. A rectangular block measures 6″ by $7\frac{1}{2}$″ by 9″. Find (*a*) its volume, (*b*) its total surface area.

9. If 1 inch = 2·54 centimetres, express 1 square inch in square centimetres correct to 2 decimal places, and 1 cubic inch in cubic centimetres correct to 2 decimal places.

Paper 28 (*on Chaps. I–XII*)

1. Simplify:

(i) 2 (7·9 – 2·8) + 4 (3·2 – 2·3).

(ii) 7 + 3·4 (7 – 5) + (5 – 2) (2·1 + 3·9).

2. Divide 43·104 by 0·00768.

3. Simplify: (i) $13\frac{1}{6} - (5\frac{1}{2} - \frac{2}{3} \times \frac{1}{4})$.

(ii) $13\frac{1}{6} - (5\frac{1}{2} - \frac{2}{3}) \times \frac{1}{4}$.

4. Express £3·799 in £ *s.* *d.*, to the nearest penny.

5. A boy spends $\frac{2}{3}$ of his pocket money in the first week of term and $\frac{1}{4}$ of the remainder in the second week. If he then has 1*s.* 3*d.* left, how much had he at first?

6. A pound of paraffin is made into 9 candles, each of which lasts for 4 hours. How long would each last if there were 12 to the pound?

7. What is $\frac{3}{16}$ of £17. 12*s.* 2*d.*, to the nearest penny?

8. How many cubic centimetres of a metal would weigh 0·4536 kilogram if 1 c.c. weighs 7·2 gms.?

9. The velocity of a shell on leaving the muzzle of a 13·5 in. gun is 2016 ft. per second. Express this in miles per hour (to nearest unit).

Paper 29 (*on Chaps. I–XII*)

1. The ancient Egyptians used to express all fractions with unit numerator, e.g.: $\frac{1}{12}+\frac{1}{76}+\frac{1}{114}$. Reduce this to a single fraction in its lowest terms.

2. Add the difference between 5·2 and 3·47 to the difference between 4·6 and 5·3.

3. Multiply 3·1678 by 0·045.

4. Find the value of (i) $\frac{\text{5 tons 4 cwts.}}{\text{4 tons}}$, (ii) $\frac{\text{5 tons 4 cwts.}}{4}$ and explain the meaning of the two operations.

5. A man starts a holiday with £5. He spends $\frac{1}{4}$ of it in railway fares, $\frac{1}{3}$ of the remainder in boating, $\frac{1}{2}$ of what was then left in hotel expenses. After this he bought as many curios as he could pay for at 8*s.* each. How many did he buy and what money remained over?

6. Find the value of 0·4685 ton in cwts., qrs., lbs. to the nearest lb.

7. Find the area in square yards of a courtyard 56 feet long and 38 feet wide. [*See next page*

8. If 12 navvies can fill 12 trucks in 12 minutes, how many will it take to fill 40 trucks in a quarter of an hour?

9. The length of a man's step is 2 ft. 8 in. How many steps does he take in a minute when walking at 4 mi./hr.?

Paper 30 (*on Chaps. I–XII*)

1. Divide 0·00234 by 0·000036.

2. Write down the values of

$$(\cdot 2)^2; \qquad 1\cdot 2 \times \cdot 04; \qquad \frac{1\cdot 11 \times \cdot 11}{30}.$$

3. Find the value of

(i) $2\frac{1}{10} \times 3\frac{3}{7} \times 6\frac{1}{9}$. (ii) $5\frac{7}{8} \div 2\frac{13}{32}$. (iii) $(5\frac{1}{3} + 3\frac{1}{5})\left(1 - \frac{3\frac{1}{5}}{5\frac{1}{3}}\right)$.

4. Express 4 cwts. 3 qrs. 13 lbs. as a decimal of a ton correct to 4 places.

5. The weight of an eight-day clock falls 4 feet in a week. In how many minutes does it fall $\frac{1}{4}''$?

6. If a man travels 540 miles in 24 days, walking 6 hours a day, how many miles will he travel in 3 days, walking 8 hours a day?

7. How many pieces 3·5 cm. long can be cut off from a rod of length 5·75 metres, and what length remains?

8. How many litres are contained in a tank 1·2 metres square and 85 centimetres deep? How many c.c.?

Paper 31 (*on Chaps. I–XII*)

1. Find the value of:

(i) $\cdot 08 + \cdot 00016$. (ii) $\cdot 08 - \cdot 00016$.
(iii) $\cdot 08 \times \cdot 00016$. (iv) $\cdot 08 \div \cdot 00016$.

2. Simplify $23\cdot 478 \times \cdot 0689$.

3. Add together $3\frac{7}{15}$ and $2\frac{11}{12}$. Subtract your answer from $9\frac{21}{80}$.

4. Simplify $\dfrac{4\frac{2}{3} \times 8 \times 5\frac{1}{7}}{28\frac{1}{2} \times \frac{8}{57}}$.

5. Express 1·714 tons in tons, cwts., qrs. and lbs. correct to the nearest 1/4 lb.

6. If 3 men reap 2 acres in 1 day, how many men will be required to reap 60 acres in 5 days?

7. How many cubic feet are contained in a room 11 ft. 6 in. high, 18 ft. 9 in. broad, and 22 ft. 3 in. long? Give your answer to the nearest cubic foot.

8. Which is the faster and by how many yards per minute, a torpedo boat doing 30 knots, or a motor car travelling at the rate of 35 miles per hour? (A knot is a speed of one nautical mile [6080 ft.] per hour.)

9. There are two rulers, A and B. The length that on A is marked into 100 divisions is on B divided into 180. If they be put side by side with the zero marks together, where will the 63rd mark on A fall?

Paper 32 (*on Chaps. I–XII*)

1. Simplify:

(i) $3\frac{1}{7} - 2\frac{1}{2} + 1\frac{1}{14} - 1\frac{11}{56}$, (ii) $4\frac{3}{11} \times 3\frac{1}{7} \div 13\frac{3}{7}$.

2. Simplify $\left(1 - \frac{2}{3} \times \frac{5}{6}\right) \div 1\frac{1}{3}$.

3. Find correct to 2 places of decimals the value of:

(i) 3·7 − 2·54. (ii) 65·2 × 41·5.

(iii) 5·7 × ·0041. (iv) 4·7362 ÷ ·0435.

4. Divide 56·32 by 12·7 correct to two decimal places.

5. Distinguish clearly between $\dfrac{12\cdot5 \text{ tons}}{5 \text{ tons}}$ and $\dfrac{12\cdot5 \text{ tons}}{5}$. Give the value of each.

6. If 24 boys consume 480 sheets of paper in a week, for how many weeks should 960,000 sheets of paper be sufficient for 432 boys?

[*See next page*

7. Find, in square decimetres, the area of the floor of a room which is 6 m. 24 cm. by 4 m. 50 cm.

8. Astronomers use a plate of glass ruled with parallel equidistant lines of which there are 20,000 to the inch. Find the distance between the lines, in mm. (1″ = 2·54 cm.)

9. A man leaves home at 11.45 a.m. and walks at 4 miles an hour until 2.30 p.m. If he stays 40 minutes for lunch and then walks home at $3\frac{1}{2}$ miles an hour, at what time does he get back?

Paper 33 (*on Chaps. I–XII*)

1. Multiply 64·22 by 0·00374.

2. Find the value of $\frac{12{\cdot}32 - 7{\cdot}56}{20{\cdot}35 + 3{\cdot}45}$.

3. Find the value of $8\frac{1}{3} - 2\frac{2}{5} - 3\frac{7}{10}$.

4. Simplify $$\frac{\frac{3}{4} + \frac{2}{5}}{\frac{3}{4} - \frac{2}{5}} \div 1\frac{9}{14}.$$

5. Find the value of (i) $\frac{£15.\ 18s.\ 4d.}{£5}$, (ii) $\frac{£15.\ 18s.\ 4d.}{5}$, and explain in words the difference between them.

6. Find the cost of feeding a boy for 261 days at 1*s*. 7·092*d*. per day.

7. If a cu. dm. of air weighs 1·3 gm., find in Kgm. the weight of air in a room 10 m. long, 8 m. broad, and 350 cm. high.

8. How much silver, at 5*s*. 6*d*. an ounce troy, can be bought for £36. 8*s*. 9*d*.?

9. If 48 men dig a trench containing 1575 cub. yds. in 261 hours, how long will it take 1740 men to dig a trench containing 73,500 cub. yds.?

Paper 34 (*on Chaps. I–XII*)

1. Divide 23·451 by 21·2 correct to 2 places of decimals.

2. Find the product of 10·34 and ·0273.

3. Express as a decimal correct to two decimal places

(*a*) $1\frac{3}{8} - \frac{21}{74} + \frac{38}{115}$, (*b*) $\frac{2{\cdot}5 \times 1{\cdot}007}{({\cdot}02)^2}$.

4. Simplify:

(i) $\frac{4\frac{3}{4} - 1\frac{2}{7} - 2\frac{1}{5}}{1\frac{1}{2}}$, (ii) $\frac{35}{84}$ of $\frac{36}{15}$,

(iii) $(4\frac{5}{8} + 3\frac{3}{8})(4\frac{5}{8} - 3\frac{3}{8})$, (iv) $\frac{29}{36} \div 3\frac{67}{84}$.

5. What is the value of an oak log 24 ft. long and of square section 30″ × 30″ at 1/9¼ per cu. ft.?

6. The price of a mining share is £$1\frac{7}{32}$; what is this in pounds, shillings and pence?

7. If a train travels uniformly at 42 miles per hour, how long does it take to travel 110 miles?

8. If 9 tons are carried 220 miles by rail for £1. 13*s*., what should be charged for the conveyance of 8 tons over a distance of 77 miles? (To the nearest penny.)

Paper 35 (*on Chaps. I–XII*)

1. Find the square of 0·367.

2. Multiply ·00625 by 1·2 and divide 7·5 by 62·5.

3. Find the value of (*a*) $4\frac{5}{13} \times 2\frac{15}{38}$. (*b*) $6\frac{3}{4} \div 3\frac{3}{8}$.
(*c*) 8·27 × 0·135. (*d*) 17·79 ÷ 12·3, to 2 places.

4. Simplify (*a*) $3\frac{3}{5} + 2\frac{1}{2} - 3\frac{1}{3} - 1\frac{2}{15}$.

(*b*) $\frac{1\frac{3}{8} + \frac{11}{12}}{2\frac{11}{15} - 1\frac{9}{20}}$. (*c*) $13\frac{2}{9} - \frac{3}{7} \times 4\frac{1}{12}$.

[*See next page*

5. Find the value per day of an income of £1000 a year.

6. How many lengths of 15·3 metres can be cut from a kilometre of wire, and how many metres will be left over?

7. If 8 fires burning 9 hours a day consume $2\frac{1}{4}$ tons of coal in 25 days, how many fires can be kept burning 10 hours a day for 15 days by 21 tons 12 cwts.?

8. I wish to form a rough estimate of the acreage of a rectangular field by stepping along two adjacent sides. If my steps average 33 inches, and the two sides are 57 and 159 steps, show that the field contains about $1\frac{1}{2}$ acres.

Paper 36 (*on Chaps. I–XII*)

1. Add one-third to two-sevenths; subtract one-fourth from two-fifths; and find the product of the two results.

2. Simplify $$\frac{\frac{1}{2}+\frac{1}{3}}{\frac{1}{2}-\frac{1}{3}} \times \frac{\frac{1}{2}\times\frac{1}{3}}{\frac{1}{2}\div\frac{1}{3}}.$$

3. Express $\frac{1}{80}$ as a decimal and ·025 as a vulgar fraction.

4. Divide 3·42 by ·023, correct to two places.

5. From Euston to Oban is 504 miles. A tourist return ticket costs £3. Find the cost per mile to the nearest tenth of a penny.

6. If the wages of 11 labourers for 15 weeks be £103. 2*s.* 6*d.*, in what time will the wages of 13 labourers amount to £170. 12*s.* 6*d.*?

7. A pane of glass measures 146 cm. by 94 cm. and is 0·33 cm. thick. The glass weighs 2·61 grams per c.c. Find the weight of the pane, to the nearest kilo.

8. A destroyer steams 33·64 n. miles per hour; how far does it go in 4·5 hours at this rate?

9. A tradesman makes a profit of £1. 12*s.* 8*d.* by selling one hundredweight of butter at 1*s.* 2*d.* per pound. How much did the butter cost him per pound?

Paper 37 (*on Chaps. I–XII*)

1. Simplify $\frac{4}{15} + \frac{5}{14} - \frac{1}{42}$.

2. Find the value of $\dfrac{\frac{3}{8} - \frac{2}{5} \times \frac{5}{8}}{2\frac{3}{4} - \frac{4}{3} \div 5\frac{1}{3}}$.

3. Find the value of $4725 \times 0{\cdot}00752$.

4. Simplify:
$\cdot 3 \times \cdot 4$, $\cdot 78 \times \cdot 02$, $1{\cdot}324 \div \cdot 4$, $5{\cdot}238 \div \cdot 09$, $1 \div (\cdot 2)^2$.

5. A water-cart which contained 360 gallons travels at the rate of $2\frac{1}{2}$ miles an hour, and is emptied after going 1/4 mile; at what rate per second does the water flow out?

6. If an 8*d.* loaf weighed 48 oz. when wheat was at 39*s.* per quarter, what should a $6\frac{1}{2}$*d.* loaf weigh when wheat is at 30*s.* per quarter?

7. During 1912–13, 1,277,000 lbs. of chocolate was manufactured for the Navy at a total cost of £35,048. How much is this per 100 lbs., to the nearest penny?

8. The area of Montenegro is 3486 sq. miles, and it has a population of 225,000. How many is this to the acre?

9. By altering the pitch of her propeller the builders of a small steam yacht increased her speed from 14 knots to $15\frac{1}{2}$ knots. What was the difference in her times over a nautical mile? (To the nearest 1/5 second.)

Paper 38 (*on Chaps. I–XII*)

1. Simplify $\frac{5}{8}$ of $1\frac{1}{3} + 2\frac{1}{4}$ of $3\frac{1}{5}$.

2. Find the value of $(8\frac{2}{7} - 4\frac{1}{2}) \div (13 - 4\frac{2}{7})$.

3. Divide 32·8532 by 439·4 correct to 4 places.

4. A sheet of paper is 8·5 inches long and 6·25 inches wide. If a strip 1/2 inch wide be cut off all the way round, what is the area of the remainder?

[*See next page*

5. A is ·875 times as old as C, and C is 1·08 times as old as B; B is 25; how old is A? (In years and months.)

6. A car at Brooklands covered 50 miles in 27 mins. 2·4 secs.; how many miles was this per hour? Give your answer correct to the nearest mile.

7. A man walks 3 miles in an hour; how many yards does he cover in a minute? If this man starts at 10.20 a.m. for a place 13 miles distant, and walks uniformly as above except for two rests of 15 minutes each, find the time at which he will reach his destination.

8. If 5 tons of coal last a household 6 months when using 3 fires, how many tons will be required to supply the same household for 5 months if they have four fires?

9. A tramway company takes 2,127,314 penny fares, 357,299 twopenny fares, and 87,465 threepenny fares. Give the total receipts in £ *s.* *d.*

Paper 39 (*on Chaps. I–XII*)

1. Simplify $1\frac{2}{3} + 1\frac{3}{4} - 1\frac{4}{5}$.

2. Simplify
$$\frac{2 - \frac{1}{3} \text{ of } \frac{2}{5}}{\frac{1}{3} \text{ of } \left(2 + \frac{2}{5}\right)}.$$

3. Multiply:

(i) 4·27 by 2·3. (ii) 4·38 by 23. (iii) 4·49 by ·23.

4. A nautical mile = 2028 yards, and a yard = 91·44 cm. Express 6 nautical miles in kilometres, to the nearest metre.

5. £114. 3*s.* 8*d.* is to be divided equally among 87 men. What is each man's share, to the nearest penny?

6. How many lengths of 1·72 inches can be cut off a stick 2 feet long? And how much will be left?

7. 1270 cubic yards of water are pumped into a skating pond with vertical sides. The area of the pond is 8760 sq. feet. Find, to the nearest inch, the depth of water.

8. The daily rations allowed a horse are: oats 22 lbs., hay 9 lbs., straw 4 lbs. Calculate to the nearest shilling the weekly cost per 100 horses, the prices being: oats 17*s*. 6*d*. per quarter of 312 lbs., hay 75*s*. per ton, straw 28*s*. per ton.

9. A man who has paid £30/13/4 as a life insurance premium is allowed to deduct £1/7/0 from his income tax. How much in the £ is this, to the nearest $\frac{1}{10}d$.?

Paper 40 (*on Chaps. I–XII*)

1. Divide 14·852 by 71·866 to two places.

2. Simplify $(3\frac{1}{3} + 7\frac{1}{4})$ of $\frac{2}{5} + 9\frac{3}{4}$.

3. Simplify $$\frac{\frac{3}{4}+\frac{2}{5}}{\frac{3}{4}-\frac{2}{5}} \times \frac{1\frac{1}{3}}{\frac{2}{3}+\frac{3}{7}}.$$

4. One pint = 568·2 c.c. How many pints could be taken from a cask containing 100,000 c.c., and how many c.c. would be left over?

5. What is the cost of 100 fathoms of chain cable at £140/2/10 per $12\frac{1}{2}$ fathoms?

6. Six men can do a piece of work in 14 days of 10 hours. Prove that 7 men can do the same work in 15 days of 8 hours.

7. What is the angle between the hands of a clock at 20 minutes past 7?

8. A bicyclist left home at 9 a.m. and reached his destination, 60 miles off, at 6 p.m. He took one hour for lunch, half an hour for tea and 20 minutes for odd stops. Find his average rate of riding when in the saddle, to the nearest tenth of a mile per hour.

9. A garrison have rations calculated to last them 4 months. How long will they be able to hold out if they are put on half rations at the end of two months?

Paper 41 (*on Chaps. I–XII*)

1. Simplify $1 - \frac{1}{2} + \frac{2}{3} + \frac{3}{4} + \frac{4}{5} - \frac{13}{6}$.

2. Simplify $\dfrac{\frac{1}{5} + \frac{1}{6}}{\frac{1}{4} - \frac{1}{5}} \times \dfrac{\frac{1}{5} - \frac{1}{6}}{\frac{1}{4} + \frac{1}{5}}$.

3. Multiply 47·8 by 3·142, (1) correct to the nearest integer; (2) correct to two decimal places.

4. Find the weight of a steel plate measuring 12·5 ft. by 3 ft. and weighing 7·65 lbs. per sq. ft.

5. A tennis-court is in the middle of a grass plot. The dimensions of the tennis-court are 72 ft. by 36 ft., and all round the court is left a border of grass which is 12 ft. wide. What would be the cost of erecting wire-netting around this plot if the netting was 12 ft. high and cost 14*s*. 9*d*. per roll of 50 yds., the width of the roll being 6 ft.?

6. A batsman makes scores of 3, 0, 7, 2 (not out), 15, 30, 0, 127, 43. What must he make in his next innings to bring his average up to 30?

7. A piece of plate glass, measuring 25″ by 16″, costs 2/1. What is the cost per square yard?

8. I bought 2000 francs for £79/10/6. At this rate what is the value of a franc in pence (to 2 places), and of £1 in francs (to nearest centime)?

9. On a rubber estate in the Malay States in 1909 there were harvested 545,319 pounds of rubber from 1274 acres. The average yield per tree was 3 lbs. Find, to the nearest whole number, the average number of trees per acre.

Paper 42 (*on Chaps. I–XII*)

1. Simplify $\frac{1}{4} + \frac{3}{4}$ of $\frac{7}{9} - \frac{1}{6}$.

2. Express as a simple fraction $\dfrac{\frac{1}{2} - \frac{3}{16} + \frac{5}{8} \text{ of } 3\frac{1}{5}}{\frac{1}{2} \times \frac{3}{16} - \frac{1}{4} \times \frac{3}{32}}$.

3. Find the value of:

(i) 91·008 ÷ 379·2. (ii) ·06828467 ÷ ·6781.

4. A debtor is able to pay only 14*s*. 9*d*. out of every pound he owes. What will a creditor receive to whom he owes £93. 10*s*.?

5. An open rectangular tank has the following internal dimensions—21 metres long, 5 decimetres broad, and 7 decimetres deep. Find (1) its volume in c.c.; (2) the total surface area of inside of tank; (3) the number of litres of water it would hold.

6. A colliery makes three contracts, one to supply 3500 tons of coal at 15*s*. 9*d*. a ton, another to supply 1800 tons at 16*s*. 3*d*. a ton, and the third to supply 1280 tons at 16*s*. 6*d*. a ton. Find to the nearest penny the average price per ton for the whole quantity.

7. If 600 men can dig a trench 5 ft. 6 in. broad, 4 ft. deep, and 135 yards long, in 1/2 an hour, what length of trench 10 ft. broad and 8 ft. deep can 2500 men dig in 6 hours?

8. A ton of ordinary coal in an open fire lasts 6 weeks and costs 27*s*. One-and-a-half tons of anthracite coal, at 45*s*. a ton, last 30 weeks; but a special closed stove must be bought, and this costs £4. 10*s*. In how many weeks does the saving in coal pay the cost of the stove?

9. In 1914–15, income tax was at the rate of 9*d*. in the £ for 8 months and at twice this rate for the remaining 4 months. How much in the £ does this amount to taken over the whole year? What would a man pay whose income is £200, a rebate of £160 being allowed? What fraction of his income does he pay?

Paper 43 (*on Chaps. I–XII*)

1. Find (i) the sum of $10\frac{5}{6}$, $5\frac{3}{10}$ and $8\frac{1}{24}$;

(ii) the difference between $5\frac{3}{10}$ and $8\frac{1}{24}$;

(iii) the product of $\frac{3}{5}$, $\frac{7}{8}$, $\frac{15}{28}$ and $3\frac{5}{9}$

[*See next page*

2. Find the value of $\dfrac{\frac{1}{3} + \frac{2}{3} \text{ of } \frac{4}{11}}{1 - 3\frac{1}{9} \div 7\frac{1}{3}}$.

3. Find the value of $1{\cdot}21 \times {\cdot}121 \times {\cdot}0121$ to 3 places.

4. A brass boiler tube 6′ 7″ long costs £1/0/7. What is the cost of 100′ of tube?

5. If 22 men dig a trench 420 yards long, 5 wide, and 3 deep in 350 days of 9 hours each, in how many days of 11 hours each will 252 men dig a trench 210 yards long, 3 wide, and 2 deep?

6. A cistern measuring 4 ft. by 3 ft. is filled with water to a depth of 18 in. Find the number of cubic feet of water. If this water is transferred to a cistern which measures 6 ft. by 6 ft. what will be its depth?

7. The diameters of moon and earth are 2160 and 7920 miles. Find the ratio in the form $1 : n$ to 2 places.

8. If a cubic foot of water weighs 1000 ounces, and a gallon of water weighs 10 lbs. avoirdupois, find the number of cubic inches in the imperial bushel of 8 gallons.

9. The average London working man has been estimated to make, amongst others, the following weekly purchases: 0·6 lb. of tea at 1*s.* 4·2*d.* per lb.; $5\frac{1}{3}$ lbs. of sugar at 2*d.* per lb.; 12 oz. cheese at 7·1*d.* per lb.; 17 lbs. of potatoes at 1·29*d.* per lb.; 10 pints of milk at 4*d.* per quart. Calculate to the nearest farthing the amount spent on each item.

Paper 44 (*on Chaps. I–XII*)

1. Simplify $\quad \frac{3}{4} + \frac{4}{13} \div \frac{3}{52} - 6.$

2. Simplify $\quad \dfrac{\left(1 - \frac{1}{3}\right)\left(\frac{1}{5} - \frac{1}{7}\right)}{1 - \frac{1}{3} + \frac{1}{5} - \frac{1}{7}}.$

3. Find the quotients of:

(i) $1735{\cdot}5 \div 3{\cdot}25$. (ii) $0{\cdot}00044408 \div 0{\cdot}0224$.

4. How many c.c. of iron of specific gravity 7·2 weigh 3500 gm. (to the nearest c.c.)?

5. If a piece of cardboard 18 cm. long and 14 cm. wide weighs 50·4 gm., find the area of a piece which weighs 56·45 gm.

6. The consumption of sugar in this country is stated to be 1,960,601 tons per annum. If the population is 44 millions, find to the nearest pound the average annual consumption per head.

7. A train starts at 11.56 a.m. and does a journey of 117 miles in 2 hours 35 mins. At what time does it arrive? Find the average speed in miles an hour to the nearest mile.

8. Carolina rice sells at £1/19/6 per cwt.; Java rice at 27/- per cwt. What is the ratio of the former price to the latter, in the form $n : 1$?

9. A man's salary is £800 a year, on which he pays an income tax of 9*d.* in the £. He also receives £246. 17*s.* in the year as interest on invested capital; on this he pays 1*s.* 2*d.* in the £. How much in the £ is he then paying, reckoned on his whole income (to the nearest farthing)?

Paper 45 (*on Chaps. I–XII*)

1. Add together $\frac{15}{56}, \frac{2}{35}, \frac{4}{55}, \frac{5}{143}, \frac{59}{104}$.

2. Simplify $$\frac{\left(3-\frac{4}{5}\right) \div \left(2+\frac{4}{7}\right)}{\left(2-\frac{4}{13}\right) + \left(\frac{1}{2}+\frac{1}{9}\right)}.$$

3. Divide: (i) 6·3 by 7. (ii) 6·3 by ·07.
(iii) 672·88 by 6·47. (iv) 232·379 by ·0373.

4. Find, correct to 1 place, the value of $\frac{0{\cdot}8672 \times 17{\cdot}1}{0{\cdot}807}$.

5. How far will a man walk between 9.15 a.m. and 10.20 a.m., if he can walk $1\frac{1}{4}$ miles in 25 minutes?

[*See next page*

6. What is the cost of a cubic yard of sawdust at 3/6 per 100 cubic feet?

7. How many pieces 2·5 cm. long can be cut from a roll of wire 72·432 m. long? Express the remainder in mm.

8. Household salt costs $3\frac{1}{2}d.$ per 7 lb. bag, and $6\frac{1}{2}d.$ per 18 lb. bar. Find the ratio of the former to the latter price.

9. A train which travels at the rate of $48\frac{3}{4}$ miles an hour leaves London at noon, meets another train from York at 2.30 p.m., and arrives at York at 3.52 p.m. At what rate does the second train travel if it left York at 12.27 p.m.?

10. Under the Customs Tariff for the United Kingdom the revenue in 1908–09 received in respect of currants was £117,795; of prunes, £10,011; of raisins, £240,705; and of tea, £6,046,210. The taxes which produced these amounts were 2/- per cwt. for currants, 7/- per cwt. for prunes and for raisins, and $5d.$ per lb. for tea. Find the number of tons of each article on which the taxes were paid that year.

Paper 46 (*on Chaps. I–XII*)

1. Simplify (i) $3\frac{1}{7} \times \left(5\frac{1}{4} - 3\frac{1}{2}\right) \times \left(5\frac{1}{4} + 3\frac{1}{2}\right)$.

(ii) $$\frac{\frac{1}{3} \text{ of } \left(\frac{1}{2} + \frac{5}{6}\right) + \frac{4}{7} \times 2\frac{1}{10}}{4\frac{3}{4} + \frac{2}{3} \text{ of } 5\frac{2}{5}}.$$

2. Find the value of $\frac{0\cdot069 \times 74\cdot63}{8\cdot95}$ (to 2 places).

3. Divide 0·8607535 by 0·000974 to the nearest integer.

4. Marmalade costs $4\frac{1}{2}d.$ per lb., and 21/- per cwt. What is the ratio of the retail to the wholesale price?

5. What is the cost of 435 lbs. of rice at 39/6 per cwt.?

6. H.M.S. *Dreadnought* on her trials covered 168 nautical miles in 8 hours; what was her average speed in knots (i.e. nautical miles per hour)? She consumed $17\frac{1}{2}$ tons of coal per hour; what was her (1) total consumption, and (2) consumption per nautical mile on this run?

7. A steel plate measuring 6 feet by 2 feet weighs 7·65 lbs. per sq. ft. and costs 7/0. An open tank is made of these plates: the base is 6′ by 6′ and the depth is 2′. Neglecting the overlap needed for riveting plates together, find (i) the weight of the empty tank, (ii) the cost of the steel plate used.

8. Fourteen Jersey pounds are equivalent to 15 lbs. 2 oz. avoirdupois; find, to the nearest farthing, the price per lb. avoirdupois of tea which is sold at 2*s*. 6*d*. per Jersey lb.

Paper 47 (*on Chaps. I–XII*)

1. Find the value of $1\frac{3}{4}+\frac{5}{8}-\frac{3}{16}+\frac{\frac{5}{16}}{2\frac{1}{2}}$ and express your result as a decimal correct to two decimal places.

2. Simplify $$\frac{3\frac{3}{7}\times\left(6\frac{1}{3}-5\frac{5}{9}\right)}{\frac{3}{7}\text{ of }3\frac{4}{9}+19\frac{2}{5}\times\frac{1}{12\frac{3}{5}}}.$$

3. Evaluate $\frac{0\cdot678\times9\cdot01}{0\cdot0234}$ to 2 places.

4. How many times can 0·254 be subtracted from 10 and how much is left over?

5. Find the price of 13 dozen oysters at 22/9 per 100.

6. Find the area in sq. metres of all the sides and bottom of a box without a lid if the length is 1·5 metres, the breadth 0·9 metres, and the height 5·4 decimetres.

7. Find the cost of 75 lbs. of macaroni at 8/6 per 28 lbs.

8. A ship's Patent Log showed 41 miles over a distance known to be 40 miles exactly. When the reading of the log is 80, what is the real distance run (to the nearest mile)?

9. Gold, silver, and copper are melted together to form a coin in weights proportional to the numbers 32 : 5 : 3. If the coin weighs 2·30 gm., what weights of gold, silver and copper were used?

Paper 48 (*on Chaps. I–XII*)

1. Simplify $\frac{0{\cdot}498+0{\cdot}131}{5{\cdot}689}$, to 3 places.

2. Evaluate to 2 places

$$(9{\cdot}7 - 0{\cdot}03) \times 1{\cdot}103 \div 2{\cdot}36.$$

3. Simplify $\dfrac{31\frac{1}{3} - 22\frac{1}{5}}{11\frac{1}{5} - 1\frac{1}{7}} \div 1\frac{49}{88} + 2\frac{5}{12}$.

4. Find the value of (*a*) $\frac{£9.\ 15s.}{£3}$, (*b*) $\frac{£9.\ 15s.}{3}$.

5. Find to the nearest yard the difference between 5 miles and 8 kilometres; given 1 metre = 39·3708 inches.

6. A small box, whose internal dimensions are height 6 cm. and horizontal section 25 sq. cm. is filled with mercury, which is 13·6 times as heavy as water. Find the cost of the mercury at 6*s.* 6*d.* a kilogram.

7. In a West Indian Island it is estimated that there are 3 white inhabitants to every 50 blacks. If the total population is 100,000, how many of these are white?

8. 50 bottles of Perrier table water cost 22/-. If I drink 2 bottles a day, what will it cost me per year? (Returned bottles are allowed for at the rate of 7*d.* per dozen.)

9. Below are tabulated for certain towns the acreage and the population in the years 1891 and 1901:

	Population		
Town	1891	1901	Acreage
Hull	200,472	240,259	8,989
Newcastle-on-Tyne	186,300	215,328	5,355

Calculate for each town the increase in population, and also (to the nearest integer) the average number of persons per acre in 1901.

Paper 49 (*on Chaps. I–XII*)

1. Simplify $\dfrac{1\frac{1}{2}-\left(\frac{1}{3}-\frac{1}{4}\right)}{1\frac{1}{2}-\frac{1}{3}-\frac{1}{4}} \div \dfrac{1\frac{1}{2}\text{ of }\left(\frac{1}{3}-\frac{1}{4}\right)}{1\frac{1}{2}\text{ of }\frac{1}{3}-\frac{1}{4}}$.

2. What must be added to the sum of $\frac{2}{3}$ and $\frac{5}{6}$ to make 2?

3. Simplify $\dfrac{12{\cdot}5+9{\cdot}7}{0{\cdot}05-0{\cdot}013}$.

4. Evaluate 4·375 of £2. 10*s.*

5. Divide the product of 61·93 and 0·117 by 1·43.

6. Divide £608. 18*s.* 4*d.* between 3 persons, so that their shares may be in the ratio 4 : 5 : 11.

7. A rectangular tank containing water is 4′ long, 3′ broad, and 2′ deep. Find how much the water will rise if a block of lead 1′ × 6″ × 3″ is put into the tank.

8. The following prices are quoted for corned beef: 2 lb. tin, $10\frac{1}{2}$*d.*; 4 lb. tin, 1*s.* 9*d.*; 6 lb. tin, 2*s.* 11*d.* There is a note that the 2 and 4 lb. tins are nominal weights, but the 6 lb. tins contain full weights. What actual weights of beef should be contained in the 2 lb. and 4 lb. tins (to the nearest ounce)?

9. The following table is copied from the accounts of a gold mine:

	In tons	Value in dollars per ton of ore
Ore in view	30,718 ..	.. 9·73
Ore extracted	62,183 ..	.. 5·64

Find the total value in thousands of pounds of all the ore if 1000 dollars exchange for £206.

Paper 50 (*on Chaps. I–XII*)

1. Simplify $\dfrac{\frac{3}{4}-\frac{2}{3}}{\frac{4}{5}-\frac{3}{4}}+\dfrac{\frac{19}{20}-\frac{18}{19}}{\frac{18}{19}-\frac{17}{18}}+\dfrac{\frac{30}{29}-\frac{31}{30}}{\frac{28}{29}-\frac{27}{28}}$.

2. Evaluate $\dfrac{£3.\ 6s.\ 8d.}{30s.}$.

3. Simplify:

(i) $(12{\cdot}41 + 2{\cdot}09) \times 1{\cdot}2 - (4{\cdot}5 - 0{\cdot}9) \div 1{\cdot}8$.

(ii) $12{\cdot}41 + 2{\cdot}9 \times 1{\cdot}2 - 4{\cdot}5 - 0{\cdot}9 \div 1{\cdot}8$.

4. Find the value of $\dfrac{23{\cdot}51 - 8{\cdot}9}{32{\cdot}72 + 7{\cdot}33}$ correct to 3 places.

5. The death rate in 3 towns whose population is 25,600, 112,500, and 73,200, is 15, 28, and 25, per 1000 respectively. What is the death rate on all the towns collectively (to nearest whole number)?

6. Find the number of complete revolutions made by a roller in passing from one end of a lawn to the other. The lawn is 49 yds. long. The circumference of the roller is 4·12 feet.

7. How much will it cost to paper a room 18 ft. 6 in. broad by 20 ft. long and 11 ft. high with paper costing 9*d.* per square yard? The room has a door 4 ft. by 7 ft. and a window 5 ft. by 8 ft.

8. 12 bricklayers are set to build a wall, and the job is expected to take a week. When $\frac{1}{3}$ of the wall is finished, orders are given that the work must be finished in 2 days more. How many more hands must be taken on?

9. Two gases, hydrogen and oxygen, combine together and form water. The weight of hydrogen to that of oxygen is in the ratio of 1 : 8. What is the weight of hydrogen required to form sufficient water just to fill a tank, the internal dimensions of which are $4\frac{1}{2}'$ by $2'$ by $1'$, given that 1 cub. foot of water weighs 62·5 lbs.?

Paper 51 (*on Chaps. I–XII*)

1. Simplify $$\frac{\frac{5}{7} \div 3\frac{1}{13} - \frac{3}{8} \text{ of } \frac{7}{12}}{\left(1 - \frac{1}{2} + \frac{1}{8}\right) \div \left(1\frac{5}{6} - \frac{2}{3}\right)}.$$

2. Simplify:

(*a*) 75·023 + 0·56 + 149 + 1·527. (*b*) 12 − 2·894.

(*c*) 2·54 × 0·025. (*d*) 1·307 ÷ 2000.

(*e*) 0·0561 ÷ 0·003. (*f*) 209·28 ÷ 65·4.

3. Evaluate to the nearest ton 0·525 of 541 tons.

4. A room is 21 ft. long, 16 ft. 6 in. broad, and 12 ft. high. Find (1) the cost of carpeting the floor at 1*s*. 4*d*. a sq. ft., leaving a border 1 ft. 6 in. wide all round the room; (2) the cost of staining the above border at 6*d*. a sq. ft.; (3) the area of the walls; (4) the number of cu. ft. of air in the room.

5. Prize money to the value of £1848 was to be distributed among a captain, 5 junior officers and 73 men. If shares of captain, officer, and man are as 10 : 7 : 3, find the share of each individual.

6. A contractor undertakes to finish a piece of work in 30 days and employs 16 men upon it. After 12 days only one-quarter of the work is done. How many additional men must be engaged in order to carry out the contract?

7. Standard gold consists of $\frac{11}{12}$ of fine metal and $\frac{1}{12}$ of alloy. 240 ounces of standard gold are coined into 934 sovereigns and one half-sovereign. Find the intrinsic value of 1 oz. of pure gold, neglecting the value of the alloy and the cost of coining.

8. A man who gives away 1/20 of his net income finds that owing to the income tax being increased from 10*d*. to 1*s*. in the pound, his annual charities are reduced by 10*s*.; what is his gross income?

[*See next page*

9. The railway between A and B ascends at a gradient of 1 in 275 from A to a place C $16\frac{1}{4}$ miles distant from A. It descends at a gradient of 1 in 165 from C to B. If the station at B is 28 ft. lower than that at A, find the distance from A to B.

Paper 52 (*on Chaps. I–XII*)

1. Simplify $\dfrac{\frac{1}{6}-\frac{1}{7}}{\frac{1}{7}-\frac{1}{8}}$ of $\dfrac{\frac{1}{8}-\frac{1}{9}}{\frac{1}{9}-\frac{1}{10}}$ of $\dfrac{\frac{1}{10}-\frac{1}{11}}{\frac{1}{11}-\frac{1}{12}}$.

2. Find the value of

(i) $13{\cdot}64 - 2{\cdot}0008 - (12{\cdot}345 - 3{\cdot}861)$.

(ii) $36{\cdot}041 \times 0{\cdot}0135$.

(iii) $\dfrac{35{\cdot}68}{17{\cdot}82}$ (correct to 2 places of decimals).

3. What is the error involved in assuming that 3·14159 is equal to $\frac{22}{7}$? (To 4 places of decimals.)

4. A stick is broken in two pieces whose ratio is 2 : 3; the smaller fragment is again broken, in the ratio 4 : 5. Find the ratio each of the three pieces bears to the whole.

5. In Scotland the charge for driving is 1*s.* 3*d.* a mile, in France the charge is 1 franc per kilometre. Which is the dearer rate, and what would be the difference in pence for a journey of 25 miles? (25 francs = £1. 8 Km. = 5 miles.)

6. A runs a hundred yards in eleven seconds, B a hundred and ten yards in twelve seconds. If B gives A a yard start in a hundred, which wins?

7. A contractor employs 15 men, each working 8 hours a day, to do a certain piece of work in 19 days. At the end of 10 days the work has to be suspended for 2 days owing to an accident in which 4 of the men are disabled. How many more men must he engage to complete the work in the specified time, all the men now working 9 hours a day?

8. Compressed vegetable rations for 1400 men can be contained within the space of a cubic foot, allowing 25 grams per head. Find, correct to 1 place, the weight in grams of a cubic centimetre of the substance.

9. A party of tourists set out for a station 3 miles distant and go at the rate of 3 miles an hour. After going half a mile one of them has to return to the starting point; at what rate must he walk in order to reach the station at the same time as the others?

Paper 53 (*on Chaps. I–XII*)

1. Simplify $\dfrac{8 \cdot 04 - 2 \cdot 85 \times 2 \cdot 5}{4 \cdot 2 \div \cdot 35 + 8}$.

2. Divide the sum of two-thirds and five-sixths by the sum of one-quarter and three-eighths. From the result take one-fifth of seven.

3. Reduce to decimals $\frac{17}{625}$ and $\frac{15}{16}$.

4. If 5/7 of a ship be worth £4000, what is the value of 3/8 of it?

5. A room is 20 feet long, 15 wide, and 9 high. Find the cost of (1) whitewashing the ceiling at 3*d.* per square yard; (2) papering the walls with paper 21 inches wide at 3*s.* the piece of 12 yards; (3) carpeting the floor with carpet 3/4 yard wide at 3*s.* a yard.

6. The record time for the Oxford and Cambridge boat race (6800 metres) is 19 mins. 33 secs.; the speed of a given point in the Isle of Wight due to the rotation of the earth is 299 metres per sec.; express the ratio between these two speeds in the form $x : 1$ (to 3 places).

7. A traveller returning from France had in his possession on landing at Dover seven Napoleons (20 francs) and seven franc pieces; on going to a money-changer he received an exchange at the rate of 15*s.* for a Napoleon and 8*d.* for a franc piece. To what extent was he cheated, the daily papers indicating the fact that the exchange was 25·20 francs for £1?

[*See next page*

8. Between two places A and B there are two routes, the shorter of which goes up-hill for half its length and down-hill for the other half; the other route is 2 miles longer and is level. A man who can go on foot $3\frac{1}{2}$ miles an hour up-hill and $4\frac{1}{2}$ miles an hour down-hill, and cycles on the level at 10 miles an hour, finds that he can cycle by the longer route in 56 mins. Find the distance between the two places by the longer route, and find how long it will take him to walk by the shorter route.

9. A man starts to row up a river a distance of 128 miles at 6 o'clock on Monday morning. He rows 2 hours before breakfast, $3\frac{1}{2}$ hours between breakfast and dinner, $3\frac{1}{2}$ hours between dinner and tea, and 3 hours after tea, resting an hour for each meal; and does the same on following days. His rate is four miles an hour against the stream, which, however, carries him back during meals and sleep at the rate of $1\frac{1}{2}$ miles an hour; when will he reach his destination?

Paper 54 (*on Chaps. I–XII*)

1. Express the following instruction in symbols and then simplify the expression:

Divide the sum of one-fifth and one-sixth by the product of $2\frac{3}{4}$ and one-fifth: from this subtract one-quarter of the quotient obtained when the sum of one-half and one-third is divided by the sum of one-third and one-quarter.

2. Find correct to three places of decimals the value of $\frac{92}{117} + \frac{4}{105} - \frac{5}{17}$.

3. Evaluate (to 2 places) $\frac{6{\cdot}43^2 - 2{\cdot}53^2}{7{\cdot}8 \times 5{\cdot}75}$.

4. If an income of £924 pays £40/8/6 tax, at how many pence in the £ is the tax charged?

5. How many lines at intervals of ·007 of an inch can be engraved on a six-inch rule? (Count the end as one line.)

6. A field was drained by pipes, each 14 inches long, which were placed end to end along the drains. The total length of all the drains was 784 yards. The pipes cost 30*s.*

per 1000, and the labour cost 5*s.* 6*d.* for every length of 110 yards measured along the drains. Find what was the total cost, correct to the nearest shilling.

7. Divide £105. 5*s.* 4*d.* between 2 men, 3 women, and 4 children, so that a woman may have as much as 2 children, and a man as much as a woman and child together.

8. A bicyclist rides up a hill 2 miles long in 9 minutes 20 seconds, and then down a hill of the same slope 3 miles long in 10 minutes 30 seconds. How long will it take him to ride back?

9. A chronometer shows the following times at noon on the given dates: May 18th, 12 hrs. 2 mins. 3 secs.; May 31st, 12 hrs. 0 min. 51 secs. Find the average daily loss correct to the nearest tenth of a second, and also find when the chronometer will be correct.

MISCELLANEOUS EXERCISES

On Chaps. I–XII

Some of the following problems are solved most easily by algebraical methods: such methods are generally allowed in arithmetic examinations.

1. Two yard measures, one divided into 360 equal parts and the other into 288 equal parts, are placed side by side with the zero graduations in contact. What other graduations are in contact?

2. Draw a figure to show that $\frac{8}{10} \times \frac{7}{10} = \frac{56}{100}$.

3. The earth describes its path round the sun in 31556925·51 seconds. Express this interval in days, to the nearest ten-thousandth.

4. Find the difference between the above period and the average length of the year, estimated on the basis that of 400 years 303 are of 365 days, and the remainder leap years of 366.

5. Prove the following rule: Multiply the price in farthings per lb. by $2\frac{1}{3}$ and you will get the price in £ per ton.

6. Find the least number of ounces of standard gold from which an exact number of half-sovereigns can be coined, at the rate of £3. 17*s.* $10\frac{1}{2}$*d.* to the ounce.

7. If a clock takes 6 seconds to strike 6, how long does it take to strike 12?

8. A regiment marching in fours at 3 miles an hour comes to a foot-bridge which the men have to cross in single file. At what rate must they cross so as not to delay the march?

9. The pulley on the headstock of a lathe is 3 inches in diameter, and is belted to a pulley 7 inches in diameter, revolving 390 times in a minute. How many revolutions a minute will the first pulley make?

10. A regiment is marching along a straight road, 128 paces to the minute, the band in front. Each man keeps time with the beat of the drum, as he hears it; but in consequence of the fact that sound takes a second to travel 1100 feet, the men are not all in step. Find, to the nearest yard, the least length of the column in order that those in the rear may be in step with those in front.

11. The following stores were consumed at a school in the quarter ending 31st March. Find the total cost: (i) 1344 lbs. of biscuits at £1/5/11 per cwt.; (ii) 5200 lbs. of sugar at 14/10 per cwt.; (iii) 556 lbs. of tea at $8\frac{1}{2}$*d.* per lb.; (iv) 153 lbs. of coffee at £5/11/11 per cwt.; (v) 925 lbs. of chocolate at £2/18/3 per cwt.; (vi) 50 lbs. of mustard at £2/18/6 per cwt.; (vii) 30 lbs. of pepper at £3/16/0 per cwt.; (viii) 158 pints of vinegar at £1/14/6 per 100 gallons.

12. A boy, wishing to estimate the speed of a train in which he is travelling, times by his watch that it takes 1 min. 8 secs. to pass 10 of the intervals between telegraph posts. He guesses these intervals to be 80 yards each. If the intervals are really 88 yards, and the correct time was 1 min. 7 secs. find, to the nearest tenth of a mile per hour, the error in his estimate.

13. Find the value in English money of 1 yd. of silk worth 7·5 fr. a metre. (1 m. = 39·37 in. and 1 fr. = 9·38*d.*)

14. If a sovereign is worth 25 francs 41 centimes and a kilogram weighs 2·204 lbs. and lead is worth £20 per ton, how many Kgm. of lead can be purchased for 121 fr.?

15. In Guernsey 1 franc = 10*d.* The exchange between England and France is 1 franc for 9·6*d.* A man changed £100 into francs in France and then changed the proceeds back into English money in Guernsey. How many £ *s.* *d.* does he gain?

16. Assuming that, owing to higher prices, what could be bought for £1 in England would cost 24*s.* in America, find which is the better off, a clerk in England with a salary of £100 a year, or one in America with a salary of 600 dollars a year. A dollar = 4*s.* $1\frac{1}{2}$*d.*

17. Write down in order of magnitude all the proper fractions whose denominator is not greater than seven.

18. A man said that all his ground was planted, 1/3 with potatoes, 1/2 with cabbages, and 5/12 with turnips. Could he have been speaking the truth? If not, supposing the various things to have been in the proportions he mentioned, what were the actual fractions of his land taken up by them?

19. If a parcel of 12 lbs. be carried 80 miles for 2*s.* 4*d.* and the rate per mile for distances over 50 miles is $\frac{2}{3}$ of the rate for the first 50 miles, how far can a parcel of 8 lbs. be carried for 4*d.*?

20. My watch loses 4 minutes in 24 hours on Greenwich time. I put it right at 9 o'clock on Monday morning. What is the right time when it says 6 minutes to 11 on Wednesday morning?

21. A vessel of 3628 tons register, making 9 knots and burning 22 tons of coal a day, leaves London for Melbourne. What will she save by going via the Cape instead of via Suez?

Price of coal 26/- per ton.
Suez Canal dues 7/- per ton register.
London to Melbourne via Cape, 11,930 n. miles.
London to Melbourne via Suez, 11,050 n. miles.

22. An engine while driving machinery burns coal at the rate of 1 ton 12 cwts. 2 qrs. in 8 hrs. 45 mins. When the machinery is not in motion the consumption is only 4/11 of this rate. How much coal will the engine burn in 1584 hours, during 1/9 of which time machinery is at rest?

23. Firewood costs 15/5 per 100 bundles. If 3 fires are lighted for 250 days in the year, and 1 fire only for the remaining days, 1 bundle being enough for 1 fire, what does a year's firewood cost?

24. At the siege of Port Arthur it was found that a certain length of trench could be dug by the soldiers and coolies in 4 days; but that when only half the coolies were present it required 7 days to dig the same length of trench. Show that coolies did 6 times as much work as soldiers.

25. In a collection plate the value of the gold is 7/8 of that of the silver, and 32 times as great as that of the copper. The total amount is £22. 16*s*. $6\frac{3}{4}$*d*. Find the amount in each metal.

26. In a Parliamentary election $\frac{1}{20}$ of the electors do not vote, and the successful candidate, who is supported by $\frac{17}{35}$ of the whole constituency, polls 51 more votes than his opponent: find the number of votes recorded for each.

27. A man died, leaving 17 cows, and directing that his eldest son should have 1/2 of them, his second son 1/3, and his youngest son 1/9. Being unable to arrange the distribution, the sons consulted a friend. The friend brought his own cow, so making the number 18, and distributed 1/2 of the 18 to the eldest son, 1/3 to the second, and 1/9 to the third. His own cow was left over and he took it home. Were the shares the sons received larger or smaller than those the father directed, and were they in the same proportion?

28. How much too short is a "foot"-rule which makes a room 10 ft. long appear to be 10 ft. 8 in. long?

29. The distance between two wickets was marked out for 22 yards, but the yard measure was 5/12 of an inch too short; what was the actual distance?

30. A tradesman's yard measure is one inch too short; by what amount does he defraud a customer to whom he sells $17\frac{1}{4}$ yards of silk at 10*s*. 9*d*. a yard?

31. *A* barters some sugar with *B* for flour which is worth 2*s*. 3*d*. per stone, but in weighing his sugar uses a false stone weight of $13\frac{1}{2}$ lbs. *B* on discovering this says nothing, but raises the price of his flour; what value should *B* set on his flour that the exchange may be fair?

32. A rectangle of cardboard is 31·9 cm. long by 15·8 cm. broad. From this, rectangular strips 15·5 cm. by 2 cm. are to be cut. How many strips will be obtained if (1) the strips are cut with their long sides parallel to the long sides of the rectangle; (2) with their long sides parallel to the short sides of the rectangle? In both cases, find the amount of cardboard left over (in sq. cm.).

33. A rectangular piece of ground 70 ft. by 28 ft. is surrounded by wire netting and divided by wire netting into 10 equal squares. What will be the cost of the netting, its price being 6*s*. 9*d*. per roll of 50 yards and odd lengths 2*d*. a yard?

34. A piece of land is 202′ long by 108′ broad. It is required to plant a double row of bushes all round the inside of the plot, the rows to be 1′ apart, the plants in the rows to be also 1′ apart, and the outside row of plants to be 1′ from the edge. How many bushes will be required?

The ground is enclosed by palings 4′ high made of planks 8″ wide. Find the length of planking that would be required.

35. How many squares 2/3 of an inch in each side can be cut from a sheet of paper measuring $19\frac{1}{2}$ in. by 11 in., and what is the area of the remainder?

36. Instead of steel armour plates 10 inches thick, plates weighing 400 lbs. per sq. ft. are built into a ship. How much weight is thus saved for each 1000 sq. ft. of surface of this armour? Give your answer in tons to the nearest 1/10 of a ton. 1 cu. ft. of steel armour plate weighs 490 lbs.

37. A rectangular piece of ground of 780 square feet area was sold for £25,050; the cost was said to be £1565. 10*s*. 0*d*. for each foot facing the street; how many feet frontage were there? At the same rate, how much would the land cost per acre (to the nearest £)?

38. A wrought-iron rod has to withstand a pull of 20 tons. If the stress on the rod is not to exceed 10,400 lbs. per square inch, what is the least possible area of cross-section?

39. Find the cost of excavating a ditch half a mile long, $7\frac{1}{2}$ ft. wide, and 4 ft. deep at the rate of 6*d*. per cubic yard for the first foot in depth, the rate increasing 1*d*. per cubic yard for each foot in depth.

40. Two houses are built, one in 4 months and the other in 7 months. The number of workmen engaged on the first is double that employed on the second, and they work two hours a day overtime for which they are paid half as much again as for work done in the ordinary working day of 10 hours. If the sum paid for labour is £1740 in all, find how much of this sum was spent on each house.

41. A man hires a servant for three weeks (18 days) upon the condition that for every day he worked he was to receive 2*s*. 8*d*., and for every day he absented himself he should forfeit 3*s*. 4*d*., and it was found that no payment had to be made either way; how many days did he work?

42. *A* and *B* walk 100 yards against each other. If *A* walks 2 yards in one second, *B* 5 yards in two, how far will *A* have to walk when *B* has finished?

43. *A* and *B* engage to run a race of 100 yards, and *A*, being the better runner, offers either to start 4 yards behind the starting line, or to give *B* 4 yards start in front of it; which offer should *B* accept, and why?

44. Three boys start simultaneously for a race with the same leg foremost, running at the same pace; if *A* takes 5 strides in 3 seconds, *B* 12 strides in 5 seconds, and *C* 9 strides in 4 seconds, how soon will they be together again, (1) in step, (2) out of step?

45. A hare pursued by a greyhound was 86 yards before him at starting; whilst the hare ran 5 yards the dog ran 7; how far had the dog run when he caught the hare?

46. In a hundred yards race *A* can give *B* 5 yards start and *B* can give *C* 10 yards start. How much start can *A* give *C*?

47. *A* can run a quarter-mile in 56 seconds; in running a race of this distance he can give *B* 20 yards and *C* 3 seconds start. If *B* and *C* run a level quarter-mile race, which will win and by how much, assuming that each runs uniformly at the speed indicated by his performance against *A*?

48. A and B ride a 10-mile race; after B has ridden $\frac{1}{4}$ mile A leads by 10 yards, and B passes the half distance post half a minute behind. Hitherto the pace has been uniform, but at this moment B takes up the pace with which A started, while A subsides into that of B, and these rates are maintained till the end of the race. Find the result of the race and its time.

49. A cyclist rides to a certain place and returns to his starting point by the same road. Uphill he averages 8 miles an hour, and downhill 14 miles an hour. Show that for the whole distance he averages $10\frac{2}{11}$ miles an hour.

50. A man rides from A to B and back. He rides at the same speed both ways; but, owing to the clock at B being a quarter of an hour behind that at A, he appears to do the outward journey at 9 miles an hour, and the return journey at 6 miles an hour. What was the distance AB?

51. On my way home I go by train to a junction; and then either walk 3 kilometres home, or wait 15 minutes, travel 10 minutes more by train, and then walk a kilometre. I get home at the same time by the two methods. At what speed do I walk?

52. Of two vessels, A and B, A weighs 4 times as much as B and has 3 times the cubical capacity. A is filled with alcohol and B with water; it is then found that the weight of B is doubled, and that the weight of A is to the weight of B as 16 to 5; what is the ratio of the weights of equal volumes of alcohol and water?

53. The rates in a certain town are 4*s.* 9*d.* in the pound; the assessment of certain premises is raised £20, but as the rates drop 3*d.* in the pound, the actual sum payable in rates is unaltered. What was the original assessment?

54. A man who has three hours to spare travels by coach at 10 miles an hour, and walks back at $3\frac{3}{4}$ miles an hour. If he returns 4 minutes before the time how far did he travel by coach?

55. Gold is 19·365 times as heavy as water, silver 10·475 times as heavy as water. If 1 oz. of gold is worth 30 oz. of silver, what volume of silver will be of the same value as a cubic foot of gold?

56. The incomes of A and B are in the ratio of 5 to 4, their expenditures are in the ratio of 6 to 5, and their savings are in the ratio of 10 to 7. Find the ratio of A's income to B's expenditure.

57. A rate of £1125 is to be raised by a union of three parishes in which the ratepayers number 215, 160, 110 respectively. The average assessments of ratepayers in the three parishes are as $2:3:4$. Find the amount to be raised from each parish.

58. Four people wish to be conveyed as quickly as possible to a place 5 miles distant. They can walk 4 miles an hour, and have a motor car that can go 18 miles an hour but only holds two besides the driver. Two set off on foot; the car takes the other two a certain distance, sets them down to continue on foot, and then returns to bring the first two. At what point must the car set down those it took first in order that the four may arrive at the same instant?

Find also the time taken to do the journey.

59. A train after travelling an hour is detained 15 minutes, after which it proceeds at 3/4 its former rate, and arrives 24 minutes late. If the detention had taken place 5 miles farther on the train would have arrived 3 minutes sooner than it did. Find the original rate of the train and the distance travelled.

60. Three workmen, A, B, C, start on a piece of work which will take them exactly 6 weeks to finish: the estimated cost of the work is £20. 5*s*. 0*d*. At the end of the first week A catches the influenza and stops work; and half-way through the second week B also is laid up; but both A and B return at the end of the third week. The work is completed at the expected date by all three working overtime, for which a higher rate is paid, and the cost of the work is increased to £21. 11*s*. 3*d*. Find what each man receives.

61. There are two tumblers, A and B; A is half-full of wine, B of water. A teaspoonful of wine is taken out of A and poured into B; a teaspoonful of the mixture in B is then transferred to A. As the result of both operations, is the quantity of wine that has been removed from A greater or less than the quantity of water that has been removed from B?

62. Two casks contained originally 40 gallons of wine and 20 gallons of water respectively. On three successive occasions 8 gallons of liquid are drawn from each cask and placed in the other, the liquids being thoroughly mixed. Express in gallons and decimals of a gallon the quantity of wine now in each cask.

63. Three tramps meet together for a meal. The first has 5 loaves, the second 3, and the third, who has his share of the bread, pays the others 2*d*. How ought they to divide the money?

64. Two passengers have between them 345 lbs. of luggage and pay on their excess luggage 4*s*. 2*d*. and 6*s*. 8*d*. respectively. If the luggage had belonged to one of them the excess charge would have been 15/-. How much free luggage is allowed?

65. In a boat race, the length of the course being $1\frac{1}{2}$ miles, the winning boat did the distance in 7 mins. 18 secs., whilst the losing boat took 7 mins. 20 secs.; by how many yards was the race won?

PART III

CHAPTER XIII

PERCENTAGE

§129. Two armies of 80,000 and 50,000 suffer losses amounting to 4000 and 3000 respectively. The larger army has lost the greater number of men: but if we wish to discover which army has been the more heavily engaged, we should consider in each case the *ratio* of the loss to the total strength. The ratios are 4000 : 80,000 and 3000 : 50,000 respectively, which reduce to 1 : 20 and 3 : 50. These ratios might be compared in the usual way by reducing each to the form $n : 1$. Another and even more useful way is to consider how many men each army loses *per hundred* of its numbers, or what is the loss **per cent.**

A loss of 4000 out of 80,000, or 1 out of 20, is equivalent to 5 out of 100 or 5 per cent.

A loss of 3000 out of 50,000, or 3 out of 50, is equivalent to 6 out of 100 or 6 per cent.

Thus the smaller army has lost a larger **percentage** of men.

Notice that a percentage is simply a ratio of which the second term is 100.

The sign for "per cent." is %.

To Find a Given Percentage of a Given Quantity

EXERCISE XIII a (oral)

1. The population of a village is 600. (i) 30 in every hundred are men. How many men are there? (ii) 40 per cent. are women. How many women are there? (iii) 2 % are paupers. How many paupers are there? (iv) 10 % are labourers. How many labourers are there? (v) 20 % are children. How many children are there?

2. What fraction of the whole population are (1) men, (2) women, (3) paupers, (4) labourers, (5) children?

3. Express as fractions the following rates per cent.:

(1) 50 %.	(2) 25 %.	(3) 5 %.	(4) 10 %.
(5) 15 %.	(6) 75 %.	(7) $\frac{1}{2}$ %.	(8) $\frac{1}{10}$ %.
(9) $2\frac{1}{2}$ %.	(10) 100 %.	(11) 200 %.	(12) 250 %.
(13) $2\frac{1}{4}$ %.	(14) $3\frac{1}{8}$ %.	(15) $4\frac{1}{5}$ %.	(16) x %.

4. What is 5 % of £100, £20, £40, £200, £240, £1?

5. What is 3 % of £50, £150, £250, £700?

6. What is $2\frac{1}{2}$ % of £100, £200, £50, £600, £1?

7. Express as decimals the following rates per cent.:

(1) 50 %.	(2) 45 %.	(3) 20 %.	(4) 22 %.
(5) 15 %.	(6) 11 %.	(7) 125 %.	(8) 200 %.
(9) 5 %.	(10) 2 %.	(11) $2\frac{1}{2}$ %.	(12) 3·5 %.
(13) 2·75 %.	(14) 18·25 %.	(15) 3·12 %.	(16) 0·5 %.

8. Express in metres and decimals of a metre:

(1) 4 % of 100 m., 10 m., 16 m., 15 m., 25 m., 17 m., 7 m., 12 m., 5 m., 1 m.

(2) 5 % of 200 m., 12 m., 15 m., 7 m., 6 m., 2 m.

9. If 5 % of an army is lost, what percentage remains?

10. If an army lost 5 % killed in action, 15 % died of wounds and 20 % were captured, what percentage remains?

§ **130.** *Example.* A certain alloy called "invar" contains 36 % of nickel. How many pounds of nickel are there per ton?

A ton of invar contains

$\frac{36}{100}$ of 2240 lbs. of nickel

= <u>806·4 lbs.</u>

```
      2240 lbs.
        36
      ----
      6720
     13440
     -----
100)80640
     -----
     806·4 lbs.
```

Note. In simplifying fractions involving 100 or other powers of 10 (e.g. $\frac{36}{100} \times 197$) never "spoil" the 100 by cancelling; it is easier to reduce the original expression to a decimal than to reduce $\frac{9}{25} \times 197$.

Example. What is $2\frac{3}{4}$ % of £158/5/-, to the nearest penny?

This is $\frac{2\frac{3}{4}}{100}$ of £158·25

= £4·351875

= <u>£4/7/0</u> approx.

```
     £158·25
        2¾
     -------
      316·50
       79·125  (= ½)
       39·5625 (= ¼)
     --------
100)435·1875
     --------
      £4·351875
      £4·352
          20
      ------
       7,·04s.
          12
      ------
       0,·48d.
```

Note that the easiest way to multiply by $2\frac{3}{4}$ is by a sort of practice method; take twice, then $\frac{1}{2}$, then $\frac{1}{4}$.

EXERCISE XIII b

1. 7·5 % of the population of a village die during an epidemic. It was 680. How many die?

2. How much metal (to the nearest ton) will be obtained from 358 tons of ore, if the ore contains 6·1 % of metal?

3. 3·2 % of a regiment 1500 strong were lost in a campaign: how many survivors were there?

4. In a school of 150 children 34 % are girls. How many boys are there?

5. If gunpowder contains 75 % of saltpetre, 10 % of sulphur, and 15 % of charcoal, how many pounds of each are there in 2 tons of gunpowder?

6. Find correct to 3 places of decimals:

(1) 23 % of 15·62. (2) 5·25 % of 15·2.

(3) 0·035 % of 11460. (4) 0·75 % of 7·38.

7. Sea-water contains 2·9 % of salts. Find to the nearest pound the amount of salt that will be obtained by evaporating a ton of sea-water.

8. From 1 lb. (Avoir.) of bronze 48 pennies are made. Bronze contains 1 % by weight of zinc. Find the total quantity of zinc required for the bronze coinage of 1899, which amounted in value to £139,065.

9. If the length of an iron rail increases by ·00067 % for every degree F. through which it is heated, find:

(1) The increase in length of 100 miles of iron rails between temperatures of 40° and 90° F. (nearest yard).

(2) The amount of play that must be allowed at one end of an iron bridge 750 feet long, if the other end is fixed and the temperature varies from 20° F. to 120° F. (to the nearest hundredth of an inch).

Increase and Decrease per cent. Profit and Loss

EXERCISE XIII c (oral)

1. What are the numbers at the end of the year if a school of (i) 100 increase by 4 %, (ii) 200 increase by 4 %, (iii) 100 decrease by 4 %, (iv) 200 decrease by 4 %?

2. What is the selling price if an article costing (i) £100 is sold at a profit of 5 %, (ii) £100 is sold at a loss of 5 %, (iii) £50 is sold at a profit of 3 %, (iv) £50 is sold at a loss of 2 %, (v) £1 is sold at a profit of 5 %, (vi) £1 is sold at a loss of 10 %?

3. Class practice in supplying the answers to such sums as the following: At a profit of 5 %, £100 becomes £105; £50 becomes £$52\frac{1}{2}$; £1 becomes £$\frac{105}{100}$; £23 becomes £23 × 1·05. The final answer to be in the form: Given amount × ratio.

EXERCISE XIII d

1. The population of a town was 406,025 in 1900 and increased by 4 % during the year. What was it in 1901?

2. The numbers of a school were 620 in 1901 and decreased by 5 % during the year. What were the numbers in 1902?

3. A man wishes to make 5 % profit on the following purchases. What must he sell them for? (1) A bicycle costing £22; (2) a bicycle costing £15; (3) a motor car costing £320; (4) a motor car costing £460; (5) a house costing £2160.

4. If the wheat harvest at a farm was 12·5 % better in 1902 than in 1901, and in 1901 amounted to 3248 bushels, what was it in 1902?

5. A gold coin fresh from the mint weighs 0·24 oz. After being 3 yrs. in circulation it is found to have lost $2\frac{1}{2}$ % of its weight. What does it then weigh?

6. Find the selling prices in the following cases:

Buyer's price	£243	£156	£156	£114/8/-	£215/6/6	£118/19/8
Gain %	− 5	+ $2\frac{1}{2}$	− $2\frac{1}{2}$	+ $3\frac{1}{2}$	− $2\frac{1}{4}$	+ $4\frac{3}{4}$

7. A man buys 6 boxes of matches for 1*d*. What must he sell one box for to make 20 % profit, and how many does he sell for 1*d*.?

8. A man buys 4 boxes of matches for a penny and makes 60 % profit. How many does he sell for 1/-?

9. A fishmonger buys herrings at a penny each and makes 50 % profit: how many does he sell for 1*s*.?

10. A man buys 6 boxes of matches for 1*d*. How many does he sell for 1*s*. if he loses 10 % on the transaction?

To express one quantity as a percentage of another

EXERCISE XIII e (oral)

1. During a campaign 1/20 of the force is killed in action, 1/10 dies of wounds, 1/5 of disease, and 1/25 is captured. Express these fractions as percentages.

2. Express as percentages the following fractions:

(1) 1/2.	(2) 1/4.	(3) 1/10.	(4) 1/20.
(5) 1/5.	(6) 2.	(7) $2\frac{1}{2}$.	(8) 3/4.
(9) 3/5.	(10) 1/8.	(11) 3/8.	(12) 0·5.
(13) 0·6.	(14) 0·56.	(15) 0·05.	(16) 0·025.
(17) 0·015.	(18) 0·032.	(19) $\frac{x}{100}$.	(20) $\frac{x}{50}$.
(21) $\frac{x}{20}$.	(22) $\frac{x}{25}$.	(23) $\frac{x}{10}$.	(24) $\frac{x}{40}$.

3. What per cent. is

(i) £20 of £100?	(ii) £5 of £100?
(iii) £5 of £50?	(iv) 15 m. of 200 m.?
(v) 1 cm. of 1 m.?	(vi) 3 m. of 15 m.?
(vii) 12 cm. of 1 m.?	(viii) 0·5 m. of 1 m.?
(ix) 0·34 m. of 1 m.?	(x) 0·025 m. of 1 m.?

EXERCISE XIII f

1. An electrical machine was bought for £1050 and costs on an average £13 a month to keep it working. What percentage of the cost is the annual working expense?

2. If a ton of ore yields 336 lbs. of lead, what is the yield per cent.?

3. Find the profit per cent. (to 1 place of decimals) if

Buying price	£150	£200	£183	£475	£16/10/-	£3260
Profit	£9	£15	£6/8/-	£5/10/-	£2/2/6	£1065

4. Find the profit or loss per cent. in the following cases (to 1 place of decimals):

Buying price	£150	£220	£182	£156	£23/5/-	£4380
Selling price	£200	£210	£175	£167	£21/10/-	£5127

5. A sample of brandy is found to contain pure spirits, 2·24 oz.; pure water, 3·27 oz.; miscellaneous, 0·21 oz. Express these proportions as percentages (to 1 place).

6. In a ton of black powder there are 1676 lbs. of nitre, 231 of sulphur, 311 of charcoal, and 22 of water. Find the percentage by weight of each in the powder (to 1 place).

7. In 1898 the total expenditure of the army and navy was £44,283,000; in 1899 it had risen to £69,815,000. Find the increase per cent. correct to 1 place.

8. A bankrupt's liabilities are £5126 and his assets £3420. 10*s*. What percentage of his debts can he pay?

9. Tea can be bought at 1/3 a lb. retail and at $8\frac{1}{2}d.$ per lb. wholesale. How much per cent. is gained by buying wholesale and selling retail?

10. Demerara sugar costs $2\frac{1}{2}d.$ per lb. and 21/6 per cwt. What is the profit per cent. on buying at the hundredweight rate and selling at the retail rate?

11. The apparent diameter of the sun is $32' \ 32''$ on January 1st and $31' \ 32''$ on July 1st. What is the variation per cent. (to the nearest integer)?

12. A man buys eggs at 10*d*. a dozen and sells them at 10 a shilling; what is his gain per cent.?

13. I buy 5 articles for 6*d*. and sell them at the rate of 7 for a shilling. What is my gain per cent.?

Given that A is a known percentage of B, to find B

EXERCISE XIII g (oral)

1. There are 30 "scholars" in a certain school, and they form 5 % of the total number. Find the total number.

2. Find a number of which (i) 15 is 5 %, (ii) 20 is 4 %, (iii) 50 is 25 %, (iv) 64 is 4 %.

3. Of what sum of money is £18 three per cent.?

4. If a man devotes 8 % of his income to charity, and in this way spends £72 a year, what is his income?

5. 18 tons of lead are derived from ore which yields 6 %. How much ore was used?

§131. *Example.* The population of a town increases at the rate of 2·7 % each year. It is 52,320 now; what was it a year ago?

The difficulty is that we do not know the original population on which the 2·7 % is to be calculated. Notice how this difficulty is overcome.

For 2·7 % increase, 100 becomes 102·7; i.e. the original population is $\frac{100}{102 \cdot 7}$ of the final population, or $\frac{100}{102 \cdot 7} \times 52320$.

This is $\frac{52320}{1 \cdot 027} = \underline{50944}$ to the nearest unit.

```
          50944·5—
1·027)5232
      5135
       9700
       9243
        4570
        4108
         4620
         4108
          5120
```

EXERCISE XIII h

1. A man saves 15 % of his income. What is his income if he saves £114?

2. A man saves 20 % of his income and spends £500. What is his income and how much does he save?

3. A man spends 95 % of his income and saves £15. What is his income?

4. The population of a town is 21,000 and is 5 % more than it was five years ago; what was it then?

5. 3·2 % of a regiment were lost in a campaign. The survivors number 1452. What was the original strength?

6. If I deduct 5 % from a bill and pay £38, what was the original bill?

7. A man receives £340 as his salary and reckons that 15 % has been deducted. What did he expect?

8. A farmer sells his horse for £54, and reckons that he has made 8 % profit. What did he pay for the horse?

9. Find the buying price in the following cases:

Selling price in £	105	63	39	100	100	95	144	100
Gain %	+ 5	+ 5	+ 4	+ 20	+ 5	− 5	− 4	− 5

10. By selling eggs at 8 for 1*s.* a farmer makes 20 % profit. What did they cost him?

EXERCISE XIII i

Miscellaneous Straightforward Exercises on Percentage

1. A farmer buys 4 cwts. of wheat. If 3 per cent. of it is stolen, express the farmer's loss in lbs. of wheat, to the nearest pound.

2. A book marked £4 is sold for £3. 15*s*. 0*d*. How much per cent. is the reduction?

3. If there were 12 fine days in September, what percentage of the days of the month were fine?

4. A ship consumes 1021 tons of coal during a certain voyage. On her next trip she encountered heavy weather and so burnt 1357 tons. What is the increase per cent. of the coal consumption?

5. Goods bought at £245 are sold at a loss of 10 per cent. on the cost price. Find the amount of the loss.

6. On a certain day in June, one county had gained 20 points at cricket out of a possible 35; another county had gained 25 points out of a possible 40. Which had the better percentage at that time?

7. A man who earns £470 a year pays £12 in income tax. What percentage does he pay (to 1 decimal place)?

8. An article is sold for £5. 13*s*. 4*d*. at a loss of 15 per cent. What did it cost?

9. 14 per cent. of the 3750 inhabitants of a town were vaccinated last year. How many were vaccinated?

10. If the population of a town was 325,619 in 1881 and 363,715 in 1891, find the increase per cent. (to 1 place).

11. A ship carries 2480 tons of cargo; 570 tons are discharged. What percentage of the total was discharged?

12. The value of exports from the United Kingdom in 1912 was £487,000,000. The value of goods exported to India was £58,000,000 and to other British Possessions, £117,000,000. What percentage of the whole is each of these latter values?

13. If by selling my horse for £17. 4*s*. I lose 20 per cent., what did my horse cost?

14. Copper ore yields 30 per cent. of iron and 13 per cent. of copper. What weights of iron and copper could be obtained from 20 tons of the ore?

15. Of all the people in a steamer, 15 % were crew, the rest were all passengers. (1) What fraction of the whole number of people consisted of passengers? (2) If there were 460 passengers, how many persons composed the crew?

16. From a vessel containing 2·5 kilograms of liquid 0·75 kilograms are removed. What percentage is this?

17. It is proposed (Oct. 24th, 1912) under the Insurance Act to give doctors 9*s*. per annum for each man on their books. This is an increase of $35\frac{5}{7}$ % on the original proposal. What was the payment proposed originally?

18. An iron rod is "turned down" in a lathe so as to reduce its weight by 12 per cent. It now weighs 11 lbs. What did it weigh originally?

19. A man's income was £267. 17*s*. 4*d*. Out of this he paid one-seventh for rent and 75 per cent. of the remainder went to pay other expenses. How much did he save?

20. What percentage of £750 is £28. 2*s*. 6*d*.?

21. In Jersey potatoes are sold by the "cabot." The price rises from 1*s*. 7*d*. to 1*s*. 8*d*. per cabot. What is the difference in cost to a merchant who, in consequence of the rise in price, reduces his usual order of 260 cabots by 5 per cent.?

22. Bronze is made of copper, tin, and zinc, the percentages (by weight) of the two former being 90·75 and 8·41 respectively. How many pounds of bronze would be made if 1 lb. of zinc were used?

23. A draper buys 1000 shawls for £500. He sells 653 of them at 15*s*. 7*d*. each, and the rest at half that price. How much does he gain? What is his gain per cent.?

24. In a certain fleet there were 127 vessels in all, but 5·7 per cent. of these have been destroyed in a war. If 76 of the vessels that now form the fleet are torpedo boats what per cent. of the present fleet consists of torpedo boats?

25. A man buys goods at the rate of £24 per cwt., and sells 2 tons 14 cwts. 3 qrs. for £1500. How much has he gained or lost per cent. on the outlay?

26. In the following ships, built for the Japanese Navy, weights were as follows:

	Battleship *Asahi*	Destroyer *Akutsuki*
Displacement in tons with normal coal	14525	366
Weight of machinery	1437 tons	155 tons
„ „ hull	5714 „	129·5 „

Find for each ship, what percentage of the displacement is made up of (i) machinery, (ii) hull (correct to 1 place).

27. Sea-water contains $2\frac{1}{2}$ per cent. of salt. What weight of water would be required to yield half a ton of salt?

28. The entrance fee to an exhibition being reduced by one-quarter, the daily attendance is increased by 30 per cent. What is the percentage decrease on the daily receipts?

29. Supposing that a diminution of 30 % of the customs duty on tobacco led to an increase of 40 % on the amount consumed, what would be the percentage increase (or diminution) of the revenue derived from this source?

30. A manufacturer's list price is 60 per cent. above the cost of manufacture; he allows a trade discount of 20 per cent. from the list price. What is his profit per cent.?

31. The report of the Egyptian Consul-General on the trade of Port Said states that during the year 1911 the total number of vessels that passed through the Suez Canal was 4969, and that of these 62·2 per cent. were British ships, and 13·4 per cent. German ships. The net tonnage of all the ships passing through the canal was 18,324,794 tons. Find, to the nearest ton, the average net tonnage of the ships that passed through the canal, and, to the nearest whole number, the actual number of British and German ships that made use of the canal.

32. For a certain examination 1892 girls and 1370 boys entered. Of these 249 girls and 370 boys "obtained honours." Of the rest 1002 girls and 648 boys "satisfied the examiners," whilst the remainder failed to satisfy the examiners. Calculate to the nearest integer in each case:

(*a*) the percentage of boys who failed;

(*b*) the percentage of girls who obtained honours;

(*c*) the percentage of the total number of candidates who satisfied the examiners, but did not obtain honours.

33. A man's expenses for the years 1910 and 1911 are tabulated below under certain heads:

	1910			1911		
	£	*s.*	*d.*	£	*s.*	*d.*
Rent, rates, taxes	93	13	0	96	15	7
Household expenses ..	201	3	8	211	19	3
Dress	59	4	7	68	13	8
Personal expenses	54	6	8	80	8	6
Sundries	65	3	3	50	13	9

Calculate, to the nearest integer in each case, the increase or decrease for 1911 as a percentage of the amount in 1910, (*a*) in household expenses, (*b*) in the total.

34. The quantity of jute imported into Great Britain for the year 1910 was 296,700 tons, valued at £4,670,000. That for 1911 was 300,900 tons, valued at £5,996,100. Find, to the nearest tenth in each case, the percentage increase during 1911 (i) in the quantity of jute imported, (ii) in the value of the jute imported, (iii) in the value per ton.

35. A man buys tobacco at 5*d.* an ounce and receives with each ounce two coupons. In exchange for 2520 of these coupons he can obtain an article valued at 42*s.* 6*d.* Find in lbs. what weight of tobacco he must buy in order to obtain the article, and express the value of the article as a percentage of the amount he pays for this tobacco.

36. An agent receives a regular salary of £150 a year, and, in addition, a commission of 10 per cent. on the selling price of all goods that he sells, but he has to pay his own expenses. During the year (365 days) his average daily expenses amount to 9*s*. 4*d*. and he pays in addition £51. 17*s*. in railway fares. If he sells goods for £5271. 15*s*. find how much he is to the good on the year.

*HARDER SUMS ON PERCENTAGE

§132. *Example.* By selling a horse for £50. 8*s*. a farmer makes 5 % profit. What must he sell it for to make 10 % profit?

To give 5 % *profit,* £100 *must become* £105, i.e. the buying price is $\frac{100}{105}$ of the selling price.

If the selling price is £50/8/-, the buying price is $\frac{100}{105}$ of £50/8/-

$$= £\frac{100}{105} \times 50{\cdot}4 = £\frac{5040}{105} = £48.$$

$\therefore$ the farmer gave £48 for the horse.

To realise 10 % profit, the horse must be sold for

$$£48 \times \frac{110}{100} = £\frac{528}{10} = £52{\cdot}8 = £52/16/\text{-}.$$

EXERCISE XIII j

1. A dealer buys two pictures from an artist at £135 and £90, and sells them again at profits of 20 per cent. and 25 per cent. respectively on the purchase price. Find his percentage of profit on the whole transaction.

2. I buy a number of articles at 4 for 3*d*., and sell them at 3 for 4*d*. What is my gain per cent.?

3. How much per cent. is gained by buying eggs at 15 for a shilling, and selling them at 11*d*. a dozen?

4. A man alternately gains and loses 10 per cent. per annum of what he has at the beginning of each year. Show that this is equivalent to losing 1 per cent. every two years.

* May be omitted at a first reading.

5. A dealer gains 5 per cent. by selling a horse for £28. How much money would he gain if he sold it for £30?

6. A man makes a profit of 44 per cent. by selling apples at 10 for a shilling. How must he sell them to make a profit of 8 per cent.?

7. A man sells goods, which cost him £900, through an agent. The agent keeps 10 per cent. of the selling price himself, but this still leaves the owner with a profit of 10 per cent. For how much did the agent sell the goods?

8. If a man sells a certain article at 8*s*. 7*d*., he makes a profit of 3 per cent. How many must he sell at 8*s*. 6*d*. in order to make £1 profit?

9. A manufacturer sells goods to a tradesman at a profit of 80 per cent. If the tradesman becomes bankrupt and pays a dividend of 11*s*. 2*d*. in the £, how much does the merchant gain or lose per cent.?

10. A portmanteau will hold 60 lbs. of clothes or 200 lbs. of books. What will it weigh when packed full of clothes and books if (1) 30 % of the volume is occupied by books and the remainder by clothes; (2) 30 per cent. of the weight is made up of books and the remainder of clothes? (Neglect the weight of the empty portmanteau.)

11. A 10-gallon barrel is full of beer which contains 8 % of alcohol. If 2 gallons are run off and the barrel is filled up with water, what is the percentage of alcohol in the mixture?

12. A motor car is made at a cost of £400. The manufacturer sells it at a profit of 5 % to the middleman, and the latter sells to a purchaser at a profit of 10 %. What does the purchaser pay for it? How much per cent. is this above the cost price?

13. What would a dishonest dealer make per cent. by using a 14 oz. weight for a pound weight?

14. A dealer professes to sell goods at cost price, but uses fraudulent weights, thereby making a profit of 12 %. What does his pound weight really weigh?

15. If 2 gallons of whisky containing 70 % of alcohol are mixed with 1 gallon containing 40 %, what is the percentage of alcohol in the mixture?

16. What per cent. above cost price must a man mark his goods so that he may take off 10 % and still gain 20 %?

17. A tradesman deducts successively discounts of 25 %, 5 %, and $2\frac{1}{2}$ %, each discount being calculated on what remains after the preceding discount is deducted. What percentage does he deduct altogether?

18. A dealer mixes 8 lbs. of tea at 1*s.* 6*d.* per lb. with 4 lbs. at 2*s.* per lb. What is the mixture worth? What does the dealer gain (1) per lb., (2) per cent., if he sells the mixture at 1*s.* 10*d.* per lb.?

19. A man buys 30 lbs. of tea at 2*s.* 6*d.* a lb., and 20 lbs. at 3*s.* 6*d.*; he mixes them and sells the mixture at 3*s.* a lb. What is his gain per cent. (to 1 place)?

20. A grocer buys 15 lbs. of tea at 1*s.* 6*d.* per lb., and 21 lbs. at 2*s.* 6*d.* per lb.; he mixes them and sells the mixture at 2*s.* 2*d.* per lb. What is his gain per cent.?

21. A spirit merchant mixes 40 gallons of whisky at 15*s.* 6*d.* per gallon with 48 gallons at 17*s.* 1*d.*, and sells the mixture so as to gain 10 per cent. At what price per gallon does he sell it?

22. Eggs bought at 21 for a shilling are mixed with others bought at 19 for a shilling, and the whole sold at 20 for a shilling. Find the gain or loss per cent. when equal numbers of each kind are bought.

23. If the length and breadth of a rectangle be each increased 20 per cent., what is the percentage increase in the area?

24. If the 3 dimensions of a solid body be each increased 20 per cent., what is the percentage increase in the volume?

25. A plan is drawn to the scale of 3 cm. to the foot and is to be reduced in scale to 1 inch to the foot. What reduction per cent. will there be in (1) lengths of lines on the plan, (2) the area of the plan?

26. A cube has an edge 1 ft. 8 in. long. If the length of the edge were diminished by 5 per cent., what would be the percentage reduction in the surface? the volume?

CHAPTER XIV

APPROXIMATION

§ 133. In arithmetic, one of the chief lessons to be learnt is absolute accuracy in dealing with figures. The mention of approximate results must not lead anyone to suppose that mistakes do not matter.

§ 134. At the same time it is essential to realise that data derived from measurements *can never be taken as absolutely exact.* Data obtained by counting may be exact; a bag of coins can be counted exactly. But length cannot be measured exactly; nor can weight, volume, time, etc.

The diameter of a ball from a ball-bearing might be ascertained by rough measurement to be 4 mm. to the nearest millimetre. With greater care we might find it to be 4·3 mm. to the nearest $\frac{1}{10}$ mm. With a micrometer screw gauge we might find it to be 4·34 mm. to the nearest $\frac{1}{100}$ mm. But 4·34 mm. is not the absolutely exact diameter any more than is 4 mm.

Similarly a weight, a period of time, etc., can never be determined with absolute accuracy, but only to a degree of accuracy depending on the measuring implements used and the skill of the experimenter.

§ 135. **Significant figures.** Imagine that we have found the diameter of a wire to be 2·47 mm. to the nearest hundredth of a millimetre. We might suppose that the accuracy of our measurement is sufficiently indicated by the number of decimal places given. This is so if the unit is given. But 2·47 mm. = ·00247 metres: thus by a change of units we have altered the number of decimal places from 2 to 5. But the measurement in its new form is neither more nor less accurate than in its old form.

For many purposes, therefore, we must fix our attention not on the number of decimal places, but on the number of **significant figures** in a result. In 2·47 mm. we have 3 significant figures. In ·00247 m. we do not regard the two zeros as significant figures; they result simply from a change of units and do not *signify* greater accuracy: they merely indicate that the figure 2 means $\frac{2}{1000}$.

Again, suppose that a railway truck is known to contain 2075 kilograms of goods (to the nearest kilogram). Here we have 4 significant figures: the zero in this case *is* significant, for we are sure that the weight is not 2175 or 2275. On the other hand, 2075 kilos = 2075000 grams: the last three zeros are not significant; they are introduced by a change of units. We have no right to say that the weight is 2075000 gm. in preference to 2075135 gm.

Broadly speaking, zeros at the beginning or end of a number are not significant figures. But we must always use our common sense. For instance if a birth certificate shows a boy's age to be 10, we know that it is not 11 or 12; the zero here gives an additional piece of information and is a significant figure. On the other hand, when we speak of the 100 years' war, we do not mean to assert that the war lasted exactly 100 years; as a matter of fact the duration was about 116 years. Accordingly the 2 zeros in this case are not significant figures.

§136. **Absolute error.** If the strength of a battalion is given as 920 correct to the nearest ten, we know that the exact number lies between 925 and 915. The **absolute error,** i.e. the difference between the exact and the approximate number, may be 4 in either direction: we may say that the exact number is 920 ± 4, the ± 4 denoting the limits of absolute error possible.

§137. **Relative error.** For many purposes, the absolute error is of less importance than the relative error, which is now to be explained. An uncertainty of ± 4 in the number of a battalion of about 900 is much less important than an uncertainty of ± 4 in a squad of 9; on the other hand, it is much more important than an uncertainty of ± 4 in an army of about 90,000. From this point of view, the important thing to know is the ratio of the absolute error to the total number; this ratio is called the **relative error.**

In the case of a battalion of 920 ± 4, the relative error is $\pm \frac{4}{920}$. This is approximately equal to $\pm \frac{4}{900}$; in fact.

in calculating relative error it is generally sufficient to take *one significant figure* of the total number.

If a length is given as 2·47 m. to the nearest ·01 m. (i.e. the nearest cm.) we know that the real length lies between 2·475 m. and 2·465 m.: the absolute error is ± ·005 m.; and the real length may be quoted as 2·47 m. ± ·005 m. The range of error is ·01 m.; accordingly it is sometimes said that such a length is determined to be 2·47 m. *to within* ·01 m.; strictly speaking we ought to say "to within ± ·005 m." In this case the relative error is $\pm \frac{\cdot 005}{2\cdot 47}$ = about ± ·002.

§ 138. Notice that absolute error is a *quantity*; relative error is a *number*.

Also remember that when working by decimal places we generally have to attend to the absolute error; but when working by significant figures we must attend to the relative error.

§ 139. **Percentage error,** or error per cent., is relative error expressed as a percentage; e.g. a relative error of $\pm \cdot 002 = \pm \frac{\cdot 2}{100} = \pm \cdot 2$ per cent.

EXERCISE XIV a (oral)

Read off the values of the following, correct to 3 sig. figs.:

1. 3·4249. **2.** 743·26. **3.** 0·52169.
4. 1·23596. **5.** 1·23546. **6.** 1·2355.
7. 263420. **8.** 17424. **9.** 71260.
10. 71209. **11.** 0·0071209. **12.** 0·06499.
13. 0·069991. **14.** 40·04. **15.** 40·09.

By common sense, decide whether or no final zeros in the following cases are significant figures:

16. 1000 mm. = 1 metre.

17. 1000 recruits joined in a single day.

18. $10^3 = 1000$.

19. The British Navy has existed for 1000 years.

20. 1 ft. = 30·25 cm., or approximately 30 cm.

21. 1 cable = 203 yds., or approximately 200 yds.

22. Sea-water is 1·025 times as heavy as fresh water, or roughly 1·0 times as heavy.

¶EXERCISE XIV b

Give the absolute error, the approximate relative error, and the approximate percentage error in results found to the following degrees of approximation:

1. 107 cm. to the nearest cm.

2. 4·39 gm. to the nearest 0·01 gm.

3. 0·52 in. to the nearest $\frac{1}{100}$ in.

4. 0·027 litres to the nearest $\frac{1}{1000}$ litres.

5. 7·35 m. to the nearest cm.

6. 0·048 gm. to the nearest mgm.

7. 1800 Km. to the nearest Km.

8. 0·095 kilos to the nearest gm.

9. 4/2½ to the nearest farthing.

10. £1/1/2 to the nearest farthing.

11. £19/3/- to the nearest shilling.

12. 2 hrs. 5 mins. to the nearest minute.

13. 5 lbs. 3 oz. to the nearest oz.

14. 10 yds. 2 ft. to the nearest ft.

Find to 1 significant figure the percentage errors in the following experimental results:

	Experimental result	Correct theoretical result
15.	2·05 cm.	2·03 cm.
16.	4·28 in.	4·25 in.
17.	16·28 gm.	16·34 gm.
18.	$\pi = 3{\cdot}15$.	$\pi = 3{\cdot}1416\ldots$
19.	25·9 c.c.	25·86 c.c.
20.	$\sqrt{2} = 1{\cdot}416$.	$\sqrt{2} = 1{\cdot}414\ldots$

21. In an experiment to find π the circumference of a cylinder was measured and taken to be 19·2 cm. The diameter of the cylinder was 6 cm. From this π was calculated*. Find the percentage error to the nearest integer. [Take 3·142 as the correct value of "π."]

22. Four boys A, B, C, D obtain in a laboratory the following results: A 3·1, B 3·22, C 3·144, D 3. If the correct result is 3·142, give the absolute error of each, the error per cent., the number of significant figures to which he has given his result, and the number of significant figures to which it is correct. Find also the average of the results to three decimal places, the absolute error of this mean, and the error per cent.

23. The earth rotates once in 23 hrs. 56 mins. 4 secs. What error per cent. is made in counting this as 24 hrs.?

24. In solving the following problem there is (i) an easy and obvious method which gives an approximate answer; (ii) a longer method which gives an exact answer. Find the error per cent. resulting from the easy method.

> A boy's watch is quite right when he leaves home at 9 o'clock on Tuesday morning, but it loses 4 seconds an hour; how much will he be late if he trusts to his watch to arrive in school at 9 a.m. on Wednesday?

25. In the cotton trade the standard make is called "79 inch, $37\frac{1}{2}$ yards, $8\frac{1}{4}$ lbs. shirting." Find what percentages of error are made in taking these quantities as equal respectively to 2 metres, 34 metres, $3\frac{3}{4}$ kilograms.

§ 140. **How many significant figures to give in a result.** A result should not be given to a greater number of significant figures than is justified by the data.

Example. To find 1/3 of a length which has been measured as 8·5 m. to the nearest ·1 m.

Wrong way	Right way
3)8·50000..	3)8·5× × × ×
2·83333..	2·8× × × ×

In dividing, one would at first be inclined to supply

* See Chap. XVI.

zeros after the 5, and obtain an indefinite number of 3's in the quotient. But really we have no right to supply zeros: the figures that follow the 5 are unknown and may be represented by crosses. The figures after the 8 in the quotient are equally unknown.

This × method gives a rough-and-ready idea of the number of figures that it is fair to set down in the answer. The most exact way however is to say that the quotient lies between $\frac{8{\cdot}55}{3}$ and $\frac{8{\cdot}45}{3}$; i.e. between 2·85 and $2{\cdot}81\dot{6}$. This again shows that the answer should not be given except as 2·8...+.

§141. *Example.* To find the sum of the following lengths, which are determined to the nearest $\frac{1}{100}$ of an inch: 3·45″, 0·21″, 6·84″, 5·26″.

Greatest possible sum	Least possible sum
3·455	3·445
0·215	0·205
6·845	6·835
5·265	5·255
15·780	15·740

The sum therefore lies somewhere between 15·78″ and 15·74″. If we had simply added together the four numbers given, we should have obtained 15·76″. But we now see that the final 6 is worthless. We cannot even give the sum as 15·8″ correct to 1 place; for it may really be nearer to 15·7″. The best form of answer is 15·76″ ± 0·02″. This means that the sum may differ from 15·76″ by anything up to 0·02″ either way.

Example. To find the difference of weights found to be 562 gm. and 431 gm. to the nearest gm.

Greatest possible difference	Least possible difference
562·5	561·5
430·5	431·5
132·0	130·0

The difference may be given as 130 gm. correct to 2 significant figures, or as (131 ± 1) gm.

Example. What is the area of a rectangle whose sides measured to the nearest 1/10 inch are 4·3″ and 2·5″?

Greatest possible area	Least possible area
4·35	4·25
2·55	2·45
8·70	8·50
2·175	1·700
·2175	·2125
11·0925	10·4125

We can only say that the area, to 1 significant figure, is 10 sq. in.

Example. 6·2 c.c. of a substance are found to weigh 14·9 gm., the determinations being reliable to the number of figures given. Calculate the weight of 1 c.c.

Greatest weight per c.c.	Least weight per c.c.
2·43...	2·37...
6·15)14·95	6·25)14·85
12 30	12 50
2 650	2 350
2 460	1 875
1900	4750
1845	4375

The weight of 1 c.c. is therefore 2·4 gm. correct to 2 significant figures.

§ 142. The "method of limits" illustrated above is, of course, too tedious to be used except in cases where very precise information is needed.

Another method is shown below, and applied to the last example. Figures represented by crosses are quite unknown; figures in italics are of doubtful value. In 12·*4*×, the *4* is doubtful, as it may be affected by carrying from the × when 6·2× is multiplied by 2. The *5* under the *4* is equally doubtful. In the last line we are left with a doubtful *1*, and cannot obtain another trustworthy figure in the quotient.

```
       2·4×
6·2×)14·9××
     12·4×
      2 5××
      2 48 ×
        1××
```

§143. These questions connected with degree of approximation are rather difficult and do not admit of solution by definite rules. The beginner can hardly be expected to judge very precisely the number of figures that are justifiable in a result; but he should at least learn caution and the habit of not putting in figures that are flagrantly absurd.

EXERCISE XIV c

¶1. What would be a reasonable degree of approximation for

(*a*) The distance of the sun from the earth? Would you state it to the nearest foot, mile, or million miles?

(*b*) The distance from John o' Groat's to Land's End?

(*c*) The length and breadth of the room?

(*d*) Your height?

(*e*) The bore of a gun?

(*f*) The diameter of a piece of wire?

(*g*) Your weight?

(*h*) The mass of the earth?

(*i*) The weight of a parcel for the post?

(*j*) The weight of a cubic centimetre of mercury?

(*k*) The cost of a battleship?

(*l*) The cost of a pound of tea?

(*m*) The cost of a ton of coal?

(*n*) The time of a hundred yards race?

(*o*) The date of the ice-age?

(*p*) The velocity of a torpedo boat?

(*q*) The length of the base line for the Ordnance Survey of England?

2. A long string is measured by finding the lengths of successive pieces, and adding them together. The pieces are measured to the nearest millimetre; they are 946, 936, 938, 982, 943, 205 millimetres. What is the length of the whole string, and what is the greatest possible absolute error?

3. A quantity of water is measured by pouring successive portions into a measuring glass graduated in cubic centimetres, and adding together all the volumes so found. To the nearest c.c. they measure as follows: 84, 90, 86, 33 c.c. What is the greatest possible absolute error in the determination of the volume?

4. The weights of a boat's crew, to the nearest pound, were 11 st. 5 lbs., 11 st. 7 lbs., 13 st., 10 st. 4 lbs. What is the possible absolute error in (1) the total weight, (2) the average weight determined from these figures?

¶**5.** An experimenter has a lath whose length he finds to be 147·6 cm., and wishes to cut off a piece long enough to reduce the length of the lath to 1 metre. How much should he try to cut off? His measurements can be trusted to a millimetre, but not further. What is the possible absolute error in the length of the resulting metre?

6. A measuring glass contains water; the reading is 147·6 c.c. By means of a pipette 100 c.c. are removed. Neither the reading of the graduated glass nor of the pipette is trustworthy beyond 0·1 c.c. What is the possible absolute error in the calculated volume of the water that is left?

7. I have a set of twenty 100-gram weights. What should be the weight (in kilograms) of the whole set? If the marking on the weights is trustworthy to the nearest milligram only, by how much may the weight of the whole set be in error?

¶**8.** 30 cubes of wood (each measuring 1 c.c.) were found to weigh 26·44 grams (to the nearest centigram). What was the average weight of a cube? and what is the absolute error to which this answer is subject? Show that the absolute error in the average is less than the absolute error in the total. (This shows why it is advantageous to take an average.)

9. A fine thread was wound 10 times round a cylinder (as the thread is wound round the handle of a cricket bat). The total length of the thread was found to be $40{\cdot}62'' \pm 0{\cdot}005''$. What is the circumference of the cylinder? and what is the greatest absolute error possible in this result?

10. A class of boys, measuring the number of inches in the circumference of a halfpenny (diameter 1 inch), obtain the following results: 3·18, 3·19, 3·13, 3·15, 3·17, 3·12, 3·13, 3·11, 3·12. Find the average of these results to a proper degree of accuracy.

11. The following results were obtained on weighing an ounce weight in grams: 28·4, 28·38, 28·37, 28·39, 28·41, 28·36, 28·36, 28·37, 28·34. (i) Find the average of these results, to a proper number of places. (ii) Find the error in the average result, given that 1 lb. = 453·593...gm.

12. The sides of a rectangle are measured, and found to be 10·4 cm. × 12·7 cm. (to the nearest millimetre). What is the possible relative error in the sides? in the area calculated from these data? To how many significant figures will it be proper to give the area?

13. Answer the questions of Ex. 12 for rectangles of the following dimensions:

(i) 4·2 in. by 3·6 in. (to the nearest 1/10″).

(ii) 465 ft. by 971 ft. (to the nearest foot).

(iii) 79·1 metres by 86·7 metres (to the nearest dm.).

14. In Exs. 12, 13 did you find any instance in which the determination of the area was trustworthy to a *greater* number of significant figures than the determination of the sides?

15. The weight of 1 c.c. of water was found by weighing a measured volume of water. In fact, 95·6 c.c. of water was found to weigh 95·43 gm.; the two measures being correct to the number of figures stated. Calculate the weight of 1 c.c. of water; also find the greatest absolute error to which this result is subject.

16. Repeat Ex. 15 for the case in which the weight of 1 c.c. of mercury is calculated from the observed result that 9·5 c.c. weighed 130·42 gm.

17. Repeat Ex. 15 for the following result: 96·4 c.c. of methylated spirits weighed 79·05 gm.

18. Find the greatest possible absolute error in the volume of a cuboid calculated from edges 4·2″, 3·7″, 2·1″, measured to the nearest 1/10″.

Contracted Addition and Subtraction

§ 144. *Example.* Find the sum of 2·4672, 0·28419, 34·635, 0·02186, correct to one place of decimals.

For safety, retain 2 more places of decimals, making 3 places in all: draw a line after the 3rd, and ignore everything to the right of that line. The sum is 37·4... to one place.

```
 2·467|2
 0·284|19
34·635|
 0·021|86
------
37·407
```

If there are many terms in the sum, it might become necessary to allow more than 2 extra places.

Example. Find to 2 places of decimals the difference between 4·1285 and 5·28347.

Retain 4 places. The difference is 1·15... to 2 places.

```
5·2834|7
4·1285|
------
1·1549
```

Note that if we had only taken one extra place, we should have obtained a difference 1·155, and should have been uncertain whether to take 1·15 or 1·16 as the best approximation to one place.

A sum or difference cannot be calculated to more places than are given in the terms. If it were required to find to 4 places the difference between two measured lengths 4·872 m. and 0·2615 m., the process might be shown thus. The × after 4·872 denotes that we have no precise information as to the figure in that place. We have no right to replace the × by a zero. Accordingly, we do not know what figure to write in the difference under the 5, and we cannot obtain an answer trustworthy to 4 places.

```
4·872×
0·2615
------
4·611×
```

EXERCISE XIV d

Find the following sums:

1. 25·679 + 0·00273 + 165·295 + 1·6675 to 1 place.

2. 0·02497 + 0·003982 + 0·1493 to 2 places.

3. 17·992 + 2·4221 + 0·09997 to the nearest integer.

4. $\frac{1}{3}$ of 2·3 + $\frac{1}{7}$ of 12·5 + $\frac{1}{5}$ of 0·0072, to 1 place.

Calculate:

5. 9·2468 – 1·6524 to 1 place.

6. 5·6381 – 0·79243 to 2 places.

Contracted Multiplication

§145. The need of rules for contracted multiplication and division has disappeared to a great extent now that logarithms are generally used for approximate work. As the rules are still sometimes taught, they are given below.

Example. Calculate 42·605 × 0·31624 to 1 place.

Reduce the multiplier to standard form. Allowing 2 extra places, mark off 3 places of decimals in the multiplicand, and multiply by the left-hand figure of the multiplier. Retain 3 columns of decimals. After performing each line of the multiplication cut off the right-hand figure of the multiplicand. The process will be understood if the full-length process is compared with the contracted process.

42·605 × 0·31624 = 4·2605 × 3·1624.

4·2605	
3·1624	
12·781	5
426	05
255	630
8	5210
1	70420
13·473	40520

4·,2,6,0,5
3·1624
12·780
426
252
8
13·466

Product = <u>13·5.</u>

If the result is required to a given number of *significant figures*, we must begin by estimating roughly the product, to 1 or 2 significant figures. We shall then be able to judge how many places of decimals must be retained.

Example. Find the product of 2·73065 and 0·0094738 to 3 significant figures.

The product is roughly 2·7 × ·01 = ·027; therefore 3 significant figures mean 4 places of decimals. We must therefore retain 6 columns of decimals.

0·009,4,7,3,8
2·73065
0·018946
6629
282
0·025857

Product = <u>0·0259.</u>

EXERCISE XIV e

Find the value of:

1. 42·605 × 0·31624 to 2 places.
2. 0·7138 × 27·96 to 1 place.
3. 0·3482 × 0·9277 to 2 places.
4. 8·295 × 76·82 (nearest integer).
5. 76·82 × 61·18 (nearest integer).
6. 79·76 × 18·302 to 3 significant figures.
7. 0·2415 × 0·09658 to 2 significant figures.
8. 625·2 × 0·3365 to 4 significant figures.
9. $(3{\cdot}752)^2$ to 2 places.
10. $(0{\cdot}2391)^2$ to 3 significant figures.
11. $(27{\cdot}23)^2$ to 3 significant figures.
12. $(939{\cdot}7)^2$ to the nearest 1000.
13. $(8135)^2$ to the nearest hundred-thousand.
14. $(6742)^2$ to 3 significant figures.
15. $(316{\cdot}23)^2$ to 3 significant figures.

Contracted Division

§146. *Example.* Find 0·25638 ÷ 50·287 correct to 2 significant figures.

Reduce divisor to standard form. Retain 2 more figures in the divisor than the number of significant figures required in the answer; in this case retain 4. After each partial division cut off the right-hand figure of the divisor. Stop when 3 significant figures of the quotient have been obtained.

$$\frac{0{\cdot}25638}{50{\cdot}287} = \frac{0{\cdot}025638}{5{\cdot}0287}.$$

```
              0·00509
5·0,2,8,7)0·025638
             25140
               498
               450
```

Quotient = 0·0051 to 2 significant figures.

If the answer is required to a given number *of decimal places*, the number of significant figures required must be found by first making a rough estimate of the quotient.

EXERCISE XIV f

Calculate the following quotients correct to 3 significant figures:

1. $47{\cdot}238 \div 0{\cdot}41076$.

2. $6{\cdot}9623 \div 0{\cdot}00716423$.

3. $1 \div 0{\cdot}00917435$.

4. $0{\cdot}9243 \div 76{\cdot}816$.

5. $0{\cdot}007268439 \div 0{\cdot}84275$.

6. $\frac{1560}{247392}$.

7. $\frac{1}{4978216}$.

8. $\frac{1}{3{\cdot}141596}$.

Calculate the following quotients:

9. $17{\cdot}24695 \div 0{\cdot}428643$ to 2 places.

10. $0{\cdot}824 \div 926{\cdot}84$ to 5 places.

11. $42{\cdot}0763 \div 0{\cdot}4216$ to the nearest integer.

12. $\frac{8247}{79215}$ to the nearest hundredth.

13. $\frac{2462}{64371} + \frac{17209}{48256}$ to 3 places.

EXERCISE XIV g

Application of Contracted Methods

1. Find the cost of 5·655 metres of silver wire at 2·85 fr. per metre (to the nearest franc).

2. Find the cost of 1255·55 metres of hurdles at 6 fr. 35 c. per metre (to the nearest franc).

3. Divide £1562 between A, B and C so that A may have 0·415 of the whole, B 0·315, and C the remainder (to the nearest pound).

4. Divide £1852 between A, B and C so that A may have 0·615 of the whole, and B 0·615 of *what is left*, and C the remainder (to the nearest pound).

5. Prove by contracted multiplication that

(1) $\sqrt{2}$ lies between 1·4142 and 1·4143;

(2) the square root of 15 lies between 3·87298 and 3·87299 (multiply correct to 4 places).

6. 107,927,701 cwts. of grain at an average cost of 6·69 shillings per cwt. were imported into England in 1901. Calculate total value (to the nearest £100,000).

7. Find the area of a rectangle:

(1) 4·256″ by 7·415″ to the nearest 1/10 of a sq. in.
(2) 4 Km. 642 m. by 16 Km. 941 m. to the nearest sq. Km.
(3) 4 yds. 2 ft. 6 in. by 7 yds. 1 ft. 3 in. to the nearest sq. yd.
(4) 140 yds. 2 ft. by 65 yds. $1\frac{1}{2}$ ft. to the nearest sq. yd.
(5) 5·425 m. by 10·246 m. to the nearest sq. dm.

8. Find to the nearest inch the diameter of a circle which has a circumference of 87·97 inches* (take $\pi = 3{\cdot}1416$).

9. Find to the nearest cm. the diameter of a circle which has a circumference of 115·8 cm.*

10. An imperial gallon is 277·274 cubic inches, and a Winchester bushel is 2150·42 cubic inches. How many Winchester bushels are equal to 1000 imperial bushels? (1 bushel = 8 gallons; answer correct to the nearest integer.)

11. The velocity of light is 186,330 miles per second. The distance of the sun is 92,900,000 miles. Find, to the nearest second, how long light takes to travel from the sun to the earth.

12. The planet Mercury describes its orbit round the sun in 87·96926 days; the earth in 365·2564 days; Neptune in 60181·11 days. Find, to 2 significant figures, the number of our years in (1) Mercury's year, (2) Neptune's year.

13. The lunar month (from new moon to new moon) contains 29·53059 days; the year contains 365·2564 days. How many lunar months are there in a year? (To the nearest integer.)

14. The moon revolves round the earth in 27 days 7 hrs. 43 mins. 11·5 secs. How many revolutions does it perform in 365 days (to 3 significant figures)?

* See Chap. XVI.

15. During the years 1898 and 1911 food, drink and tobacco to the following values were imported into Great Britain and Ireland from abroad. Find, to the nearest shilling, the value of food, etc., imported per head in each of these years:

Year	Value of food, etc. imported	Population of United Kingdom
1898	£208,187,405	40,380,792
1911	£263,958,137	45,297,114

16. The population of the United Kingdom in 1911 was 45,297,114.

The Imperial Revenue was	£185,090,286
The Expenditure	£178,545,100
The National Debt	£733,072,610
The Value of Imports	£577,398,393
The Value of Exports	£454,119,298

Calculate the values per head (to the nearest 1/-).

17. What percentages (to the nearest integer) of the United Kingdom do England, Wales, Scotland, and Ireland form?

Area of England = 32,346,000 acres.
,, Wales = 4,774,000 ,,
,, Scotland = 19,456,000 ,,
,, Ireland = 20,334,000 ,,

18. By a certain census the population of Great Britain was found to be about 18,844,000; the parts employed (1) in agriculture, (2) in trades and manufactures, were respectively 1,499,000 and 3,110,000. How much per cent. of the whole population was each of these classes? (To one place.)

19. The equatorial diameter of the earth is 41,847,662 feet and the polar axis 41,707,536 feet. Find, to 2 significant figures, by what percentage of the polar axis the two measurements differ.

20. A bankrupt's liabilities are £4769. 4*s*. 7*d*. and his assets £3452. 3*s*. 6½*d*. What percentage of his debts can he pay? (To 3 significant figures.)

21. The value of the exports of British produce was, in 1899, £264,492,211; and in 1912, £487,223,439. Find the increase per cent. (To 2 significant figures.)

22. The value of the export of British coal was, in 1872, £10,442,000; and in 1912, £42,584,454. Find the increase per cent. (To 2 significant figures.)

23. The cost of the Navy increased from £24,068,000 in 1898 to £42,858,000 in 1911. The amount of national income that paid income tax was £548,229,450 and £720,640,587 in these two years. Find for each year what percentage the cost of the Navy formed of the national income. (To 2 significant figures.)

24. The amount of fish delivered at Billingsgate market was 3,889,540 cwts. in 1911 and 3,766,400 cwts. in 1912. Find to 2 significant figures the decrease per cent.

25. The following table gives the quantity and value of fish landed in the United Kingdom in 1908 (excluding salmon and shell-fish):

	Weight in tons	Value in £
England and Wales	663,520	7,739,334
Scotland	431,498	2,511,492
Ireland	34,338	261,431

Find to the nearest integer how much per cent. by weight of the total for the United Kingdom was landed in England and Wales, and how much per cent. by value. Also give, to the nearest penny, the value per cwt. of the fish landed (*a*) in England and Wales, (*b*) in Scotland, (*c*) in Ireland.

26. The figures below give the revenue received from the Post Office and from Telegraphs in the years 1908–09 and 1909–10.

	1908–09	1909–10
Post Office	£12,227,070	£12,568,617
Telegraphs	£3,831,992	£3,872,472

Express the increase for the year 1909–10 as a percentage of the revenue in 1908–09 (*a*) from each department separately, and (*b*) from the two departments combined. Give each answer to the nearest tenth of one per cent.

27. In 1913 we imported 278,465 cwts. of dead poultry, valued at £954,540. Find the average value per cwt.

28. Between 1897 and 1911 the total amount of income on which income tax was paid increased from £525,211,200 to £720,640,587; and the total salaries of government, corporation, and public company officials increased from £37,499,958 to £67,160,629. Find in each case the increase per cent. (correct to 2 significant figures).

29. In 1907 there were 10,692,555 accounts open in the Post Office Savings Bank, and the total amount deposited during the year was £44,217,288. In 1908 there were 11,918,251 accounts open, and the amount deposited was £44,770,782. Express the average amount deposited per account in 1908 as a percentage of the corresponding amount in 1907. Give your answer to the nearest whole number.

30. The following table gives the number of bushels of wheat produced in certain parts of the British Empire for the years 1906 and 1907:

	1906	1907
United Kingdom	60,618,442	56,531,198
India	305,606,933	212,941,867
Canada	123,505,691	92,581,571
Australia	66,421,359	44,655,673

Find, to the nearest integer in each case, the reduction in 1907 as a percentage of the production in 1906.

31. The following figures refer to the working of all the railways in the United Kingdom in the year 1910:

Total receipts from passengers	£43,247,345
Total number of passengers	1,307,481,246
Total number of miles travelled by all trains	423,221,538
Total locomotive expenses	£20,623,351

From these data calculate (1) the average sum paid by each passenger, correct to the nearest tenth of a penny: (2) the average locomotive expenses incurred in a journey of 150 miles, correct to the nearest shilling.

32. The figures given below represent the acreage under the respective crops mentioned and the produce in quarters of the land under each crop in the United Kingdom for the year 1912. Find the yield in quarters per acre for each crop (to 2 significant figures).

	Acreage	Produce
Wheat	1,970,542	7,175,288
Barley	1,813,559	7,275,900
Oats	4,075,054	20,600,079

33. Below are tabulated certain statistics relating to the imports of quicksilver during the years 1907 and 1908:

	1907	1908
Quantity	2,958,603 lbs.	3,270,412 lbs.
Value	£275,197	£353,396

Find, correct to within a penny, the average value per lb. in each year.

34. In 1907 the United Kingdom imported 57,314,200 cwts. of wheat, valued at £21,602,734, from foreign countries, and 39,853,800 cwts. of wheat, valued at £15,743,814, from British Possessions; also 11,726,946 cwts. of flour, valued at £5,892,172, from foreign countries, and 1,570,420 cwts. of flour, valued at £801,910, from British Possessions. Find, to within a penny, the value per cwt. of imported wheat (*a*) from foreign countries, (*b*) from British Possessions. Also find to the nearest integer what percentage by value of the total import of wheat and flour came from British Possessions.

35. In 1911 the tax on tea was 5*d.* per lb., and brought in £5,930,008, and the population was 45,365,599. Find the average consumption of tea per head of the population to the nearest ounce.

36. In 1907–8 the Boards of Guardians in England and Wales received from rates £11,928,863. This sum was raised on an assessable value of £197,443,250. Find, to the nearest halfpenny, the average rate per £. The amount received from rates was equivalent to 6*s.* 10*d.* per head of population. In addition the Boards of Guardians received £3,800,000 in grants. Find, to the nearest penny, the total amount per head of population which the Boards of Guardians received.

37. The Army Estimates for 1912–13 were £27,860,000 and the Navy Estimates £45,075,400. The corresponding figures for 1913–14 were: Army, £28,220,000; Navy, £46,309,300. What was the increase per cent., to the nearest tenth of one per cent., in each case? If the Estimates for 1914–15 showed the same percentage increase on those for 1913–14, what were they to the nearest £100,000?

38. The following table exhibits the wire mileage in telegraphs and telephones in 1902 and 1907:

	1902	1907
Telegraphs, land wires	3,659,659	5,038,981
Telegraphs, submarine cables	212,894	259,000
Telephones, cable	7,467,417	19,839,537
Telephones, subscribers' stations	3,534,036	8,406,336

Find to the nearest whole number the percentage increase in the 5 years in each of these four items, and in the total wire mileage.

39. If the percentage increase in the total wire mileage continued, what, to the nearest million miles, was the total wire mileage in 1912?

40. In 1912 the value of the total imports into the United Kingdom was £744,640,631; this sum representing £16/6/2 per head of population. Find the population to the nearest ten-thousand.

41. The following table gives certain information about the counties of Cambridge and Oxford:

	Area in acres	Population	Rateable value	Poor rate
Cambridge	566,493	215,109	£1,325,292	£58,054
Oxford	491,421	198,491	£1,268,650	£57,377

Find, to within 1, the average population per sq. mile, and to within a farthing the poor rate per £1 of rateable value for each county.

42. In 1907 the Transvaal produced 2,062,855 carats of diamonds, valued at £2,268,075; in 1908 it produced 2,022,687 carats valued at £1,549,815. Find the average value per carat in each year to within a penny, and express that for 1908 as a percentage of that for 1907.

43. The fineness of a gold coin is the fraction of pure gold in the coin. From the table given below, calculate to three decimal places the fineness of the following standard coins of Great Britain, France and Germany:

	Sovereign	20 Franc piece	20 Mark piece
Weight in grains	123·27	99·56	122·918
Weight of pure gold in grains	113·0016	89·610	110·626

44. If the value of the coins is proportional to the amount of pure gold in them, find the equivalent of one sovereign in French and in German money to the nearest centime and pfennig respectively.

CHAPTER XV

SQUARES, SQUARE ROOTS, RECIPROCALS

Graph of Squares and Square Roots

§147. To construct a graph of squares, take a table such as the following:

Number	1	2	3	....
Square	1	4	9	....

Plot the corresponding graph, taking $\frac{1}{2}$ inch to each unit along the horizontal axis, and $\frac{1}{2}$ inch to 10 units along the perpendicular axis.

Observing the points on the graph which indicate that $6^2 = 36$ and $7^2 = 49$, find from the graph the approximate value of $6{\cdot}5^2$ and verify by multiplication. In the same way find the values of $8{\cdot}3^2$, $4{\cdot}7^2$, $5{\cdot}7^2$ and verify by multiplication.

Now take the perpendicular scale to represent Numbers and the horizontal scale to represent Square Roots; and notice that the two points already found indicate that $\sqrt{36} = 6$ and $\sqrt{49} = 7$.

Ex. 1. Find from the graph the approximate values of $\sqrt{40}$, $\sqrt{85}$, $\sqrt{30}$ and verify each result by squaring.

This graph may be used to find approximate values (i) for the square of any number between 1 and 10, (ii) for the square root of any number between 1 and 100.

(The graph may be kept permanently before the class for checking results roughly.)

TABLE OF SQUARES

§ 148. The square of any number may be found by multiplication, but labour is saved by using printed tables of squares.

For instance, by multiplication we find that $29^2 = 841$. Searching for 29 in the left-hand column of the table we find the figures 8410 in the next column.

Now 29^2 is rather less than 30^2, or 900; this enables us to fix the position of the decimal point; thus $29^2 = 841{\cdot}0$.

Ex. 2. Find from the table the value of each of the following and verify by multiplication:

(i) 52^2, (ii) 73^2, (iii) 97^2.

The *figures* in the value of 29^2 are the same as those in the value of $2{\cdot}9^2$; and $2{\cdot}9^2$ is a little short of 3^2, or 9. Thus $2{\cdot}9^2 = 8{\cdot}41$.

Similarly, 290 lies between 200 and 300; therefore 290^2 lies between 40,000 and 90,000, and since the *figures* are the same as before, $290^2 = 84{,}100$.

Again, $0{\cdot}29^2$ lies between $0{\cdot}2^2$ and $0{\cdot}3^2$, i.e. between 0·04 and 0·09. Therefore $0{\cdot}29^2 = 0{\cdot}0841$.

Ex. 3. Find from the table the value of:

(i) $5{\cdot}2^2$, (ii) $6{\cdot}4^2$, (iii) 730^2,
(iv) 880^2, (v) 9700^2, (vi) $0{\cdot}52^2$,
(vii) $0{\cdot}64^2$, (viii) $0{\cdot}073^2$, (ix) $0{\cdot}88^2$.

§ 149. To find from the table the square of a number of three figures, e.g. 6·24, take the number opposite to 62 in the column headed by 4; this is 3894, which gives the figures in the required square. $6{\cdot}24^2$ is approximately 36, and therefore $= 38{\cdot}94$. By multiplication it is found that $6{\cdot}24^2 = 38{\cdot}9376$, so that the answer obtained from the table is *correct to* 4 *significant figures*.

RULE. **In using a table of squares, first find the figures from the table and then insert the decimal point.**

Ex. 4. Find from the table the value of:

(i) $6{\cdot}25^2$, (ii) $6{\cdot}26^2$, (iii) $2{\cdot}47^2$,
(iv) $7{\cdot}21^2$, (v) $9{\cdot}85^2$, (vi) $62{\cdot}5^2$,
(vii) 626^2, (viii) $0{\cdot}247^2$, (ix) $72{\cdot}1^2$.

§150. To find from the table the square of a number of 4 figures, e.g. 6·247. The table gives $6{\cdot}24^2 = 38{\cdot}94$, and the **difference columns** on the right give 9 in the column headed 7; this means that $6{\cdot}247^2$ is 38·94 with a 9 added to the last figure; that is 39·03. Now $6{\cdot}247^2 = 39{\cdot}025009$ by multiplication, i.e. 39·03 correct to 4 significant figures.

It will be found very conducive to speed and accuracy if the printed table be laid on the left-hand side, and the left hand used in searching therein, the right hand being left free to write. Run the little finger of the left hand down the first column until it is immediately under the first two figures 62; shift it to the right until it is in the column headed by the third figure 4 (it now indicates 3894); keeping it in this position, use the first finger of the same hand to determine the appropriate difference (9) for the fourth figure 7. Write down the result 3903 before moving the left hand.

EXERCISE XV a

1. Find correct to 4 significant figures the squares of:

(i)	2·476.	(ii)	5·374.	(iii)	4·162.
(iv)	0·03917.	(v)	24·76.	(vi)	53·74.
(vii)	65·54.	(viii)	0·4015.	(ix)	247·6.
(x)	537·4.	(xi)	2·315.	(xii)	0·4009.
(xiii)	0·2476.	(xiv)	0·5374.	(xv)	721·7.
(xvi)	0·1008.	(xvii)	0·02476.	(xviii)	0·05374.

Plenty of oral practice in the use of the table should be given at this stage, each pupil writing down the answer to every question.

2. Find from the table the area of a square whose side is:

(i)	24·3 inches.	(ii)	173·1 yards.
(iii)	5·73 feet.	(iv)	29·37 inches.
(v)	42·74 cm.	(vi)	0·9983 m.

3. Find by means of the table the value of $a^2 + b^2$ if

(i) $a = 58{\cdot}2$, $b = 64{\cdot}7$.

(ii) $a = 79{\cdot}21$, $b = 56{\cdot}39$.

(iii) $a = 4{\cdot}631$, $b = 7{\cdot}174$.

4. Find by means of the table the value of $a^2 + b^2 - c^2$ if

(i) $a = 7{\cdot}5$, $b = 6{\cdot}3$, $c = 2{\cdot}9$.

(ii) $a = 6{\cdot}32$, $b = 2{\cdot}94$, $c = 6{\cdot}34$.

(iii) $a = 17{\cdot}3$, $b = 23{\cdot}7$, $c = 7{\cdot}9$.

Square Root

§ 151. Since

$$6^2 = 36, \quad \therefore \sqrt{36} = 6.$$
$$0{\cdot}6^2 = 0{\cdot}36, \quad \therefore \sqrt{0{\cdot}36} = 0{\cdot}6.$$
$$60^2 = 3600, \quad \therefore \sqrt{3600} = 60.$$
$$600^2 = 360000, \quad \therefore \sqrt{360000} = 600.$$

etc., etc.

But what is $\sqrt{360}$, $\sqrt{3{\cdot}6}$, $\sqrt{36000}$, etc.?

Now $19^2 = 361$, $\therefore \sqrt{360} = 19$ nearly,

and $1{\cdot}9^2 = 3{\cdot}61$, $\therefore \sqrt{3{\cdot}6} = 1{\cdot}9$ nearly,

and $190^2 = 36100$, $\therefore \sqrt{36100} = 190$ nearly, etc.

We see therefore this peculiarity about square root: given the *figures* of a number, there are two quite different series of figures which may form its square root.

§ 152. If the square roots above be written in order of magnitude, it will be observed that in this instance when the given number has

$$\sqrt{3{\cdot}6} = 1{\cdot}9 \text{ nearly}$$
$$\sqrt{36} = 6{\cdot}0$$
$$\sqrt{360} = 19{\cdot}0 \text{ nearly}$$
$$\sqrt{3600} = 60$$
$$\sqrt{36000} = 190 \text{ nearly}$$
$$\sqrt{360000} = 600$$

1 or 2 figures before the decimal point, the square root has 1

3 ,, 4 ,, ,, ,, ,, ,, 2

5 ,, 6 ,, ,, ,, ,, ,, 3

In fact, any number between 10 and 100 has 2 figures before the decimal point; its square however, lying between 100 and 10000, has either 3 or 4 figures before the point.

This suggests that, if we mark off pairs of figures from the decimal point in the given number, for each pair or group there is one figure before the decimal point in the square root. The last group marked off may consist of a pair or a single figure. Thus $\sqrt{36'00'00'00} = 6000$, and $\sqrt{3'60'00'00} = 1900$ nearly.

A similar rule is suggested in the case of the square root of numbers less than 1, viz.: for every pair of zeros after the decimal point in the given number, there is *one* zero in the square root.

$\sqrt{\cdot 36} = \cdot 6$
$\sqrt{\cdot 03'6} = \cdot 19$ nearly
$\sqrt{\cdot 00'36} = \cdot 06$
$\sqrt{\cdot 00'03'6} = \cdot 019$ nearly
$\sqrt{\cdot 00'00'36} = \cdot 006$
etc., etc.

§ 153. The first steps in finding a square root are to determine

(i) the position of the decimal point,

(ii) the first significant figure in the answer.

When these have been determined, the *figures* in the square root are not further affected by the position of the decimal point in the number. For instance, $\sqrt{4\cdot 785}$, $\sqrt{478\cdot 5}$, $\sqrt{47850}$ all consist of the same series of figures.

§ 154. **To find a first approximation to a square root,** e.g. $\sqrt{4407\cdot 952}$.

Beginning at the decimal point mark off pairs of figures thus: 44′07′·95′2.

Consider the most important pair, that on the extreme left; the number is somewhat greater than 44′00.

Now the perfect square immediately below 44 is 36, and $\sqrt{44'00}$ is somewhat greater than $\sqrt{36'00}$ or 60, but somewhat less than $\sqrt{4900}$ or 70.

A first approximation to $\sqrt{4407\cdot 952}$ is therefore 60, and we may write $\sqrt{4407\cdot 952} = 60 + \ldots$.

Now take the case of a number less than 1; and to illustrate the chief peculiarity of square root we will take the same series of figures as before, e.g. $\sqrt{0\cdot 04407952}$. Exactly as before, mark off pairs of figures, beginning from the decimal point thus: 0·04′40′79′52.

Taking the most important pair and neglecting the rest, we see that $\sqrt{0\cdot 04407952}$ is a little greater than $\sqrt{0\cdot 04}$ or 0·2; we may write $\sqrt{0\cdot 04407952} = 0\cdot 2 + \ldots$. (What was $\sqrt{4407\cdot 952}$?)

EXERCISE XV b (oral)

What is the value of:

1. 5^2.
2. $\sqrt{25}$.
3. 7^2.
4. $\sqrt{49}$.
5. $\sqrt{144}$.
6. $\sqrt{1}$.
7. $\sqrt{100}$.
8. $\sqrt{10000}$.
9. 20^2.
10. $\sqrt{900}$.
11. $\sqrt{300^2}$.
12. $\sqrt{160000}$.
13. $\sqrt{14400}$.
14. $\sqrt{1210000}$.
15. $\sqrt{3600}$.
16. $\sqrt{360000}$.
17. $0{\cdot}2^2$.
18. $\sqrt{0{\cdot}09}$.
19. $\sqrt{0{\cdot}01}$.
20. $0{\cdot}4^2$.
21. $\sqrt{0{\cdot}25}$.
22. $0{\cdot}02^2$.
23. $\sqrt{0{\cdot}0009}$.
24. $\sqrt{0{\cdot}0036}$.
25. $\sqrt{0{\cdot}49}$.
26. $\sqrt{0{\cdot}0049}$.
27. $1{\cdot}1^2$.
28. $\sqrt{1{\cdot}44}$.
29. $\sqrt{1{\cdot}69}$.
30. $\sqrt{1{\cdot}96}$.
31. $\sqrt{0{\cdot}0144}$.
32. $\sqrt{0{\cdot}000009}$.
33. $\sqrt{0{\cdot}0016}$.
34. $\sqrt{640000}$.
35. $\sqrt{0{\cdot}0004}$.
36. $\sqrt{2500}$.
37. $\sqrt{0{\cdot}000025}$.
38. $\sqrt{1440000}$.
39. $\sqrt{0{\cdot}000049}$.
40. $\sqrt{810000}$.
41. $\sqrt{0{\cdot}000121}$.
42. $\sqrt{0{\cdot}000016}$.
43. $\sqrt{2250000}$.
44. $\sqrt{0{\cdot}000001}$.
45. $\sqrt{0{\cdot}000064}$?

EXERCISE XV c (oral)

Determine the first significant figure in the answer and the position of the decimal point in each of the square roots in Exercise xv d, e.g. $\sqrt{1521} = 30 + \ldots$

$$\sqrt{0{\cdot}001369} = 0{\cdot}03 + \ldots$$

Rule for Square Root

§ 155. The following **geometrical illustration** may help the reader to understand the rule laid down in the subsequent paragraphs.

Suppose that a square (AKFH in the figure) is known to contain 1190·25 sq. cm. It is required to find the length of its side, i.e. $\sqrt{1190{\cdot}25}$ cm. Now 1190·25 lies between 900 and 1600. Therefore the length of the side is between 30 and 40 cm. If AB = 30 cm., the square ABCD contains 900 sq. cm. Thus the remaining (shaded) area contains

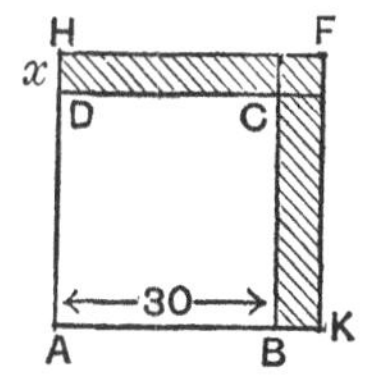

$$1190{\cdot}25 - 900 \text{ or } 290{\cdot}25 \text{ sq. cm.}$$

Now this shaded area consists of *two* rectangles and the square CF. The length of each rectangle is 30 cm.; let the breadth (BK or DH) be x cm. Then the area of the two rectangles is $60x$ sq. cm. and they are together by far the larger part of the shaded area.

Thus $60x$ = something rather less than 290·25,

$\therefore x$ = about 4.

Note that this number 4 is found by dividing the remainder 290·25 by **twice** the first approximation 30. We have then $\sqrt{1190 \cdot 25} = 34 + \ldots$

§156. **The rule for extracting a square root** is shown in the following examples.

Example. To find $\sqrt{1190 \cdot 25}$.

Mark off pairs from the decimal point and determine the first figure and the position of the decimal point as described above. We have $\sqrt{1190 \cdot 25} = 30 + \ldots$, and we ignore the decimal point in the rest of the calculation. Write 3 above 11 and 9 ($= 3^2$) below. Subtract 9 from 11, giving remainder 2. Bring down the next pair 90. *Double the answer* (3) *already obtained* and write the result (6) on the left, leaving room for another figure after it. 6 is called the **trial divisor.** Divide 29 (omitting the last figure) by 6. Write the result 4 after the 3 in the answer and also after the 6. Multiply the 64 by the 4 of the answer and write the result 256 below 290. Subtract; and bringing down the next pair obtain 3425. Double the answer (34) already obtained and write the result 68 on the left, leaving room for another figure as before. 68 is the new trial divisor.

```
   3|  |
  11|90|25
   9
   2
```

```
   3|  |
  11|90|25
   9
6 | 290
```

```
      3| 4| 5
     11|90|25
      9
64  | 290
    | 256
685 | 3425
    | 3425
```

Divide 342 (omitting the 5) by 68, and write the result (5) after the 34 of the answer and also after the 68. Multiply 685 by the last figure 5 of the answer and we get 3425 and there is no remainder.

Thus we find $\sqrt{1190 \cdot 25} = \underline{34 \cdot 5}$ exactly.

If you prefer to retain the decimal point throughout, it must be inserted immediately after bringing down the first pair following the decimal point in the given number.

```
        3| 4|· 5
       11|90|·25
        9
  64 |  290
     |  256
 685 |   34·25
     |   34·25
```

Example. To find $\sqrt{0{\cdot}0005671}$.

Mark off pairs from the decimal point thus: 0·00|05|67|1 and determine the first significant figure in the answer. Thus $\sqrt{0{\cdot}0005671} = 0{\cdot}02 + \ldots$, and we ignore the decimal point in the rest of the calculation.

Write down the significant figures, dividing off the groups as they appear above. Write 2 as the first figure in the answer and 4 ($= 2^2$) below 5. Subtract 4 from 5 and write 1 below. Bring down the next pair 67, obtaining 167. Double the answer 2 already obtained and write the trial divisor 4 on the left. Divide 16 (omitting 7) by 4. The result is 4. If, however, 44 is written on the left and 4 added to the answer, the product is 176, which is greater than 167. *Thus* 4 *is too large and we must take* 3 *as the next figure in the square root.*

```
    2|  |
    5|67|1
    4
 4 |167
```

```
     2| 4|
     5|67|1
     4
 44 |167
    |176
```

We now have 23 in the answer and 43 as the trial divisor on the left.

Multiply 43 by 3 and subtract the result (129) from 167. The remainder is 38. Bring down the next *pair* 10, obtaining 3810.

```
     2| 3|
     5|67|1
     4
 43 |167
    |129
    | 3810
```

Double the answer (23) already obtained and write the result 46 on the left. Divide 381 by 46. The result is 8, which is the next figure in the answer and must be written also after the 46 if the process is to be continued.

If the square root is required to 2 *significant figures only,* we have

$$\sqrt{0{\cdot}0005671} = \underline{0{\cdot}024.}$$

```
      2| 3|8
      5|67|1
      4
  43 |167
     |129
 468 | 3810
     | 3744
```

But the process may be continued indefinitely, and the square root obtained to any degree of accuracy required.

As before, the decimal point *may* be retained during the calculation.

```
         ·0| 2| 3|8
        ___________
        ·00|05|67|1
              4
        43|167
          |129
       468| 3810
          | 3744
```

Example. To find $\sqrt{4407{\cdot}952}$ correct to 4 sig. figs.

$$\sqrt{44'07'{\cdot}95'2} = 60{\cdot} + \ldots$$

```
           6| 6| 3|9 2
          ____________
          44|07|95|2
          36
    126 | 807
        | 756
   1323 |  5195
        |  3969
  13269 |  122620
        |  119421
 132782 |    319900
```

$\therefore \sqrt{4407{\cdot}952} = \underline{66{\cdot}39}$ to 4 sig. figs.

EXERCISE XV d

Calculate the square root of:

1. 841.	**2.** 1521.	**3.** 8649.
4. 60516.	**5.** 142884.	**6.** 540225.
7. 667489.	**8.** 11664.	**9.** 167281.
10. 649636.	**11.** 1002001.	**12.** 32400.
13. 10404.	**14.** 817216.	**15.** 49126081.
16. 4036081.	**17.** 82609921.	**18.** 76282756.

Find the square root of:

19. 32·49.	**20.** 1·4161.	**21.** 12·8164.
22. 3782·25.	**23.** 8649·372004.	**24.** 100·2001.
25. 0·001369.	**26.** 0·00002809.	**27.** 0·017161.
28. 0·00034225.	**29.** 0·00822649.	**30.** 0·49112064.

Calculate, to 3 significant figures, the square root of:

31. $\sqrt{200}$. **32.** $\sqrt{20}$. **33.** $\sqrt{2}$.
34. $\sqrt{30}$. **35.** $\sqrt{3}$. **36.** $\sqrt{5}$.
37. $\sqrt{13}$. **38.** $\sqrt{130}$. **39.** $\sqrt{41\cdot 9}$.
40. $\sqrt{4\cdot 19}$. **41.** $\sqrt{0\cdot 419}$. **42.** $\sqrt{0\cdot 728}$.
43. $\sqrt{0\cdot 0728}$. **44.** $\sqrt{0\cdot 00035}$. **45.** $\sqrt{\frac{2}{3}}$.
46. $\sqrt{6\frac{2}{3}}$. **47.** $\sqrt{\frac{1}{3}}$. **48.** $\sqrt{3\frac{1}{3}}$.
49. $\sqrt{1\frac{1}{2}}$. **50.** $\sqrt{4\frac{2}{5}}$.

TABLE OF SQUARE ROOTS

§ 157. The labour of calculating square roots may be avoided by using tables. It should however be borne in mind that square roots such as $\sqrt{4631}$ and $\sqrt{463\cdot 1}$ do not consist of the same figures. Before seeking these figures from the tables, the approximate answers should be determined as above (see § 154).

We have $\sqrt{4631} = 60 + \dots$ and $\sqrt{463\cdot 1} = 20 + \dots$

On searching in the tables for the figures of the answer we find 2152 or 6805. (In some books of tables these appear on different pages.)

We therefore have $\sqrt{4631} = 68\cdot 05$, and $\sqrt{463\cdot 1} = 21\cdot 52$ correct to 4 significant figures.

The rule for the use of square root table is then:

Mark off pairs from the decimal point; determine the first figure and the position of the decimal point in the answer; find the figures from the table.

As the tables are constructed only for numbers of 4 figures it is sometimes necessary to approximate.

Example. To find the value of $\sqrt{0\cdot 00728462}$.

The given number is 0·007285 correct to 4 significant figures and we search for these in the table.

Thus $\sqrt{0\cdot 00728462} = 0\cdot 08535$ correct to 4 significant figures.

EXERCISE XV e

Use square root table to find the value of each of the roots in Exercise xv d.

Applications of Square Root

EXERCISE XV f

1. Calculate (i) in feet and decimals of a foot, correct to 1 place, (ii) in feet and inches, correct to the nearest inch, the side of a square whose area is

(*a*) 169 sq. ft. (*b*) 170 sq. ft. (*c*) 2 sq. ft.

2. Calculate in yards, to 3 significant figures, the side of a square field whose area is

(*a*) 1 acre. (*b*) 2 acres. (*c*) 3 acres.
(*d*) 4 acres. (*e*) 100 acres. (*f*) $\frac{1}{2}$ a sq. mile.

3. Two squares of sheet lead, of sides 2 ft. and 3 ft., are melted together to make a single square of the same thickness. Find to the nearest inch the side of the resulting square.

4. A battalion of 1000 men is to be arranged in a solid square, how many men must stand out to make this arrangement possible? The arrangement would also be made possible by the addition of a certain number of men; what is the least number needed for this purpose?

5. Find to the nearest inch the edge of a cube whose total surface is

(*a*) 24 sq. ft. (*b*) 36 sq. ft. (*c*) 37 sq. ft.

6. A beam of timber, of square section, is 10 feet long and contains 3 cubic feet; find (i) the area of its section, in sq. inches, (ii) its thickness, to the nearest inch.

7. A square organ pipe 16 feet long contains 12 cubic feet of air; what are its internal dimensions?

8. A gentleman offers to subscribe to a charity as many guineas as the total number of subscribers, on condition that each subscriber undertakes to subscribe as much as he himself: 3600 guineas are needed; how many subscribers must come forward?

9. How many yards of fencing are required to enclose a square field of $2\frac{1}{2}$ acres? 10 acres? 40 acres?

10. Find the value of $\left(1\frac{3}{4}\right)^2$ and of $\sqrt{1\frac{9}{16}}$.

11. Find the value of

(*a*) $\sqrt{1\frac{25}{144}}$. (*b*) $\sqrt{1\frac{64}{225}}$. (*c*) $\sqrt{4\frac{25}{36}}$. (*d*) $\sqrt{9\frac{49}{64}}$.

12. Find correct to 3 significant figures $\sqrt{\cdot 3} + \sqrt{\cdot 03}$.

13. Show that $3 + \sqrt{2}$ is the square root of $11 + 6\sqrt{2}$ (work to 3 significant figures).

14. Show that $\sqrt{3} - \sqrt{2}$ is the square root of $5 - 2\sqrt{6}$.

15. Find the difference between $(\sqrt{3} + \sqrt{2})^2$ and $3 + 2$.

16. Find the difference between $(\sqrt{6} - \sqrt{3})^2$ and $6 - 3$.

17. Find the difference between $\sqrt{3} \times \sqrt{2}$ and $\sqrt{6}$.

18. Find the difference between $\sqrt{6 - 3}$ and $\sqrt{6} - \sqrt{3}$.

EXERCISE XV g

The Theorem of Pythagoras asserts that in any right-angled triangle the area of the square described on the hypotenuse is equal to the sum of the areas of the squares described on the other two sides; i.e. area **P** = area **Q** + area **R**, or $c^2 = a^2 + b^2$.

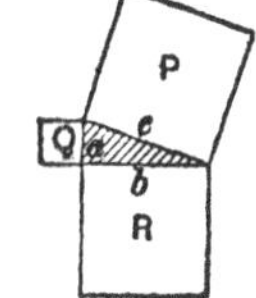

Use tables of squares and square roots to find the length of:

1. c if $a = 5{\cdot}4$ cm. and $b = 10{\cdot}8$ cm.

2. c if $a = 1{\cdot}43$ in. and $b = 3{\cdot}57$ in.

3. c if $a = 21{\cdot}45$ cm. and $b = 17{\cdot}53$ cm.

4. a if $c = 13{\cdot}4$ in. and $b = 10{\cdot}5$ in.

5. a if $c = 53{\cdot}71$ cm. and $b = 41{\cdot}62$ cm.

6. b if $c = 9{\cdot}63$ cm. and $a = 0{\cdot}874$ cm.

If a triangle PQR has $PQ = PR$ and PN is perpendicular to QR, find the length of:

7. PN if $PQ = 8$ cm. and $QR = 4$ cm.

8. PQ if $PN = 7{\cdot}3$ cm. and $QR = 3{\cdot}5$ cm.

9. QR if $PQ = 12{\cdot}6$ cm. and $PN = 4{\cdot}7$ cm.

10. PN if $PQ = 141{\cdot}3$ cm. and $QR = 122{\cdot}4$ cm.

11. Find the height of an isosceles triangle whose equal sides are each 17·45 cm. long and whose base = 13·19 cm.

12. Find the side of an isosceles triangle whose base is 6·42 in. and whose height is 7·53 in.

13. Find the diagonal of a square whose side is 41·54 cm.

14. Find the diagonal of a square whose area is 252·8 sq. cm.

15. Find the diagonal of a rectangle whose sides are 2·56 cm. and 4·73 cm.

16. Find the diagonal of the rectangle whose sides are 13·54 cm. and 23·64 cm.

17. Find the diagonals of the faces of a rectangular block whose edges are 2, 3 and 5 inches long. Find also the diagonal of the block, i.e. the distance from a top corner to the opposite bottom corner.

18. Repeat Question 17, if the edges are 3·54, 4·62, 7·40 cm.

19. Find the diagonal of a cube whose edge is 19·4 cm.

20. How long is a ladder whose foot is 6 ft. 7 in. from a vertical wall and whose top just reaches a window in the wall 11 ft. 2 in. above the ground?

21. The middle point of a straight piece of elastic 12 in. long is pulled 2·56 in. to one side. To what length is the elastic stretched?

22. The distance between two points on a mountain slope when measured on the map seems to be 980 feet. If one point is 275 feet above the other, what is their actual distance apart?

Reciprocals

§ 158. Two numbers whose product is 1 are said to be **reciprocals** of one another.

Thus 3 and $\frac{1}{3}$, $\frac{2}{5}$ and $\frac{5}{2}$, $3\frac{1}{4}$ and $\frac{4}{13}$, 2·7 and $\frac{10}{27}$ are pairs of reciprocals.

Ex. 5. Express as decimals the reciprocals of 2, 20, 200, 0·2, 0·02, 0·002; 8, 800, 0·008; 2·5, 25, 250, 0·025.

If the *figures* in the given number remain the same, the figures in the reciprocal remain the same. These figures correct to 4 significant figures are given in the table of reciprocals, which are used in the same way as tables of squares and square roots when a number consisting of only 3 figures is involved.

Example. To find the values of $\frac{1}{2{\cdot}35}$, $\frac{1}{235}$ and $\frac{1}{0{\cdot}0235}$.

Corresponding to the numbers 235 we find 4255 in the reciprocal table.

Now

$$\frac{1}{2{\cdot}35} = \text{about } \frac{1}{2} \text{ or } 0{\cdot}5, \quad \therefore \frac{1}{2{\cdot}35} = 0{\cdot}4255 \text{ to 4 sig. figs.,}$$

$$\frac{1}{235} = \text{about } \frac{1}{200} \text{ or } 0{\cdot}005, \quad \therefore \frac{1}{235} = {\cdot}004255 \text{ to 4 sig. figs.,}$$

$$\frac{1}{0{\cdot}0235} = \text{about } \frac{1}{0{\cdot}02} \text{ or } 50, \quad \therefore \frac{1}{0{\cdot}0235} = 42{\cdot}55 \text{ to 4 sig. figs.}$$

Ex. 6. Use the table of reciprocals to find the value of:

(i) $\frac{1}{5{\cdot}47}$, (ii) $\frac{1}{547}$, (iii) $\frac{1}{0{\cdot}0547}$, (iv) $\frac{1}{54700}$,

(v) $\frac{1}{8{\cdot}63}$, (vi) $\frac{1}{86{\cdot}3}$, (vii) $\frac{1}{863000}$, (viii) $\frac{1}{0{\cdot}863}$,

(ix) $\frac{1}{20{\cdot}4}$, (x) $\frac{1}{0{\cdot}0316}$, (xi) $\frac{1}{1240}$, (xii) $\frac{1}{0{\cdot}903}$.

§ 159. Since the product of a number and its reciprocal is always equal to 1, if the number *increase* its reciprocal must *decrease*.

For example $\frac{1}{4} = 0{\cdot}25$, and $\frac{1}{5} = 0{\cdot}20$.

Also $\frac{1}{2{\cdot}35} = 0{\cdot}4255$, and $\frac{1}{2{\cdot}36} = 0{\cdot}4237$.

Now 2·357 is *greater* than 2·35,

$\therefore \frac{1}{2{\cdot}357}$ is *less* than 0·4255.

The number in the difference column corresponding to the 4th figure **7** is found to be 13. This must therefore be *subtracted* from the last 2 figures of 0·4255 and we get

$$\frac{1}{2{\cdot}357} = 0{\cdot}4242 \text{ to 4 sig. figs.}$$

In this respect the Table of Reciprocals differs from the Tables of Squares and Square Roots, and the word "Subtract" is usually printed over the difference columns, or a note is added to the same effect.

Example. To find the reciprocal of 38·62.

$$\frac{1}{38{\cdot}62} = \text{about } \frac{1}{40} \text{ or } 0{\cdot}025.$$

From the table we find 2591 corresponding to 386; and, in the difference columns, 1 corresponding to the fourth figure 2.

$$\text{Thus } \frac{1}{38{\cdot}62} = 0{\cdot}02590 \text{ correct to 4 sig. figs.}$$

(Note that the final 0 is a significant figure. But, as a matter of fact, by division we find that $\frac{1}{38{\cdot}62} = 0{\cdot}02589$ to 4 significant figures. This illustrates the limitation of 4-figure tables; the 4th figure cannot be trusted.)

EXERCISE XV h

Using the table of reciprocals find the value to 4 significant figures of:

1. $\frac{1}{2{\cdot}961}$.	**2.** $\frac{1}{7{\cdot}358}$.	**3.** $\frac{1}{9{\cdot}717}$.	**4.** $\frac{1}{5{\cdot}467}$.
5. $\frac{1}{7{\cdot}045}$.	**6.** $\frac{1}{31{\cdot}63}$.	**7.** $\frac{1}{246{\cdot}2}$.	**8.** $\frac{1}{6947}$.
9. $\frac{1}{200{\cdot}6}$.	**10.** $\frac{1}{1111}$.	**11.** $\frac{1}{0{\cdot}5003}$.	**12.** $\frac{1}{0{\cdot}4997}$.
13. $\frac{1}{0{\cdot}7246}$.	**14.** $\frac{1}{0{\cdot}04738}$.	**15.** $\frac{1}{0{\cdot}02306}$.	**16.** $\frac{1}{623{\cdot}7}$.
17. $\frac{1}{39{\cdot}47}$.	**18.** $\frac{1}{0{\cdot}3663}$.	**19.** $\frac{1}{31478}$.	**20.** $\frac{1}{13{\cdot}59}$.

§ 160. It is often possible to shorten the calculation of difficult fractions by using a table of reciprocals.

Example 1. To find the value of $\frac{3}{2{\cdot}493}$.

$$\frac{3}{2{\cdot}493} = 3 \times \frac{1}{2{\cdot}493} = 3 \times 0{\cdot}4011 = 1{\cdot}2033.$$

The reciprocal is not exactly 0·4011 and by division we find that it should be 0·40112...; and this when multiplied by 3 gives 1·20336, which, to 4 significant figures, is 1·2034. Thus the 5th figure in the answer above is incorrect. But an answer derived from 4-figure tables should never be given to more than 4 significant figures; thus

$$\frac{3}{2{\cdot}493} = 3 \times 0{\cdot}4011 = \underline{1{\cdot}203.}$$

Example 2. To find the value of $\frac{1}{\sqrt{2{\cdot}631}} + \sqrt{\frac{1}{17{\cdot}3} + \frac{1}{9{\cdot}64}}$.

$$\begin{aligned}\frac{1}{\sqrt{2{\cdot}631}} + \sqrt{\frac{1}{17{\cdot}3} + \frac{1}{9{\cdot}64}} &= \frac{1}{1{\cdot}622} + \sqrt{0{\cdot}05780 + 0{\cdot}1037} \\ &= 0{\cdot}6166 + \sqrt{0{\cdot}1615} \\ &= 0{\cdot}6166 + 0{\cdot}4018 \\ &= 1{\cdot}018.\end{aligned}$$

If we calculate $\frac{1}{\sqrt{2{\cdot}631}}$ by using reciprocal tables first and square root tables afterwards, we find a different result. For $\frac{1}{\sqrt{2{\cdot}631}} = \sqrt{\frac{1}{2{\cdot}631}} = \sqrt{0{\cdot}3801} = 0{\cdot}6165.$

This illustrates the fact that results differing in the 4th significant figure may be obtained if the order of operations is varied. Accordingly, when 4-figure tables are used, it is well not to give results to more than 3 significant figures.

In the present case, the value is 1·02.

Example 3. If $\frac{1}{x^2} = \frac{1}{a^2} + \frac{1}{b^2}$, find x when

$$a = 23{\cdot}7 \text{ and } b = 29{\cdot}7.$$

$$\frac{1}{x^2} = \frac{1}{23{\cdot}7^2} + \frac{1}{29{\cdot}7^2}$$

$= \frac{1}{561{\cdot}7} + \frac{1}{882{\cdot}1}$, from table of squares,

$= 0{\cdot}001781 + 0{\cdot}001134$, from table of reciprocals,

$= 0{\cdot}002915.$

$\therefore\ x^2 = 343{\cdot}0$, from table of reciprocals,

$\therefore\ x = 18{\cdot}52$, from table of sq. roots,

$= 18{\cdot}5$ to 3 significant figures.

EXERCISE XV i

Find the value to 3 significant figures of:

1. $\frac{20}{16{\cdot}31}$. **2.** $\frac{104}{27{\cdot}63}$. **3.** $\frac{2}{13{\cdot}9} + \frac{3}{17{\cdot}2}$.

4. $\frac{35}{23{\cdot}51} - \frac{20}{25{\cdot}47}$. **5.** $\left(\frac{1}{4{\cdot}984}\right)^2$. **6.** $\frac{1}{21{\cdot}46^2}$.

7. $\frac{1}{71{\cdot}39^2}$. **8.** $\frac{1}{0{\cdot}2564^2}$. **9.** $\frac{23}{1{\cdot}247^2}$.

10. $\frac{1}{14{\cdot}3^2} + \frac{1}{12{\cdot}7^2}$. **11.** $\frac{1}{\sqrt{90{\cdot}4}}$. **12.** $\frac{1}{\sqrt{0{\cdot}0315}}$.

13. $\frac{231}{\sqrt{1771}}$. **14.** $\sqrt{\frac{10{\cdot}7}{19{\cdot}37}}$. **15.** $\sqrt{\frac{215}{42{\cdot}73}}$.

If $\frac{1}{x} = \frac{1}{a} + \frac{1}{b}$, find

16. x if $a = 4{\cdot}372$ and $b = 5{\cdot}938$.

17. a if $x = 37{\cdot}6$ and $b = 44{\cdot}7$.

18. a if $x = 0{\cdot}153$ and $b = 0{\cdot}248$.

If $\frac{1}{x^2} = \frac{1}{a^2} + \frac{1}{b^2}$, find

19. x when $a = 2{\cdot}19$ and $b = 1{\cdot}37$.

20. a when $x = 11{\cdot}63$ and $b = 15{\cdot}72$.

21. x when $a = 0{\cdot}754$ and $b = 0{\cdot}372$.

22. a when $x = 0{\cdot}0315$ and $b = 0{\cdot}0785$.

EXERCISE XV j

1. One metre = 39·37 inches. Express (i) 1 inch in cm., (ii) 1 yd. in metres, (iii) 4·7 in. in mm.

2. Given that 1 lb. = 453·593 grams, find the number of lbs. in a kilogram.

3. Express 5 cub. inches as a decimal of a cubic foot.

4. Express 111 sq. yards as a decimal of an acre.

5. If £1 = 25·35 francs, express 20 francs in shillings and pence to the nearest penny.

6. A cubic foot of water weighs 62·43 lbs. How many cubic feet are there in 100 lbs. of water?

7. How many c.c. of mercury (sp. gr. 13·6) weigh 1000 gm.?

8. How many c.c. of zinc (sp. gr. 7·19) weigh 20 gm.?

9. How many c.c. of tin (sp. gr. 7·29) weigh 99 gm.?

10. If a 500 gm. weight is made of pure copper (sp. gr. 8·79) what is its volume in c.c.?

11. If a lump of lead (sp. gr. 11·4) weighs 21 gm., what is its volume?

12. If 500 gm. of copper are mixed with 250 gm. of zinc to form brass, what is the volume of the brass, and what is its sp. gravity?

13. If 900 gm. of copper are mixed with 100 gm. of tin to form bronze, what is the sp. gravity of bronze?

A cubic foot of water weighs 1000 *ounces.*

14. How many cubic feet of mercury weigh 1000 lbs.?

15. How many cub. ft. of lead weigh 500 lbs.?

16. How many cub. ft. of copper weigh a ton?

CHAPTER XVI

CIRCLE AND CYLINDER

CIRCUMFERENCE OF CIRCLE

§161. In books on geometry it is shown that, if a regular hexagon (6 sides) is inscribed in any circle (see figure), each of its sides is equal to the radius. Thus the perimeter of the hexagon is 6 times the radius or 3 times the diameter.

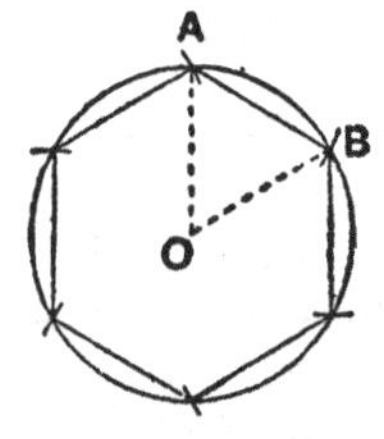

Now, since the arc AB is curved, it is longer than the straight line AB. Therefore **the circumference of a circle is rather more than 3 times the diameter.** This fact should be remembered and continually used as a rough check in more accurate calculations.

§162. The ratio (circumference : diameter) which we have just seen to be rather greater than 3 can be calculated by higher mathematics, but the following experiments provide approximate values.

Experiment 1. Take any cylindrical object, whose diameter can easily be measured, e.g. a tin; measure the diameter of the end in inches, to the nearest $\frac{1}{100}$ of an inch, by laying your rule across the widest part. Do this 3 times and take the average of your measurements.

Diameter of circle = ... in.
.... in.
.... in.
.... in.

Diameter of circle (average) = ... in. (correct to 2 places).

Now find the circumference as follows. Wrap a piece of paper *tightly* round the cylinder, making the ends overlap; with a pin prick through two thicknesses of paper: unwrap the paper and measure the distance between the pin-pricks. This is the circumference of the cylinder, and of the circular end.

Circumference of circle = ... in.

$$\frac{\text{Circumference}}{\text{Diameter}} = \frac{\ldots\ldots}{\ldots\ldots} = \ldots\ldots \text{ (correct to 2 places).}$$

Experiment 2. Make a small scratch on the end of a cylinder* at the edge. Put the cylinder on a metre rule, with the scratch coinciding with some convenient graduation (say 10 cm.). Now let the cylinder roll along the rule until it has made three complete turns, and read the graduation that now coincides with the scratch.

Distance rolled in 3 turns = ...cm.
Distance rolled in 1 turn = ...cm.
∴ Circumference of cylinder = ...cm.
Diameter of cylinder = ...cm.

$$\therefore \frac{\text{Circumference}}{\text{Diameter}} = \frac{\ldots\ldots\text{cm.}}{\ldots\ldots\text{cm.}} = \ldots\ldots \text{ (correct to 2 places).}$$

§ 163. If you have worked carefully, these two experiments will have led to nearly the same result. Theory shows that the ratio is 3·1415926.... This number is so important that a special symbol is used to denote it—the Greek letter π. $3\frac{1}{7}$, or $\frac{22}{7}$ differs from this so little that it will generally be a sufficiently accurate approximation to use.

Ex. 1. Calculate the average of the two values of π you have found.

Finally, then,

$$\frac{\text{circumference}}{\text{diameter}} = \pi,$$

$$\therefore \textbf{circumference} = \textbf{diameter} \times \pi$$
$$= \text{radius} \times 2\pi$$
$$= 2\pi r \text{ units, if radius} = r \text{ units.}$$

Inversely $$\text{radius} = \frac{\text{circumference}}{2\pi}.$$

* A pulley on an axle serves better still.

§164. *Example.* To find the circumference of a circle whose diameter is 12 inches.

$$\text{circum.} = \text{diam.} \times \pi$$
$$= 12 \times \frac{22}{7} \text{ in.} = \frac{264}{7} \text{ in.} = \underline{37 \cdot 7} \ldots \text{in.}$$

Example. To find the circumference of a circle whose radius is 3·49 cm.

$$\text{circum.} = 2\pi r$$
$$= 2 \times \frac{22}{7} \times 3 \cdot 49 \text{ cm.}$$
$$= \frac{153 \cdot 56}{7} \text{ cm.} = \underline{21 \cdot 9} \ldots \text{cm.}$$

$$\begin{array}{r} 349 \\ 44 \\ \hline 1396 \\ 1396 \\ \hline 15356 \end{array}$$

Example. To find the radius of a circle whose circumference is 14 feet.

$$\text{radius} = \frac{\text{circum.}}{2\pi}$$
$$= \frac{14}{2 \times \frac{22}{7}} \text{ ft.} = \underline{2 \cdot 23} \text{ feet.}$$

EXERCISE XVI a

Take $\pi = \frac{22}{7}$, *and give results to* 3 *sig. figs.*

1. Find the length of the circumference of a circle, of radius (i) 7 m.; (ii) 1′ 9″; (iii) 1·75 cm.; (iv) $3\frac{1}{2}$ yds.; (v) 8·4 Km.; (vi) 0·14 mm.; (vii) 2 miles 50 yds.

2. Find the circumference of a circle of diameter (i) 28 cm.; (ii) 14 miles; (iii) 4′ 1″; (iv) 3 miles 61 yds.

3. Find the radius of a circle the length of whose circumference is (i) 88 m.; (ii) 11″; (iii) $\frac{1}{4}$ mile; (iv) 1·21 cm.; (v) 2 miles 88 yds.; (vi) 18·7 Km.

4. What is the diameter of a circle whose circumference is (i) 220′; (ii) 10′ 1″; (iii) 0·0154 cm.; (iv) 17 miles?

5. The crank of a reciprocating engine is 2′ 11″ long; find the distance travelled by the crank-pin in a complete revolution.

6. A rope is wound six times round a capstan of diameter 2′ 4″, and 10′ of rope is left free; find its total length.

7. A bicycle wheel makes 720 revolutions in travelling a mile; find its diameter.

8. Find the number of revolutions per minute made by the 7 foot wheel of a locomotive when travelling at the rate of $\frac{1}{4}$ mile in 20 seconds.

9. A motor car is going at the rate of 15 miles per hour. How many revolutions will the wheel make in a minute if its diameter is 30 inches?

10. What is the circumference of a circle of radius 26 cm.?

11. Find the diameter of a circle, the length of whose circumference is 108 inches.

12. A square and a circle have the same perimeter; what is the length of the radius of the circle if the side of the square is $3\frac{1}{4}$ inches?

13. A circular ring is made from 2″ of wire; what will its diameter be?

14. The earth's equatorial diameter is 7920 miles; find the length of the equator.

15. A rope is passed over a pulley of radius 5″ and hangs vertically on each side with its ends at the same height; what will be the difference in height of the ends when the pulley has made three revolutions, assuming that no slipping takes place?

16. What is the number of revolutions per minute made by the driving wheel of a locomotive, when travelling at 90 kilometres per hour, the diameter of the wheel being 2·3 metres?

17. The driving wheel of a locomotive engine 5 ft. in diameter turned 2500 times in going 7 miles; find (to the nearest yard) what distance is lost owing to the slipping of the wheel on the rail.

Area of Circle

§165. As you are not yet able to find by pure mathematics the rule for the area of a circle, you must proceed by measurement.

The figure shows that the area of the circle is *less than four times* the square on the radius. But the shaded quarter of the circle is obviously greater than half the square on the radius; therefore the area of the whole circle is *greater than twice* the square on the radius. Thus the **area of a circle is about 3 times the square on the radius.** Use this fact continually as a rough check.

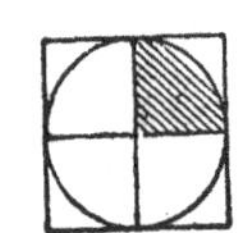

To get a closer approximation squared paper may be used, as follows:

Experiment 3. Take a piece of squared paper ruled with small squares $\frac{1}{10}''$ each way. How many small squares go to the square inch? Draw on the squared paper a circle whose radius is 2 inches. Count the number of small squares inside the circle; you will notice that the inch lines mark off blocks of 100, and thus make the counting easier. Where the circle cuts through a small square you have fractions of a small square; if there is more than half a small square inside the circle, reckon it as one; if there is less than half, leave it out; this will give you a sufficiently close approximation.

Number of small squares inside = ...

Divide by 100, and find the area in square inches.

Area of circle = ... sq. in.

Area of square on radius = ... sq. in.

$$\frac{\text{area of circle}}{\text{area of square on radius}} = \frac{\ldots\text{sq. in.}}{\ldots\text{sq. in.}} = \ldots \text{ (correct to 2 places).}$$

You may be surprised to find that the answer is very near to π; in fact it can be proved that, exactly,

$$\frac{\text{area of circle}}{\text{area of square on radius}} = \pi,$$

or **area of circle = area of square on radius × π.**

If the radius is r units, the square on the radius contains r^2 sq. units.

$$\therefore \text{ area of circle} = \pi r^2 \text{ sq. units.}$$

§ 166. *Example.* To find the area of a circle whose diameter is 42 cm.

$$\text{Area} = \pi r^2 \text{ sq. cm.}$$
$$= \frac{22}{7} \times (21)^2 \text{ sq. cm.} = \underline{1386} \text{ sq. cm.}$$

Example. To find the circumference of a circle whose area is 600 sq. inches.

If $r'' =$ the radius, $r^2 = \frac{600}{\pi}$, $\therefore r = \sqrt{\frac{600}{\pi}}$.

$$\text{And circumference} = 2\pi r = 2\pi\sqrt{\frac{600}{\pi}}$$
$$= 2\sqrt{\frac{\pi^2 \times 600}{\pi}} = 2\sqrt{\pi \times 600}$$
$$= 2\sqrt{3{\cdot}142 \times 600} = 2\sqrt{1885{\cdot}2}$$
$$= 2 \times 43{\cdot}42 = \underline{86{\cdot}8} \text{ inches.}$$

EXERCISE XVI b

Take $\pi = \frac{22}{7}$, *and give results to* 3 *sig. figs.*

1. Find the area of a circle of radius (i) 7 m.; (ii) 1′ 2″; (iii) 3·5 cm.; (iv) 0·21 Km.

2. What is the area of a circle whose diameter is (i) 28 mm.; (ii) 7′; (iii) 0·98 cm.?

3. What is the radius of a circle of area (i) 154 cm.^{2}; (ii) 6·16 m.^{2}; (iii) 616 sq. in.; (iv) 15·4 acres; (v) 1540 acres?

4. Find the circumference of a circle whose area is (i) 38·5 mm.^{2}; (ii) $346\frac{1}{2}$ in.^{2}

5. 504 coins, 1″ in diameter, are cut from a sheet of silver 7″ wide and 6′ long; what area of metal remains over?

6. What area is commanded by a gun of range 7 Km., which can be trained through 180°?

7. What is the radius of a disc of area 5544 in.^{2}?

8. Find the area of a circle of diameter 2·77 cm.

9. The radius of a piston is 1′ 8″ and of the piston-rod 4·5″; find the areas exposed to steam on each side.

10. Four goats are tethered at the four corners of a square field of side 200 yds., so that they just cannot reach one another; what area will be left ungrazed?

11. Find the area of one side of a washer of outer radius 1·92 cm. and inner radius 1·17 cm.

SURFACE OF CYLINDER

§ 167. Objects of cylindrical shape are so common that it is useful to be able to calculate the surface and volume of a cylinder. Examples of solid cylinders are the following: an uncut lead pencil, a coin, a Norman column, a straight piece of wire. Examples of hollow cylinders are: a cylindrical tobacco tin, a garden roller, the cylinder of an engine, a straight piece of gas-pipe. We know all about the size and shape of a cylinder if we know (i) the radius of either circular end, (ii) the distance apart of the two ends. This latter measurement we should naturally call the **length,** if the cylinder is a lead pencil, or a piece of wire or tubing. On the other hand, if the cylinder is an object which we are accustomed to see standing upright (e.g. a jam-jar) we should call this measurement the **height.** In the case of a coin we speak of the **thickness** rather than the height or length.

§ 168. The complete surface of a solid cylinder consists of the curved surface together with the flat ends. If the ends are circular (as they will be in all the examples given below) you will have no difficulty in finding their area. We will consider how to find the curved surface (or, strictly speaking, the area of the curved surface).

A rectangular sheet of paper can be rolled up into a hollow cylinder. If you do this without making the paper overlap, i.e. by bringing opposite edges of the paper just to meet, the curved surface of the cylinder formed has the same area as the sheet of paper.

Or suppose that you have such a cylinder as a length of stove-pipe. Imagine it to be slit straight down, and opened out flat. The resulting shape is a rectangle of the same area.

Now one of the dimensions of this rectangle is the height of the cylinder, and the other is the circumference of the end of the cylinder.

Thus the **area of the curved surface of a cylinder**

= area of resulting rectangle

= **circumference of end** × **height of cylinder.**

Algebraically, if the radius of end $= r$ units, and height $= h$ units,

area of curved surface $= 2\pi r \times h$ sq. units.

If you are asked to find the surface of a cylinder without further directions, you have to use your common sense in judging whether the ends are to be counted in or not.

¶**Ex. 2.** How many ends would you count in if asked to find (i) the area of paint on the outside of a closed cylindrical cistern, (ii) the area of galvanised sheet-iron required to make an open cylindrical cistern, (iii) the area of grass rolled by a garden roller of given dimensions in 10 turns?

EXERCISE XVI c

Take $\pi = \frac{22}{7}$ *and give results to 3 sig. figs.*

1. A sheet of note-paper, 8 inches by 6 inches, is rolled into a cylinder, its longer edges *just* meeting. Find the diameter of the cylinder so formed.

2. Find the curved surface of a cylinder, having given:

(i) radius 7″, height 11″,

(ii) height 14 cm., radius 6 cm.,

(iii) diameter 21 m., height 100 m.,

(iv) diameter 5′ 3″, height 6′,

(v) height 4·9 mm., radius 0·2 mm.,

(vi) height 37·1 cm., diameter 1·2 cm.,

(vii) length 360 yds., radius 0·07″,

(viii) length 1 mile, diameter 1·4″.

3. Find the radius of a cylinder, having given:

(i) curved surface 44 cm.^{2}, height 3·5 cm.,
(ii) length 7′, curved surface 121 ft.^{2},
(iii) height 4′ 1″, curved surface 616 in.^{2},
(iv) curved surface 9·9 cm.^{2}, length 4·2 m.

4. Find the diameter of a cylinder, having given:

(i) length 7 m., curved surface 44 m.^{2},
(ii) curved surface 77 ft.^{2}, height 2′ 0·5″,
(iii) length 28 yds., curved surface 9·9 in.^{2}

5. Find the height of a cylinder, having given:

(i) radius $3\frac{1}{2}$″, curved surface 132 in.^{2},
(ii) curved surface 11 ft.^{2}, diameter 1 yd. 2′ 3″.

6. Find the total surface of a solid cylinder, given:

(i) radius 1″, height 6″,
(ii) length 5 m., diameter 4 m.,
(iii) diameter 7′, height 3′ 6″,
(iv) radius 1·35 m., height 2·15 m.,
(v) height 4′ 6″, diameter 5′,
(vi) radius 0·6 cm., length 14·1 cm.

7. Find the length of a cylinder, having given:

(i) total surface 44 in.^{2}, radius 1″,
(ii) diameter 4″, total surface 1 ft.^{2} 21 in.^{2}

8. Find the cost of painting the outside surface of a funnel of height 15′ and diameter 5′ 3″ at 4*d.* per ft.^{2}

9. What area of sheet-iron would be required to make a cylindrical boiler 14′ long and 6′ diameter?

10. An open tunnel 200 yds. long has a cross-section in the form of an exact semicircle of radius 11′ 6″; find the area of steel required to line its whole surface.

11. A water-tube boiler contains 10 pairs of tubes each 6′ 9″ long; what must be their external diameter in order that their combined heat-receiving surface may be 160 ft.^{2}?

VOLUME OF CYLINDER

§ 169. The volume of a cylinder must be calculated if it is required to find the volume of water in a mile of water-pipe; or the volume of steam in the cylinder of a steam engine; or the weight of a solid cylinder made of material weighing so much per cubic inch.

A cylinder is a solid of uniform cross-section; therefore **the volume is the product of the area of cross-section × the height.** As the cross-section is a circle, its area is πr^2 sq. units. Therefore the volume of a cylinder is $\pi r^2 h$ units of volume.

Example. How many gallons will a water-butt hold, if it is 3′ 6″ deep and has a diameter throughout of 19″? (1 cub. foot = $6\frac{1}{4}$ galls.)

The capacity of the water-butt

$$= \pi \times \left(\frac{19}{2}\right)^2 \times 42 \text{ cu. in.}$$

$$= \frac{22}{7} \times \frac{361}{4} \times 42 \text{ cu. in.}$$

$$= \frac{22 \times 361 \times 42}{7 \times 4 \times 1728} \text{ cu. ft.}$$

$$= \frac{22 \times 361 \times \overset{\overset{2}{\cancel{6}}}{\cancel{42}} \times 25}{\cancel{7} \times 4 \times \underset{576}{\cancel{1728}} \times 4} \text{ galls.}$$

$$= \frac{397100}{9216} \text{ galls.}$$

$= \underline{43 \text{ gallons}}$ about.

$$\begin{array}{r} 576 \\ 16 \\ \hline 576 \\ 3456 \\ \hline 9216 \end{array}$$

$$\begin{array}{r} 43 \\ 9{\cdot}216\overline{)397{\cdot}1} \\ 368\ 64 \\ \hline 28\ 460 \\ 27\ 648 \\ \hline 812 \end{array}$$

Example. What is the volume of lead in a pipe 150 feet long whose external circumference is $7\frac{3}{4}$″, if the metal is $\frac{1}{4}$″ thick?

The volume of lead is found by subtracting the volume of the hollow inside from the volume of the pipe reckoned as solid. Let the inner and outer radii be r, R in. respectively.

Then $R = \dfrac{7\frac{3}{4}}{2\pi}$ in. $= \dfrac{31}{8\pi}$ in. $= \dfrac{31 \times 7}{8 \times 22}$ in.

$= \dfrac{217}{176}$ in.

$= 1{\cdot}233..$in.

$r = (1{\cdot}233 - 0{\cdot}25)'' = 0{\cdot}983..$in.

```
     1·233
1·76)2·17
     1 76
      410
      352
       580
       528
        520
```

Volume of lead

$= (\pi R^2h - \pi r^2h)$ cu. in.

$= \pi h\,(R^2 - r^2)$ cu. in.

$= \dfrac{22}{7} \times 150 \times 12 \times (1{\cdot}233^2 - 0{\cdot}983^2)$ cu. in.

$= \dfrac{22}{7} \times 1800 \times (1{\cdot}520 - 0{\cdot}966)$ cu. in.

(using tables)

$= \dfrac{22 \times 1800}{7} \times 0{\cdot}554$ cu. in.

$= \underline{3130}$ cu. in. about.

```
 0·554
 1800
 554
 443·2
 997·2
    3 1/7
2991·6
 142·5
3134·1
```

Examples of this type are generally best worked by logarithms.

EXERCISE XVI d

Take $\pi = \dfrac{22}{7}$ *or* 3·142 *according to circumstances and give results to* 3 *sig. figs.*

1. Find the volume of a cylinder, having given:

(i) area of base = 52 in.², height = 1′ 2″,

(ii) area of end = 0·7 cm.², length = 1·2 m.

2. Find the volume of a cylinder, having given:

(i) radius 7″, height 3″,

(ii) radius 2 cm., height 14 cm.,

(iii) height 1′ 9″, radius 6″,

(iv) radius 4·9 m., height 20 m.,

(v) diameter 0·42 cm., length 1 Km.,

(vi) length 14 yds., radius 1″,

(vii) diameter 2′ 11″, thickness 1/2″.

3. Find the radius of a cylinder, having given:

(i) height 7″, volume 88 in.^{3},

(ii) thickness 1/7″, volume 550 in.^{3},

(iii) volume 0·32076 cm.^{3}, height 14 mm.

4. Find the diameter of a cylinder of height 14 m. and volume 11 m.^{3}

5. What is the volume of a cylinder of radius 4″ whose curved surface is 1 ft.^{2}?

6. What is the internal volume of a gun 5′ 3″ long and calibre 4″?

7. How many cub. ft. of earth must be dug out in sinking a well 14′ in diameter and 50′ deep?

8. Find the volume of metal in 30′ of hollow shafting of external diameter 1′ $2\frac{1}{2}$″ and internal diameter $7\frac{1}{2}$″.

9. Find the weight of a solid glass rod of 0·4″ diameter and 1′ long, if a cubic foot of glass weighs 130 lbs.

10. How many tons of material must have been removed in constructing the Severn tunnel (7650 yds. long), assuming it to be cylindrical (diameter 14′) and a cubic foot of the material to weigh 350 lbs.?

11. What volume of metal is required to make a cylinder of external diameter 1″ and external length 1′, closed at both ends, the metal being $\frac{1}{4}$″ thick throughout?

12. Find the volume of metal in 12 ft. of hollow shafting, the external diameter being 2 ft. 6 in. and the internal diameter 1 ft. 2 in.

13. The internal diameter of an iron pipe is 12 inches, the metal is half an inch thick, and the length of the pipe is 6 feet. What is its weight if 1 cubic inch of iron weighs 4·4 oz.?

14. The external and internal diameters of a hollow steel shaft are 15″ and 7″ respectively. Find the area of the cross-section and the weight of 20 feet of the shaft, given that 1 cubic inch of steel weighs 0·29 lb.

15. What would be the diameter of a pump which is to replace and do the work of two pumps of 8 in. and 12 in. diameter respectively, given that the stroke and speed are the same for each pump? (To the nearest half-inch.)

16. A cylindrical glass whose internal height is 10 cm. and diameter 8 cm. contains 1 cm. of water. A lead cylinder 12 cm. high and 6 cm. in diameter is placed in the glass. Will the water overflow? How if the lead cylinder is 9 cm. high?

17. An ebony snuff-box is in the shape of a shallow cylinder. The wood is 3 mm. thick; the external dimensions of the box are: diameter 9·5 cm., depth 2·7 cm. Calculate the weight of the box when empty; the sp. gr. of ebony is 1·2. Would the empty box float?

18. A cylinder will exactly fit into a rectangular box, whose internal dimensions are: length 9 cm., breadth 4 cm., height 4 cm. What is the volume of the cylinder? What is the difference between its volume and the volume of the box?

19. Find the volume of a cylindrical tree trunk 10 feet long and of diameter 14 inches. If it were planed down so as to form a prism whose right section is an equilateral triangle, what would be the volume of the new post?

What percentage of the whole would be wasted?

20. Find the number of cubic centimetres of wood in a right cylinder of 4 cm. diameter and 6 cm. height.

A second cylinder is twice as high, but its diameter is only one-half that of the first. Find its volume.

Find also the ratio of the areas of the complete surfaces of the two cylinders.

21. A lead pencil consists of a hollow cylinder of wood with a solid cylinder of graphite fitted into it. The diameter of the pencil is 7 mm.; the diameter of the graphite 1 mm.; the length of the pencil 10 cm. Calculate the volume of the wood and of the graphite and the weight of the whole pencil if the specific gravity of the wood is 0·6 and of the graphite 2·3.

22. Find the volumes of the solids shown in plan and elevation in figs. I—O.

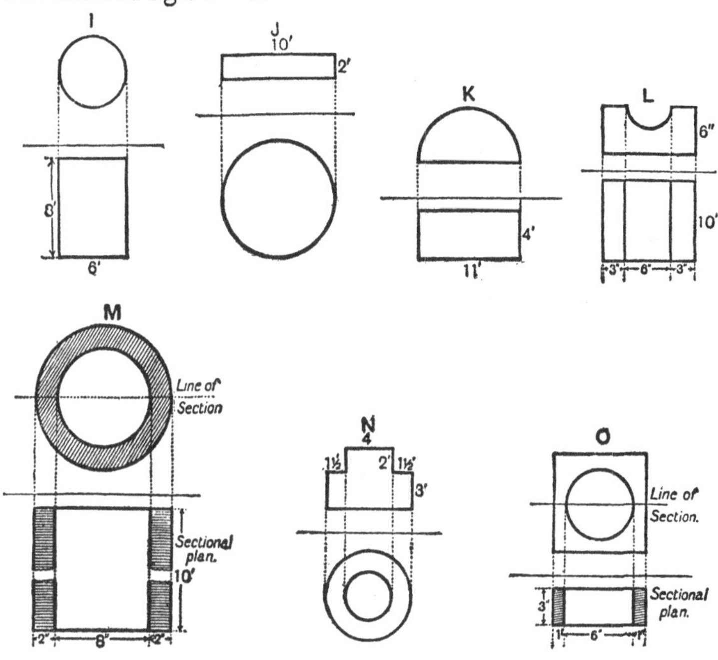

EXERCISE XVI e

Circles and Cylinders

Give results to 3 sig. figs.

1. The diameter of a wheel of a motor car is 2′ 4″; at how many miles an hour is the car travelling when a cyclometer on that wheel clicks 120 times every minute?

2. What is the capacity in gallons of a cylindrical tank 5′ long and 2·8′ in diameter? (1 c. ft. of water weighs 62·5 lbs.: 1 gal. of water weighs 10 lbs.)

3. Find the area of the annular space between two concentric circles of radii 1′ 3″ and 8″.

4. Find the weight of a metre of lead tubing 8 cm. in external diameter and 1 cm. thick. (Sp. gr. of lead 11·4.)

5. It is said of the Asiatic elephant that twice round the base of his foot is a measure of his height; what would be the diameter of the foot of an elephant 11′ high?

6. What weight of glycerine (sp. gr. 1·25) can be held in a beaker of depth 14 cm. and internal radius 4 cm.?

7. Find the total surface of a semi-cylindrical solid of radius 1′ 2″ and length 6′.

8. Find the volume of a solid formed of a cylindrical bar 6″ in diameter, with a disc 2″ thick and 1′ 8″ in diameter at each end, the length over all being 4′ 1″.

9. How often will a bicycle wheel of 28″ diameter rotate in travelling half a mile?

10. A cylindrical hole of radius 1′ 6″ is cut from end to end of a rectangular block of wood 4′ by 4′ by 7′; what volume of wood remains?

11. A locomotive boiler has 404 tubes of length 21′ and internal diameter 2″; find the heat-receiving surface in sq. ft.

12. What is the radius of a wheel which makes 7200 revolutions in travelling 10 miles?

13. A well when full to a depth of 5 m. contains 17325 litres; what is its diameter?

14. A rectangular block of metal 1′ 4″ by 11″ by 8″ is melted down and cast into a cylinder 2′ 4″ long; what will be the diameter?

15. Find the capacity in cub. ft. of a gasholder of diameter 300′ and height 180′.

16. Taking the earth's equatorial diameter to be 7927 miles, find the speed due to rotation only at a point on the equator, in miles per hour.

17. Find the weight of a steel cylinder of height 15 cm. and radius 2·4 cm. (Sp. gr. of steel = 7·79.)

18. A roller 2′ 2″ in diameter and 3′ wide makes 24 revolutions in passing from end to end of a lawn; find in sq. yds. the area rolled when it has passed from end to end 36 times.

19. The cross-section of the Simplon tunnel is equivalent to $\frac{3}{4}$ of a circle of radius 4·3 m. and the tunnel is 20 Km. long; find the volume in cubic metres of the material removed in cutting it.

20. A cub. in. of coin bronze weighs 5 oz. and 3 pennies weigh 1 oz. The diameter of a penny is 1·2″; what is its average thickness (to 2 significant figures)?

21. What is the area of a semicircle of radius 2″?

22. Find the cost of lining the curved surface and one end of a cylindrical tank of internal radius 2·5 m. and internal depth 4 m., with lead 5 cm. thick, the sp. gr. of lead being 11·4, and its price £21 for 1000 Kgm.

23. What must be the period of rotation of a capstan of diameter 4′ 6″, in order that ropes may be coiled on it at the rate of 8′ per minute?

24. If the diameter of a circle is doubled, how are the circumference and area altered?

25. A pipe 1·5 cm. in diameter fills a tank in $3\frac{1}{4}$ mins.; how long will it take a pipe 3 cm. in diameter to fill it?

26. A circle and a square have the same area; which has the greater perimeter?

27. A linen collar forms a cylinder 16 in. in circumference and 3 in. high. Find its curved surface.

28. How much metal is needed to make a telescope tube (without ends) 5 ft. long and 4 in. in diameter?

29. The height of a cylinder is 3 in. and its curved surface 15 sq. in. Find its total surface and its volume.

30. How much ink will a fountain pen hold, the depth of the reservoir being 3·5 cm. and the internal diameter 7 mm.? How often can it be filled up from a $\frac{1}{2}$-litre bottle?

31. Find which contains the greater amount of tobacco; a cigarette of length 10 cm. and diameter 1 cm., or a pipe whose cylindrical interior is 3 cm. deep and 2 cm. in diameter. What area of paper is consumed in a year by a person who smokes 10 such cigarettes a day and throws away $\frac{3}{10}$ of each?

32. From a cylindrical tank 10 ft. in diameter water is being discharged so that the surface falls 1 ft. in $16\frac{1}{2}$ minutes. How many cubic feet (to 3 significant figures) are being discharged per hour?

33. A cylindrical water-pipe bifurcates into two pipes of half its diameter. Show that the rate of flow is increased.

34. A certain quantity of water is being measured by pouring it into a measuring-glass of known diameter and reading the height. If a mistake of 1 mm. is made in reading the height, what error will this cause in the calculated volume if the internal diameter of the glass is (1) 4 cm., (2) 1 cm., (3) 1 mm.?

CHAPTER XVII

USE OF LOGARITHMS

§170. This chapter contains graduated exercises in the use of logarithms. For the theory of logarithms the reader is referred to Godfrey and Siddons' *Elementary Algebra*, Vol. II, or to the chapter on logarithms in Price's *Examples in Numerical Trigonometry*, both published by the Cambridge University Press.

At first the pupil is strongly recommended to use powers of 10, as in the examples below. When facility in the use of logarithms has been obtained, the shortened form shown on p. 349 may be used.

It should be remembered that when 4-figure tables are used, the 4th figure is liable to error.

Positive Indices

§171. *Example.*
$$\begin{aligned} 3{\cdot}563 \times 1{\cdot}212 &= 10^{0{\cdot}5518} \times 10^{0{\cdot}0835} \\ &= 10^{0{\cdot}6353} \\ &= \underline{4{\cdot}318}. \end{aligned}$$

Example.
$$\frac{3{\cdot}563}{1{\cdot}212} = \frac{10^{0{\cdot}5518}}{10^{0{\cdot}0835}} = 10^{0{\cdot}4683} = \underline{2{\cdot}940}.$$

Example.
$$\begin{aligned} (12{\cdot}12)^3 &= (10^{1{\cdot}0835})^3 \\ &= 10^{3{\cdot}2505} \\ &= \underline{1780}. \end{aligned}$$

Example.
$$\begin{aligned} \sqrt[3]{356{\cdot}3} &= (10^{2{\cdot}5518})^{\frac{1}{3}} \\ &= 10^{0{\cdot}8506} = \underline{7{\cdot}089}. \end{aligned}$$

EXERCISE XVII a

(The object of the following Exercise with easy checks is to give the beginner confidence in his work. Logarithms should not, of course, be used in actual practice for such calculations.)

Use logarithms to simplify the following and verify by ordinary multiplication and division or by using square and square root tables.

1. $2 \times 1{\cdot}417$. **2.** $3 \times 2{\cdot}379$. **3.** $2{\cdot}5 \times 3{\cdot}624$.
4. $4 \times 1{\cdot}767$. **5.** $1{\cdot}5 \times 4{\cdot}634$. **6.** $476{\cdot}2 \times 2$.
7. $72{\cdot}59 \times 3$. **8.** $5{\cdot}672 \times 50$. **9.** $6{\cdot}24 \times 25$.
10. $4{\cdot}644 \times 400$. **11.** $4{\cdot}794 \div 2$. **12.** $6{\cdot}315 \div 5$.
13. $9{\cdot}637 \div 2{\cdot}5$. **14.** $574{\cdot}4 \div 2$. **15.** $47{\cdot}67 \div 4$.
16. $96430 \div 6000$. **17.** $17{\cdot}342 \div 7$. **18.** $2{\cdot}54 \times 2{\cdot}54$.
19. $(1{\cdot}319)^2$. **20.** $(1{\cdot}116)^4$. **21.** $(1{\cdot}371)^4$.
22. $(4{\cdot}625)^2$. **23.** $(72{\cdot}54)^2$. **24.** $(625{\cdot}4)^2$.
25. $(1{\cdot}732)^4$. **26.** $(2)^{\frac{1}{2}}$. **27.** $(8{\cdot}475)^{\frac{1}{2}}$.
28. $\sqrt{2{\cdot}963}$. **29.** $\sqrt{200}$. **30.** $\sqrt{41{\cdot}92}$.
31. $\sqrt{419{\cdot}2}$ **32.** $\sqrt{150}$. **33.** $\sqrt[4]{150}$.
34. $(27)^{\frac{1}{3}}$. **35.** $\sqrt[3]{1024}$. **36.** $\sqrt[5]{32}$.

EXERCISE XVII b

Use logarithms to find the value of:

1. $4{\cdot}651 \times 1{\cdot}843$. **2.** $3{\cdot}794 \times 2{\cdot}156$. **3.** $7{\cdot}342 \times 1{\cdot}112$.
4. $5{\cdot}321 \times 1{\cdot}632$. **5.** $3{\cdot}007 \times 1{\cdot}729$. **6.** $2{\cdot}419 \times 3{\cdot}069$.
7. $2{\cdot}572 \times 14{\cdot}63$. **8.** $31{\cdot}79 \times 43{\cdot}72$. **9.** $727{\cdot}6 \times 1{\cdot}571$.
10. $6297 \times 23{\cdot}67$. **11.** $97{\cdot}64 \times 94{\cdot}67$. **12.** $212{\cdot}5 \times 409{\cdot}5$.
13. $21{\cdot}35 \times 37{\cdot}67$. **14.** $61{\cdot}12 \times 43150$. **15.** 73050×4350.
16. $3{\cdot}295 \div 2{\cdot}34$. **17.** $7{\cdot}694 \div 2{\cdot}795$. **18.** $5{\cdot}407 \div 1{\cdot}436$.
19. $8{\cdot}371 \div 2{\cdot}634$. **20.** $14{\cdot}63 \div 2{\cdot}572$. **21.** $43{\cdot}72 \div 31{\cdot}79$.
22. $727{\cdot}6 \div 1{\cdot}571$. **23.** $6297 \div 23{\cdot}67$. **24.** $57{\cdot}93 \div 41{\cdot}52$.
25. $7296 \div 47$. **26.** $312400 \div 76{\cdot}53$. **27.** $49{\cdot}35 \div 17{\cdot}21$.
28. $13{\cdot}69 \div 13{\cdot}61$. **29.** $200{\cdot}5 \div 20{\cdot}5$. **30.** $493{\cdot}7 \div 92{\cdot}41$.
31. $(2{\cdot}732)^3$. **32.** $(13{\cdot}91)^3$. **33.** $(9{\cdot}675)^5$.
34. $(2007)^{\frac{1}{3}}$. **35.** $\sqrt[3]{100}$. **36.** $\sqrt[3]{21{\cdot}08}$.

§ 172. We next give examples of a rather more complicated form.

Example.

$13{\cdot}25 \times 12{\cdot}43 \times 10{\cdot}04 = 1653$
$= \underline{1650}$ to 3 sig. figs.*

No.	Log.
13·25	1·1222
12·43	1·0944
10·04	1·0017
	3·2183

Example.

$\dfrac{87{\cdot}2 \times 64}{123{\cdot}7 \times 20{\cdot}09} = 2{\cdot}247$
$= \underline{2{\cdot}25}$ to 3 sig. figs.

No.	Log.	
87·2	1·9405	
64	1·8062	
Numerator		3·7467
123·7	2·0923	
20·09	1·3029	
Denominator		3·3952
		0·3515

Note that the two logarithms required for the final subtraction are conveniently written in a third column.

Example.

$37{\cdot}39 \times 3{\cdot}042 + \dfrac{37{\cdot}39}{3{\cdot}042}$
$= A + B$ say
$= 113{\cdot}7 + 12{\cdot}28$
$= 125{\cdot}98$
$= \underline{126}$ to 3 sig. figs.

No.	Log.	
37·39	1·5727	
3·042	0·4832	
A	2·0559	by addition.
B	1·0895	by subtraction.

In every case the framework for the calculation should be constructed with all items in the first column before the tables are used.

EXERCISE XVII c

Find the value to 3 significant figures of:

1. $3{\cdot}773 \times 6{\cdot}512 \times 8{\cdot}786$. **2.** $17{\cdot}53 \times 24{\cdot}69 \times 26{\cdot}04$.
3. $6{\cdot}41 \times 7{\cdot}435 \times 4{\cdot}914$. **4.** $86{\cdot}74 \times 13{\cdot}76 \times 21{\cdot}92$.
5. $213{\cdot}6 \times 111{\cdot}5 \times 31{\cdot}7$. **6.** $1{\cdot}754 \times 9{\cdot}863 \times 11{\cdot}78$.
7. $12{\cdot}87 \times 1{\cdot}629 \times 30{\cdot}27 \times 1{\cdot}014$.

* See § 160, Example 2.

8. $1{\cdot}23 \times 2{\cdot}34 \times 3{\cdot}45 \times 4{\cdot}56$.

9. $\dfrac{76 \times 18{\cdot}4}{17{\cdot}4}$. **10.** $\dfrac{19{\cdot}3 \times 23{\cdot}6}{14{\cdot}2 \times 8{\cdot}43}$. **11.** $\dfrac{38{\cdot}56 \times 40{\cdot}2}{31{\cdot}99 \times 5{\cdot}267}$.

12. $\dfrac{48{\cdot}74 \times 54{\cdot}16}{6{\cdot}255 \times 46{\cdot}18}$. **13.** $\dfrac{11{\cdot}57 \times 12{\cdot}09}{10{\cdot}41 \times 11{\cdot}32}$. **14.** $\dfrac{7{\cdot}694 \times 231{\cdot}5}{11{\cdot}64 \times 9{\cdot}634}$.

15. $27{\cdot}34 \times 8{\cdot}37 + 27{\cdot}34 \div 8{\cdot}37$.

16. $27{\cdot}68 \times 7{\cdot}624 - \dfrac{27{\cdot}68}{7{\cdot}624}$.

17. $5{\cdot}147 \times 5{\cdot}341 + 3{\cdot}792 \times 6{\cdot}913$.

18. $72{\cdot}35 \times 31{\cdot}94 - 212{\cdot}5 \div 1{\cdot}672$.

19. $31{\cdot}64 \div 2{\cdot}69 + 43{\cdot}94 \div 15{\cdot}17$.

20. $15{\cdot}61 \div 3{\cdot}246 - 273{\cdot}4 \div 62{\cdot}34$.

21. $2{\cdot}43 \times 6{\cdot}94 \times 3{\cdot}61 - 2{\cdot}41 \times 6{\cdot}92 \times 3{\cdot}59$.

22. $100{\cdot}3 \times 27{\cdot}4 \times 19{\cdot}3 - 99{\cdot}8 \times 26{\cdot}9 \times 18{\cdot}8$.

23. $\left(\dfrac{7{\cdot}95}{4{\cdot}321}\right)^2$. **24.** $\left(\dfrac{84{\cdot}32}{7{\cdot}63}\right)^3$. **25.** $\left(\dfrac{93{\cdot}61}{7{\cdot}3 \times 11{\cdot}4}\right)^3$.

26. $\left(\dfrac{41{\cdot}2 \times 39{\cdot}7}{625{\cdot}2}\right)^3$. **27.** $\left(\dfrac{97{\cdot}64}{29{\cdot}49}\right)^{\frac{1}{3}}$. **28.** $\sqrt[3]{\dfrac{3{\cdot}495}{2{\cdot}751}}$.

29. $\sqrt{\dfrac{93{\cdot}41 \times 23{\cdot}72}{324{\cdot}5}}$. **30.** $\sqrt[3]{\dfrac{23{\cdot}91}{4{\cdot}69 \times 4{\cdot}96}}$.

31. $\sqrt[3]{\dfrac{3 \times 72{\cdot}5}{4 \times 3{\cdot}142}}$. **32.** $\sqrt[3]{\dfrac{14{\cdot}96 \times 7{\cdot}35}{11{\cdot}41 \times 9{\cdot}24}}$.

EXERCISE XVII d

Positive Indices only

Take $\pi = 3{\cdot}1416 = 10^{0{\cdot}4971}$. *Give results to* 3 *sig. figs.*

1. The volume of a cylinder is given by the formula $V = \pi r^2 h$, where h is the height and r the base-radius.

Find V when (i) $h = 2{\cdot}4$ in., $r = 1{\cdot}45$ in.; (ii) $h = 4{\cdot}6$ cm., $r = 7{\cdot}6$ cm.; (iii) $h = 13$ cm., $r = 25{\cdot}7$ cm.

2. The volume of a sphere is given by $\frac{4}{3}\pi r^3$, where r is the radius.

Find V when $r =$ (i) $2{\cdot}5$ in., (ii) $17{\cdot}4$ cm., (iii) $31{\cdot}65$ cm.

3. The radius r of a sphere whose volume V is known is given by the formula $r = \sqrt[3]{\dfrac{3V}{4\pi}}$. Find r when

$V =$ (i) 1000 c.c., (ii) 5 cub. ft., (iii) $14{\cdot}5$ cub. in.

4. The volume of a hollow cylinder is given by the formula $V = \pi l (R + r)(R - r)$, where l is the length, R the external and r the internal radius. Find V when

(i) $l = 13$ cm., $R = 5{\cdot}3$ cm., $r = 4{\cdot}2$ cm.;
(ii) $l = 1{\cdot}45$ in., $R = 7{\cdot}6$ in., $r = 3{\cdot}7$ in.;
(iii) $l = 141{\cdot}7$ cm., $R = 12{\cdot}6$ cm., $r = 10{\cdot}7$ cm.

5. Find the number of acres in a field measuring 100 yds. each way.

6. Find the area (in acres) of a rectangular field whose length is 473 yds. and breadth 247 yds.

7. Find the number of acres in a rectangular field 447 yards long and 258 yards wide.

8. The area of a rectangular field is $3\frac{3}{4}$ acres, and the length is 156 yards. Find the breadth.

9. Calculate the volume of a cuboid (rectangular block) whose dimensions are (i) 6·72 m., 4·56 m., 3·24 m.; (ii) 4 ft. 1 in., 3 ft. 2 in., 1 ft. 4 in. (in cubic feet).

10. Find the amount of air in a room whose length is 27·4 ft., breadth 14·7 ft., and height 12·3 ft.

11. Find the volume of a cube whose edge is 1·74 dm.

12. Given that 1 in. = 2·54 cm., find (i) the number of sq. cm. in 1 sq. in., (ii) the number of cub. cm. in 1 cub. in.

13. The circumference of the earth is 40,000,000 metres; how many miles is this? (1 m. = 39·37 in.)

14. The *Royal Sovereign* cost £902,600 and was sold out of the service for £40,000. Express the selling price as a percentage of the cost correct to 3 significant figures.

15. Given that 1 lb. = 453·593 grams, find the number of lbs. in a kilogram.

16. How many planks, each $13\frac{1}{2}$ ft. long and 9 in. wide, will be required for the construction of a platform 100 ft. long and 70 ft. wide?

17. The bill for painting a wall 21′ by 11′ 9″ is £9. Find the cost per square foot to the nearest penny.

18. The frontage of a piece of building land is 57 yds., and the price is £254, at £123 per acre. Find how far back the site extends.

19. Calculate the radius of a circle whose circumference is (i) 478 miles, (ii) 27·5 feet.

20. Calculate to three significant figures the circumference of the earth, taking radius = 3963 miles.

21. Find the area of a circle of radius 5·72 cm.

22. Calculate the area of a circle whose circumference is 25,000 miles.

23. Calculate (in yards) the diameter of a circular pond whose area is 1 acre.

24. Find the radius and circumference of a circle whose area is (i) 6 sq. in., (ii) 765 sq. cm.

25. A grant of £40 is to be divided among two parishes in proportion to their populations; these are 4327 and 2916 respectively. Find, to the nearest shilling, the share of each parish.

26. The mass of the earth is 6×10^{21} tons; the mass of the earth is to that of the moon as 81·5 : 1. Find the moon's mass.

27. A cubic foot of water weighs 62·43 lbs. Find the weight of a cubic foot of mercury (sp. gr. 13·6).

28. Find the weight of a bar of aluminium, measuring 12·7 cm. by 4·6 cm. by 3·8 cm., of sp. gr. 2·56.

29. A block of lead weighs 5·7 lbs.: find the weight of a piece of copper of half the size. (Sp. gr. of lead = 11·4, of copper = 8·9.)

30. Find the weight of a cylinder of copper 5·41 cm. high and 3·48 cm. in diameter. (Sp. gr. 8·9.)

31. Find the weight in tons of a block of stone (sp. gr. 2·56) measuring 5·3 ft. by 4·7 ft. by 2·1 ft.

32. Find to the nearest cwt. the weight of a rectangular bar of metal 6 ft. 3 in. long, 4 ft. 9 in. wide and 1 ft. $1\frac{1}{2}$ in. thick, if its sp. gr. is 7·89.

33. The volume of a cone is given by the formula $\frac{1}{3}\pi r^2 h$, where h = height, r = radius of base. Find the volume of a cone of height 6·3 cm. and base-radius 3·7 cm.

Negative Indices

EXERCISE XVII e

Use logarithms to simplify the following and check by ordinary multiplication or division or by using tables of squares and square roots.

1. 0·567 × 2. **2.** 0·0345 × 3. **3.** 0·00124 × 50.

4. 0·1752 × 600. **5.** 0·0934 × 0·5. **6.** 7·123 × 0·006.

7. 0·007123 × 6. **8.** 0·2439 × 0·09. **9.** $\frac{624}{6000}$.

10. $\frac{42}{110}$. **11.** $\frac{0{\cdot}0425}{2}$. **12.** $\frac{0{\cdot}6521}{30}$.

13. $\frac{0{\cdot}00724}{500}$. **14.** $\frac{427}{0{\cdot}05}$. **15.** $\frac{2{\cdot}46}{0{\cdot}003}$.

16. $\frac{0{\cdot}0072}{0{\cdot}8}$. **17.** $\frac{0{\cdot}724}{0{\cdot}08}$. **18.** $\frac{0{\cdot}102}{0{\cdot}003}$.

19. $(0{\cdot}246)^2$. **20.** $(0{\cdot}0129)^2$. **21.** $(0{\cdot}967)^2$.

22. $(0{\cdot}00862)^4$. **23.** $(0{\cdot}0127)^{\frac{1}{2}}$. **24.** $(0{\cdot}0006351)^{\frac{1}{2}}$.

25. $\sqrt{0{\cdot}0952}$. **26.** $\sqrt[4]{0{\cdot}000654}$. **27.** $\sqrt{0{\cdot}427}$.

28. $\sqrt{0{\cdot}0065}$. **29.** $\sqrt[4]{0{\cdot}246}$. **30.** $\sqrt[4]{0{\cdot}0462}$.

EXERCISE XVII f (oral)

Simplify the following, making the decimal part of the result positive:

1. $2{\cdot}4 + \bar{2}{\cdot}3$. **2.** $\bar{1}{\cdot}2 + \bar{3}{\cdot}5$. **3.** $4{\cdot}6 + \bar{1}{\cdot}2$.

4. $\bar{2}{\cdot}6 + 1{\cdot}1$. **5.** $2{\cdot}7 + \bar{1}{\cdot}5$. **6.** $3{\cdot}5 + \bar{2}{\cdot}8$.

7. $4{\cdot}2 + \bar{1}{\cdot}9$. **8.** $2{\cdot}9 + \bar{1}{\cdot}9$. **9.** $0{\cdot}5 + \bar{2}{\cdot}7$.

10. $1{\cdot}4 + \bar{2}{\cdot}9$. **11.** $1{\cdot}5 - 3{\cdot}2$. **12.** $2{\cdot}9 - 4{\cdot}6$.

13. $0{\cdot}8 - 1{\cdot}5$. **14.** $2{\cdot}1 - 3{\cdot}2$. **15.** $0{\cdot}7 - 2{\cdot}9$.

16. $5{\cdot}3 - \bar{2}{\cdot}1$. **17.** $4{\cdot}9 - \bar{4}{\cdot}9$. **18.** $0{\cdot}3 - \bar{2}{\cdot}1$.

19. $3{\cdot}4 - \bar{3}{\cdot}3$. **20.** $1{\cdot}5 - \bar{2}{\cdot}6$. **21.** $0{\cdot}7 - \bar{3}{\cdot}9$.

22. $2{\cdot}4 - \bar{3}{\cdot}8$. **23.** $\bar{1}{\cdot}3 - 2{\cdot}1$. **24.** $\bar{2}{\cdot}2 - 3{\cdot}0$.

25. $\bar{1}{\cdot}5 - 3{\cdot}7$. **26.** $\bar{2}{\cdot}6 - 3{\cdot}8$. **27.** $\bar{1}{\cdot}3 - \bar{2}{\cdot}1$.

28. $\bar{2}{\cdot}2 - \bar{3}{\cdot}0$. **29.** $\bar{1}{\cdot}5 - \bar{3}{\cdot}7$. **30.** $\bar{2}{\cdot}6 - \bar{3}{\cdot}8$.

31. $\bar{1}{\cdot}3 \times 2$. **32.** $\bar{1}{\cdot}6 \times 2$. **33.** $\bar{2}{\cdot}1 \times 5$.

34. $\bar{2}{\cdot}2 \times 5$. **35.** $\bar{1}{\cdot}6 \times 3$. **36.** $\bar{3}{\cdot}9 \times 2$.

37. $\bar{3}{\cdot}9 \times 3$. **38.** $\bar{1}{\cdot}5 \times 10$. **39.** $\bar{2}{\cdot}7 \times 10$.

40. $\bar{1}{\cdot}9 \times 9$. **41.** $\bar{3}{\cdot}6 \div 3$. **42.** $\bar{2}{\cdot}8 \div 2$.

43. $\bar{3}{\cdot}8 \div 2$. **44.** $\bar{1}{\cdot}8 \div 2$. **45.** $\bar{1}{\cdot}7 \div 3$.

46. $\bar{3}{\cdot}6 \div 4$. **47.** $\bar{4}{\cdot}5 \div 5$. **48.** $\bar{6}{\cdot}4 \div 7$.

49. $\bar{4}{\cdot}4 \div 6$. **50.** $\bar{2}{\cdot}0 \div 10$. **51.** $\bar{2}{\cdot}4 \div 10$.

52. $\bar{9}{\cdot}6 \div 12$. **53.** $\bar{5}{\cdot}4 \div 2$. **54.** $\bar{5}{\cdot}2 \div 3$.

EXERCISE XVII g

Use logarithms to find the value of:

1. $27{\cdot}68 \times 0{\cdot}7624$. **2.** $0{\cdot}0872 \times 0{\cdot}868$.

3. $0{\cdot}02572 \times 0{\cdot}1463$. **4.** $0{\cdot}02367 \times 0{\cdot}6297$.

5. $0{\cdot}7276 \times 0{\cdot}01571$. **6.** $0{\cdot}4372 \times 0{\cdot}003179$.

7. $946{\cdot}7 \times 0{\cdot}09764$. **8.** $3{\cdot}244 \times 0{\cdot}3154$.

9. $0{\cdot}06325 \times 0{\cdot}4309$. **10.** $0{\cdot}3849 \times 0{\cdot}7830$.

Also of:

11. $\dfrac{0{\cdot}001463}{0{\cdot}2572}$. **12.** $\dfrac{0{\cdot}7276}{0{\cdot}01571}$. **13.** $\dfrac{0{\cdot}002367}{0{\cdot}6297}$. **14.** $\dfrac{0{\cdot}4372}{0{\cdot}003179}$.

15. $\dfrac{0{\cdot}09764}{0{\cdot}9467}$. **16.** $\dfrac{3{\cdot}244}{0{\cdot}3154}$. **17.** $\dfrac{1{\cdot}24}{0{\cdot}0243}$. **18.** $\dfrac{0{\cdot}259}{0{\cdot}0765}$.

19. $\dfrac{0{\cdot}2694}{0{\cdot}0357}$. **20.** $\dfrac{0{\cdot}07293}{0{\cdot}09254}$.

EXERCISE XVII h

Use logarithms to find the value of:

1. $(\cdot 0178)^2$.	**2.** $(\cdot 2657)^3$.	**3.** $(\cdot 4635)^2$.
4. $(\cdot 07931)^2$.	**5.** $(\cdot 007432)^2$.	**6.** $(\cdot 3946)^3$.
7. $(\cdot 8143)^3$.	**8.** $(\cdot 7462)^4$.	**9.** $(\cdot 06925)^3$.
10. $(\cdot 8888)^5$.	**11.** $(\cdot 6)^7$.	**12.** $(\cdot 5492)^8$.
13. $(\cdot 3065)^6$.	**14.** $(\cdot 5604)^5$.	**15.** $(\cdot 3564)^4$.
16. $\sqrt[2]{\cdot 2222}$.	**17.** $(\cdot 04371)^{\frac{1}{2}}$.	**18.** $\sqrt{\cdot 007642}$.
19. $\sqrt{\cdot 0005321}$.	**20.** $\sqrt[3]{\cdot 7243}$.	**21.** $(\cdot 03271)^{\frac{1}{3}}$.
22. $(\cdot 002121)^{\frac{1}{3}}$.	**23.** $\sqrt[3]{\cdot 0007729}$.	**24.** $\sqrt[4]{\cdot 3569}$.
25. $\sqrt[4]{\cdot 06473}$.	**26.** $(\cdot 00576)^{\frac{1}{4}}$.	**27.** $(\cdot 0004962)^{\frac{1}{4}}$.
28. $\sqrt[4]{\cdot 00007994}$.	**29.** $\sqrt[5]{\cdot 2543}$.	**30.** $(\cdot 007647)^{\frac{1}{5}}$.
31. $(\cdot 1572)^{\frac{3}{4}}$.	**32.** $\sqrt[3]{(\cdot 0012)^2}$.	**33.** $(\cdot 04632)^{\frac{2}{7}}$.
34. $(\cdot 7691)^{\frac{5}{3}}$.	**35.** $(\cdot 6947)^{\frac{3}{5}}$.	**36.** $(\cdot 1193)^{\frac{3}{2}}$.
37. $\left(\frac{743\cdot 9}{\cdot 9472}\right)^2$.	**38.** $\left(\frac{\cdot 0432}{\cdot 000693}\right)^{\frac{1}{2}}$.	**39.** $\left(\frac{\cdot 0124}{\cdot 3251}\right)^{\frac{1}{3}}$.
40. $\left(\frac{\cdot 5764}{\cdot 0245}\right)^{\frac{2}{3}}$.	**41.** $\left(\frac{0\cdot 5678}{0\cdot 6789}\right)^{\frac{3}{5}}$.	

§ 173. The following Example shows convenient arrangements of logarithmic computation. See also p. 349.

Example.

$$\frac{17\cdot 53 \times (0\cdot 15)^3}{3\cdot 95 \times \sqrt{19\cdot 4}} = 0\cdot 003400$$

$= \underline{0\cdot 00340}$ to 3 sig. figs.

The two logarithms which are used in the final operation are conveniently distinguished by an * to assist the eye in subtracting.

No.	Log.	Log.
17·53		1·2437
0·15	$\bar{1}\cdot 1761$	
$(0\cdot 15)^3$		$\bar{3}\cdot 5283$
num.		$\bar{2}\cdot 7720$*
3·95		0·5966
19·4	1·2878	
$\sqrt{19\cdot 4}$		0·6439
denom.		1·2405*
		$\bar{3}\cdot 5315$

EXERCISE XVII i

Find the value of:

1. $\dfrac{3{\cdot}547 \times 0{\cdot}631}{0{\cdot}0547}$.

2. $\dfrac{72{\cdot}63 \times 4{\cdot}735}{0{\cdot}7157}$.

3. $\dfrac{0{\cdot}5729 \times 0{\cdot}00397}{3{\cdot}754}$.

4. $\dfrac{7{\cdot}592 \times 0{\cdot}03492}{0{\cdot}5947}$.

5. $\dfrac{1{\cdot}357}{0{\cdot}01234 \times 1{\cdot}64}$.

6. $\dfrac{43{\cdot}27 \times (0{\cdot}012)^2}{0{\cdot}00263}$.

7. $\dfrac{2{\cdot}493}{0{\cdot}5615 \times 0{\cdot}0695}$.

8. $\dfrac{0{\cdot}4372}{1{\cdot}978 \times 0{\cdot}8463}$.

9. $\dfrac{7{\cdot}918}{3{\cdot}465 \times 0{\cdot}0987}$.

10. $\dfrac{1{\cdot}007}{1{\cdot}017 \times 1{\cdot}027}$.

11. $\dfrac{49{\cdot}69 \times 0{\cdot}6503}{7{\cdot}059 \times 0{\cdot}9845}$.

12. $\dfrac{0{\cdot}7384 \times 0{\cdot}0469}{1{\cdot}047 \times 0{\cdot}6982}$.

13. $\dfrac{495 \times 0{\cdot}243}{0{\cdot}4378 \times 1{\cdot}798}$.

14. $\dfrac{21{\cdot}42 \times 367{\cdot}8 \times 1{\cdot}066}{522{\cdot}9 \times 7{\cdot}624}$.

15. $\dfrac{87{\cdot}65 \times 9{\cdot}378 \times 0{\cdot}9742}{10{\cdot}99 \times 0{\cdot}009164}$.

16. $\dfrac{27{\cdot}4 \times 37{\cdot}6 \times 0{\cdot}02}{35{\cdot}7 \times 8{\cdot}625 \times 0{\cdot}007}$.

17. $\dfrac{{\cdot}6492 \times (2{\cdot}973)^2}{{\cdot}319 \times {\cdot}02465}$.

18. $\left(\dfrac{{\cdot}001257 \times {\cdot}1423}{{\cdot}7649 \times {\cdot}0135}\right)^{\frac{1}{2}}$

19. $\left\{\dfrac{{\cdot}1357 \times {\cdot}7652}{2{\cdot}431 \times {\cdot}7934}\right\}^{\frac{1}{3}}$.

20. $\dfrac{3{\cdot}1416}{{\cdot}325 \times ({\cdot}076)^2}$.

21. $\dfrac{(\ 3215)^3}{{\cdot}127 \times ({\cdot}05672)^2}$.

22. $\sqrt{\dfrac{12{\cdot}72 \times 5{\cdot}327}{84}}$.

23. $\dfrac{(11{\cdot}2)^{\frac{1}{3}} \times (0{\cdot}625)^{\frac{2}{5}}}{224}$.

24. $\sqrt{\dfrac{6{\cdot}77 \times 5{\cdot}57}{8{\cdot}69 \times 9{\cdot}89}}$.

25. $\dfrac{(0{\cdot}0327)^{\frac{2}{3}} \times (0{\cdot}015)^{\frac{1}{5}}}{0{\cdot}0027}$.

26. $\dfrac{(0{\cdot}3456)^{\frac{2}{3}} \times 34{\cdot}56}{(0{\cdot}3456)^{\frac{1}{5}}}$.

27. $\sqrt[5]{\dfrac{12{\cdot}99 \times 36{\cdot}02}{3{\cdot}571}}$.

28. $\dfrac{(27{\cdot}9)^{\frac{3}{5}} \times (0{\cdot}089)^{\frac{2}{3}}}{(0{\cdot}1753)^{\frac{1}{2}}}$.

29. $\dfrac{2{\cdot}32 \times 4{\cdot}769 \times (2{\cdot}35)^2}{1{\cdot}732 \times ({\cdot}1469)^3}$.

30. $\dfrac{(19{\cdot}4)^3 \times ({\cdot}0375)^{\frac{1}{3}}}{({\cdot}72)^{\frac{1}{5}} \times \sqrt{3606} \times 100}$.

31. $\dfrac{\sqrt[3]{8{\cdot}476} \times 4{\cdot}73 \times (1{\cdot}351)^{\frac{2}{3}}}{23{\cdot}67 \times \sqrt[3]{{\cdot}9834} \times (1{\cdot}732)^5}$.

32. $\pi\,(0{\cdot}095^2 - 0{\cdot}067^2) \times 18{\cdot}3$.

33. $\frac{4}{3}\pi\,(1{\cdot}75^3 - 0{\cdot}75^3)$.

34. $1000\sqrt{\dfrac{(0{\cdot}625)^5}{0{\cdot}57 \times 48}}$.

If the sides of a triangle are a, b, c, and the area Δ, and if $2s = a + b + c$, then $\Delta^2 = s(s-a)(s-b)(s-c)$. Find s and Δ in each of the following cases:

35. $a = 254,\quad b = 372,\quad c = 137.$

36. $a = 2{\cdot}322,\quad b = {\cdot}314,\quad c = 2{\cdot}590.$

37. $a = {\cdot}0147,\quad b = {\cdot}2563,\quad c = {\cdot}2432.$

Find the value of $ab + cd$, where

38. $a = 4{\cdot}72,\quad b = 5{\cdot}41,\quad c = 7{\cdot}24,\quad d = 1{\cdot}26.$

39. $a = {\cdot}7926,\quad b = {\cdot}0724,\quad c = {\cdot}00523,\quad d = 4{\cdot}297.$

40. $a = {\cdot}4361,\quad b = 9{\cdot}427,\quad c = {\cdot}0537,\quad d = 72{\cdot}49.$

EXERCISE XVII j

1. Find the volume of a sheet of metal 41·3 inches long, 25·7 inches wide and 0·375 inches thick.

2. Find the weight of a metal plate 4 ft. 7 in. long, 3 ft. 10 in. wide and $\frac{5}{8}$ in. thick if a cubic foot of the metal weighs 477 lbs.

3. At a certain point in its course a river has an average depth of 15·62 feet and it is 38·4 yards wide. If it flows at the rate of 2·634 miles per hour, find the volume of water in cubic yards which passes every minute.

4. A railway cutting is 18·23 ft. wide at the bottom and 37·21 ft. wide at the top. It is 28·3 ft. high and 450·2 yards long. Find the weight in tons of the material excavated, given that a cubic foot weighed on an average 321 lbs.

5. Given that a cubic foot of water weighs 1000 oz., find the weight of water in tons which fell upon a field of 17·84 acres on a day when the rainfall was 0·67 inch.

6. Find to the nearest hundredth of a ton the weight of a block of metal 5·24 feet by 3·65 feet by 1·29 feet if its specific gravity is 7·8 (see p. 188).

7. Find the volume and total surface area of a prism whose cross-section is a right-angled triangle whose shorter sides are a in. and b in. and whose height is h in. when

(1) $a = 3{\cdot}142,\quad b = 5{\cdot}316,\quad h = 25{\cdot}62.$

(2) $a = 28{\cdot}32,\quad b = 31{\cdot}64,\quad h = 34{\cdot}63.$

8. Two cubes whose edges are 3·46 in. and 5·72 in. are melted and recast in the form of a cube. What is the length of its edge?

9. A block of hewn sandstone, found in some ancient ruins, measured 20 ft. 3 in. in length, its cross-section being $10' \ 6'' \times 9' \ 5''$. Find the weight of the block to the nearest ton, given that 1 cu. ft. of the sandstone weighs 135 lbs.

10. Given that 1 metre = 39·37 in., convert:

(i) 35·7 m. into inches. (ii) 527 m. into yards.
(iii) 1 Km. into miles. (iv) 1 in. into cm.
(v) 1 yard into metres. (vi) 4·7 in. into mm.

11. Given that 1 cu. foot of water weighs 1000 oz. and that 1 in. = 2·54 cm., find the number of lbs. in 1 kilogram.

12. "A pint of pure water weighs a pound and a quarter"; a litre of water weighs 1 kilogram; how many pints are there in a litre?

13. Find the volume of a cylinder whose diameter is 4·234 in. and whose height is 28·32 in. Find also the area of its curved surface.

14. The volume of a cylinder 14·31 inches high is 420·2 cub. in. Find its diameter.

15. The curved surface of a cylinder 17·21 cm. high is 331·2 sq. cm. Find its radius.

16. A cylinder made of metal of sp. gr. 7·82 weighs 567 gm. and is 13 cm. high. Find its radius.

17. Find to the nearest pound the weight of a 5-mile length of copper wire 0·03 in. in diameter. The weight of a cubic foot of copper is 560 lbs.

18. A kilometre of telephone wire weighs 28,000 gm. and the sp. gr. of copper is 8·9. Calculate the area and diameter of the cross-section of the wire.

19. Find the height of a tin whose diameter is equal to its height if it contains 1 cu. foot.

20. The external diameter of the barrel of an air-gun is found to be 1·62 cm. and the diameter of the bore is 0·51 cm. Find the area of metal exposed in a cross-section of the barrel.

21. The bore of a pipe made of metal 0·42 in. thick is 31·34 in. Find the weight in tons of 1 mile of piping if 1 cu. ft. of metal weighs 482·3 lbs.

22. What is the weight of a new pencil 7·26 in. long if the diameter of the pencil is 0·35 in. and the diameter of the lead is 0·13 in. A cubic inch of the wood weighs 0·34 oz. and of lead weighs 1·36 oz.

23. In a narrow cylindrical glass tube there is a column of mercury 47 mm. long weighing 0·5855 gm. Find its diameter if the sp. gr. of mercury is 13·59.

24. Find the weight in tons of a hollow steel shaft 20 feet long whose external diameter is 20 inches and internal diameter 16 inches, if the weight of 1 cu. in. of steel is 0·288 lb.

25. A packet of 17 candles weighs 6 lbs.: each candle is 28 in. long and $\frac{3}{4}$ in. in diameter. Given that 1 cu. ft. of water weighs 62·4 lbs., what is the sp. gr. of the wax?

The volume of a sphere is $\frac{4}{3}\pi r^3$ *where r is the radius.*

26. A 12-inch cube is melted down and made into a sphere. Find its radius.

27. Three spheres of radii 4·21, 5·36, 7·38 inches are melted and recast as a sphere. Find its radius.

28. What is the weight of a ball of lead 2·314 in. in diameter if a cubic foot of lead weighs 698·2 lbs.?

29. What is the diameter of a ball weighing 1 ton if 1 cu. foot of the substance weighs 428 lbs.?

30. Find the weight in ounces of an ivory billiard ball $2\frac{1}{16}$ inches in diameter if the sp. gr. of ivory is 1·82.

31. What is the weight (in tons) of a hollow steel sphere of diameter 7·232 in. if the metal is 0·6 in. thick and 1 cu. foot of the metal weighs 431·2 lbs.?

32. A hollow sphere of external diameter 10 inches and made of metal 1 in. thick is melted down and recast as a solid sphere. Find the diameter of the solid sphere.

33. Find in lbs. the weight of a spherical shell whose internal diameter is 10 inches, external diameter 12 inches and sp. gr. 11·2.

34. From a cubic inch of lead 15,300 shot are made. Find the radius of a shot correct to two significant figures.

35. Brass is a mixture of two parts (by weight) of copper to one part of zinc. The specific gravity of copper is 8·788, of zinc 7·190; find that of brass.

36. Bronze contains 90 per cent. (by weight) of copper and 10 per cent. of tin. Sp. gr. of copper is 8·788, of tin 7·291; find that of bronze.

Sound travels in still air at 32° *F. at the rate of* 1090 *ft. per second, and* 1·1 *ft. per second faster for every degree Fahrenheit increase in temperature.*

37. A gun is fired at a fort from a distance of a mile; there is no wind, and the temperature is 57° F. The shell travels at 1900 ft. per second, and bursts when it hits the fort. How much sooner is the explosion of the shell heard at the fort than that of the gun?

38. Sound travels in water about 4·26 times as fast as in air. How many seconds earlier would the sound of a torpedo exploded under water two miles away reach you by water than by air, the air being at freezing point?

39. The total weight in lbs. which can be raised by a spherical balloon filled with hydrogen is approximately $0{\cdot}524D^3\ (A - G)$, where D is the diameter of the balloon in feet, A is the weight in lbs. of a cubic foot of air, and G the weight in lbs. of a cubic foot of hydrogen. Find, to the nearest cwt., the weight which could be raised by a spherical balloon 50 feet in diameter if a pound of air occupies 13·1 cubic feet and a pound of hydrogen 180 cubic feet.

40. The volume of a Rugby football is given by the formula $\frac{4}{3}\pi ab^2$, where $2a$ is the distance between the ends and b the mid-sectional radius. Calculate the volume of a ball in which $a = 6{\cdot}28''$, $b = 4{\cdot}23''$.

41. A pendulum of length l feet is known to make a complete swing in $2\pi\sqrt{\frac{l}{g}}$ seconds; where $g = 32$. Find the time of a complete swing for a pendulum 1 metre long.

42. The number of ton-inches of twisting moment that a hollow steel shaft can safely endure is given by the formula

$$T = \frac{7\pi}{16}\left(\frac{D^4 - d^4}{D}\right)$$

where D is the external and d the internal diameter in inches. Find the possible twisting moment (T) when the external diameter is $6\frac{5}{8}$ in. and the internal diameter is $3\frac{5}{16}$ in.

43. The formula $c = 1350d^2\sqrt{\frac{d}{0{\cdot}5L}}$ may be used for finding the number of cu. ft. of gas which pass through a gas main per hour; c denoting the number of cu. ft., d the diameter of the main in inches, and L its length in yards. If the diameter of the main is $2\frac{1}{2}$ inches and its length 1 mile, how many cu. ft. are discharged per hour?

CHAPTER XVIII

INTEREST

§174. If a business man borrows money, he usually has to pay for the use of it. He pays a certain percentage per year; this is called **interest.** Thus if he borrows £400 at 5 per cent. interest for 3 years, he pays £4 × 5 each year, or £4 × 5 × 3 (= £60) for the 3 years. Finally he has to repay the £400 borrowed.

He hopes to be able to pay this interest by trading with the borrowed money and earning profit at a rate higher than 5 per cent.

A business man who *lends* money usually receives interest for its use.

The sum lent or borrowed is called the **Principal.** As a rule the interest is paid periodically (e.g. every year or every half-year). But if the interest is saved up and added to the Principal at the end of the time, the total is called the **Amount.** Thus the amount of £400 at 5 per cent. **simple interest** after 3 years is £400 + £60 = £460.

§175. In simple interest, the same interest is payable each year. But the transaction is sometimes one of **compound interest.** In this case, the lender says to the borrower "I lend you £400 at 5 per cent. compound interest. At the end of the first year you will owe me £20 interest. Instead of paying me this, you shall continue to owe it; but you shall pay interest on this £20 as well as the £400. In the second year you shall pay interest on £420": this works out to £21 (for 5 % is 1*s.* in the £). At the end of the second year the sum owed will be £441. For the third year interest is charged on £441; this will be £22. 1*s.* 0*d.*; and the amount owed at the end of 3 years will be £463. 1*s.* 0*d.*

Comparing this with the amount for the same rate and time at simple interest, viz. £460, we see that a sum mounts up more rapidly at compound than at simple interest (see graph on p. 371).

Simple Interest

§ 176. Given the Principal, Rate per cent. and Time, to find the Simple Interest.

EXERCISE XVIII a (oral)

What is the Simple Interest in the following cases?

	Principal £	Rate %	No. of years		Principal £	Rate %	No. of years
1.	100	5	1	**11.**	50	3	2
2.	100	5	2	**12.**	1	5	1
3.	100	3	4	**13.**	3	5	1
4.	100	$2\frac{1}{2}$	2	**14.**	1	5	2
5.	200	4	1	**15.**	2	5	3
6.	300	3	2	**16.**	10	5	4
7.	400	4	5	**17.**	20	8	3
8.	200	$2\frac{1}{2}$	2	**18.**	150	3	4
9.	300	4	$\frac{1}{2}$	**19.**	250	4	5
10.	200	4	$2\frac{1}{2}$				

§ 177. *Example.* Find the simple interest on £564. 14*s.* 6*d.* at $3\frac{3}{8}$ % for 4 years 2 months.

20)14·5*s.*

100)£564·725

5·64725 = no. of hundreds

$3\frac{3}{8}$

16·94175

1·41181 (= $\frac{1}{4}$)

·70591 (= $\frac{1}{8}$)

£19·05947 = int. for 1 year

4

£76·23788 = int. for 4 years

3·17658 = int. for 2 months

£79·414,46

20

8,·28*s.*

12

3·36*d.*

Notice:

(1) How we multiply by $3\frac{3}{8}$.

(2) How we deal with 2 months (= $\frac{1}{6}$ of a year).

(3) That we want 3 places of decimals in the final answer, and are therefore safe if we keep 5 throughout the calculation.

Interest to nearest penny = £79. 8*s.* 3*d.*

EXERCISE XVIII b

Find the simple interest in the following cases, the answer to be correct to the nearest penny.

	Principal £	s.	d.	Rate %	No. of years
1.	250	0	0	4	5
2.	180	0	0	5	3
3.	120	0	0	5	$1\frac{1}{4}$
4.	160	0	0	$3\frac{1}{2}$	2
5.	210	0	0	$3\frac{1}{2}$	$2\frac{1}{2}$
6.	500	0	0	$4\frac{3}{4}$	$3\frac{1}{4}$
7.	800	0	0	$3\frac{1}{2}$	2 months
8.	140	10	0	4	8
9.	215	5	0	$3\frac{1}{2}$	2
10.	714	10	0	$3\frac{3}{4}$	$2\frac{1}{2}$
11.	815	2	6	$4\frac{1}{4}$	$3\frac{1}{8}$
12.	411	17	6	$3\frac{1}{8}$	$4\frac{1}{4}$
13.	915	18	0	4	$2\frac{1}{4}$
14.	14	16	0	4	3 months
15.	12	18	6	4	8 months
16.	513	8	4	3	$5\frac{1}{2}$
17.	297	7	11	$3\frac{5}{8}$	4
18.	3457	3	4	$3\frac{5}{16}$	25
19.	771	19	$6\frac{1}{2}$	$2\frac{3}{4}$	14
20.	25	4	11	$3\frac{1}{8}$	$2\frac{1}{2}$
21.	999	9	9	$2\frac{1}{4}$	35
22.	5	14	$7\frac{1}{2}$	$4\frac{1}{2}$	100
23.	218	11	5	15	2 years 5 months
24.	1257	8	10	$7\frac{3}{16}$	5

§ 178. Given the Principal, Rate per cent. and Interest, to find the Time.

EXERCISE XVIII c (oral)

What is the number of years in the following cases?

	Principal £	Rate %	Interest £		Principal £	Rate %	Interest £
1.	100	5	20	**6.**	1000	$3\frac{1}{2}$	70
2.	50	6	18	**7.**	400	$4\frac{3}{4}$	57
3.	200	4	24	**8.**	10	8	4
4.	150	4	6	**9.**	350	6	$10\frac{1}{2}$
5.	250	2	60	**10.**	2000	3	15

§ 179. *Example.* In what time (to the nearest month) will £520 earn £50 interest at $4\frac{1}{2}$ %?

$$£\frac{520}{100} \times 4\tfrac{1}{2} \text{ interest is earned in 1 year,}$$

$$\therefore \text{ £50 interest is earned in } 50 \div \frac{520 \times 4\frac{1}{2}}{100} \text{ years}$$

$$= 50 \times \frac{200}{520 \times 9} \text{ years} = \frac{1000}{468} \text{ years.}$$

The time is 2 years 2 months.

```
468)1000(2 years
     936
      64
      12
     768(2 months
```

EXERCISE XVIII d

Find the time, to the nearest month, in the following cases:

	Principal £	s.	d.	Rate %	Interest £	s.	d.
1.	150	0	0	4	15	0	0
2.	200	0	0	$2\frac{1}{2}$	7	10	0
3.	470	0	0	3	84	12	0
4.	62	10	0	5	25	0	0
5.	53	0	0	$4\frac{1}{2}$	47	14	0
6.	640	0	0	$3\frac{1}{2}$	78	8	0
7.	275	0	0	$4\frac{3}{4}$	209	0	0
8.	182	16	0	4	20	0	0

	Principal £	s.	d.	Rate %	Interest £	s.	d.
9.	745	0	0	6	80	0	0
10.	237	0	0	5	24	10	0
11.	163	0	0	$4\frac{1}{2}$	17	5	0
12.	26	10	0	$3\frac{3}{4}$	5	10	0
13.	528	10	0	$5\frac{1}{2}$	70	5	0

Find the time, to the nearest month, in the following cases:

	Principal £	s.	d.	Rate %	Amount £	s.	d.
14.	260	0	0	$4\frac{1}{2}$	300	10	0
15.	65	0	0	$3\frac{1}{4}$	76	5	0
16.	431	10	0	5	492	15	0
17.	314	5	0	$4\frac{1}{2}$	345	5	0
18.	76	0	0	$6\frac{1}{2}$	88	2	6

§180. **Given the Principal, Interest and Time, to find the Rate per cent.**

EXERCISE XVIII e (oral)

What is the rate per cent. in the following cases?

	Principal £	No. of years	Interest £
1.	100	2	10
2.	200	1	5
3.	200	3	24
4.	200	1	9
5.	300	3	36
6.	1000	$2\frac{1}{2}$	250
7.	1000	$1\frac{1}{2}$	90
8.	500	$2\frac{1}{2}$	50
9.	500	$3\frac{1}{2}$	105
10.	250	3	30
11.	150	2	18

§ 181. *Example.* What rate per cent. will bring in £65. 10*s*. 0*d*. interest on £453 during $3\frac{1}{2}$ years?

If £453 produces £65·5 interest in $3\frac{1}{2}$ years,

$$£100 \text{ produces } £65{\cdot}5 \times \frac{100}{453} \text{ interest in } 3\tfrac{1}{2} \text{ years,}$$

and $$£100 \text{ produces } £65{\cdot}5 \times \frac{100}{453} \div 3\tfrac{1}{2} \text{ interest in 1 year.}$$

This is $$£\frac{65{\cdot}5 \times 100 \times 2}{453 \times 7} = £\frac{13100}{3171} = £4{\cdot}1\ldots$$

```
           4·1
3·171)13·1 00
      12·6 84
         4 160
         3,171
```

The rate of interest is therefore approximately 4·1 %.

EXERCISE XVIII f

Find the rate per cent., to one place of decimals, in the following cases:

	Principal £	s.	d.	No. of years	Interest £	s.	d.
1.	150	0	0	2	15	0	0
2.	50	0	0	10	12	10	0
3.	510	0	0	3	40	0	0
4.	265	0	0	4	42	10	0
5.	136	0	0	5	32	5	0
6.	347	0	0	$2\frac{1}{2}$	35	0	0
7.	613	0	0	$3\frac{1}{4}$	75	0	0
8.	25	10	0	4	5	0	0
9.	76	15	0	6	20	0	0
10.	1500	0	0	$4\frac{1}{2}$	240	5	0
11.	742	0	0	$3\frac{3}{4}$	130	10	0

Find the rate per cent., to one place of decimals, in the following cases:

	Principal £	s.	d.	No. of years	Amount £	s.	d.
12.	160	0	0	$2\frac{1}{2}$	181	0	0
13.	540	0	0	3	590	10	0
14.	122	0	0	5	153	5	0
15.	504	0	0	$3\frac{1}{4}$	678	15	0
16.	1400	0	0	$4\frac{1}{2}$	1643	0	0

§182. **Given the Rate per cent., Time and Interest, to find the Principal.**

EXERCISE XVIII g (oral)

What is the principal in the following cases?

	Rate %	No. of years	Interest in £
1.	5	1	40
2.	$2\frac{1}{2}$	1	10
3.	3	2	18
4.	4	2	24
5.	4	6	24
6.	6	3	36
7.	5	4	60
8.	$2\frac{1}{2}$	4	30
9.	$3\frac{1}{2}$	2	70
10.	$3\frac{1}{4}$	4	52

§183. *Example.* What sum (to the nearest £) will produce £45. 2*s.* 6*d.* interest in $3\frac{1}{4}$ years at $4\frac{1}{2}$ %?

£100 produces £$3\frac{1}{4} \times 4\frac{1}{2}$ in $3\frac{1}{4}$ years at $4\frac{1}{2}$ %.

$\therefore$ £$100 \times \dfrac{45\frac{1}{8}}{3\frac{1}{4} \times 4\frac{1}{2}}$ produces £$45\frac{1}{8}$ in $3\frac{1}{4}$ years at $4\frac{1}{2}$ %.

This is £$\dfrac{100 \times 45\frac{1}{8} \times 8}{13 \times 9}$

$= £\dfrac{100 \times 361}{117}$

$=$ £309 about.

$$\begin{array}{r} 308{\cdot}5 \\ 1{\cdot}17\overline{)361} \\ 351 \\ \hline 1000 \\ 936 \\ \hline 640 \\ 585 \end{array}$$

EXERCISE XVIII h

Find the principal (to the nearest £) in the following cases:

	Rate %	No. of years	Interest £	*s.*	*d.*
1.	3	5	46	10	0
2.	4	3	32	15	0
3.	6	2	84	17	6

	Rate %	No. of years	Interest £	s.	d.
4.	5	$4\frac{1}{2}$	37	5	0
5.	4	$3\frac{1}{4}$	52	2	6
6.	$3\frac{1}{2}$	3	65	10	0
7.	$4\frac{1}{4}$	6	135	15	0
8.	$3\frac{1}{4}$	$5\frac{1}{2}$	43	0	0
9.	$5\frac{1}{2}$	$4\frac{3}{4}$	64	10	0
10.	$3\frac{3}{4}$	$1\frac{1}{3}$	10	2	6

§184. Given the Rate per cent., Time and Amount, to find the Principal.

Example. What sum will amount to £700 in 3 years at 4 %?

£100 amounts to £112 in 3 years at 4 %,

$\therefore$ the principal is $\frac{100}{112}$ of the amount,

$\therefore$ if £700 is the amount,

$$£\frac{100}{112} \times 700 \text{ is the principal,}$$

$\therefore$ sum required = £625.

$$\begin{array}{r} 625 \\ 1{\cdot}12)\overline{700} \\ \underline{672} \\ 280 \\ \underline{224} \\ 560 \\ 560 \end{array}$$

EXERCISE XVIII i

Find the Principal (to the nearest £) in the following cases:

	Rate %	No. of years	Amount £	s.	d.
1.	3	4	65	0	0
2.	4	2	128	0	0
3.	6	5	637	0	0
4.	$2\frac{1}{2}$	4	43	10	0
5.	$3\frac{1}{2}$	6	176	15	0
6.	$4\frac{3}{4}$	8	417	5	0
7.	$5\frac{1}{2}$	3	84	2	6
8.	7	$2\frac{1}{2}$	53	7	6

EXERCISE XVIII j

Miscellaneous Exercises on Simple Interest

1. Find the simple interest on £564 for 3 years at $3\frac{1}{2}$ %.

2. At what rate will £76 produce £7 interest in 3 years?

3. In how many years will £436 earn £56. 10*s*. interest at $4\frac{3}{4}$ %?

4. What sum amounts to £1700 in 4 years at $2\frac{3}{4}$ %?

5. Find the amount of £526. 10*s*. at $2\frac{1}{2}$ % after 3 years.

6. In what time will £62/7/6 become £80 at $3\frac{1}{4}$ %?

7. At what rate must £534. 15*s*. be invested in order to amount to £700 after 9 years?

8. A sum is invested on a boy's 16th birthday with the object of producing £400 for him on his 21st birthday: the rate of simple interest is $2\frac{3}{4}$ %. What is the sum?

9. A man bought a house for £1200 and lets it for £70 a year rent. After 5 years he sells the house for £1068. Deducting his loss from the rents, what yearly interest on his capital has he earned?

10. What difference in amount (to nearest penny) of Simple Interest shall I obtain by lending £92. 17*s*. 6*d*. for 2 years at $5\frac{1}{2}$ per cent. or 6 per cent. per annum?

11. Find the difference between the Simple Interest on £498. 17*s*. for $4\frac{1}{2}$ years at 3 % per annum and at 5 % per annum.

12. The simple interest on a certain sum for $4\frac{1}{2}$ years at $3\frac{1}{2}$ per cent. exceeds that on the same sum for 3 years at 5 per cent. by £4. 7*s*. What is the sum?

13. Two-thirds of my capital is invested at 3 % and the rest at $4\frac{1}{2}$ %. My income is £140 per annum. What is my capital?

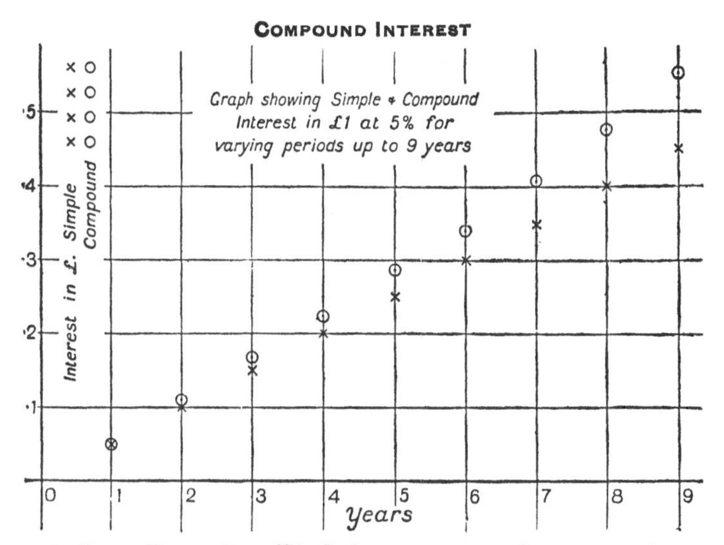

§ 185. *Example.* Find the amount and interest of £462 accumulating at $4\frac{1}{2}$ % Compound Interest for 3 years.

First, find what £1 becomes at the end of 3 years.

At the end of 1 year, £1 has become £1·045.

At the beginning of the second year the principal is £1·045. At the end of the second year this has increased to £1·045 × 1·045 or £$(1{\cdot}045)^2$. At the end of the 3rd year this has increased to £$(1{\cdot}045)^3$.

Since £1 becomes £$(1{\cdot}045)^3$ in 3 years, £462 becomes £462 × $(1{\cdot}045)^3$ in 3 years.

To obtain the answer to the nearest penny, we must work for 3 places of decimals and retain 5.

The amount is

£527·219 = £527/4/5.

The Compound Interest is £65/4/5.

Note that the Simple Interest for the same period would be £62/7/5.

```
462
  1·045
---------
462·
 18 48
  2 310
---------
482·79
  1·045
---------
482·79
 19 31 1 6
  2 41 3 9 5
------------
504·51,5,5,5
  1·04 5
------------
504·51 5 5 5
 20 18 0 6 0
  2 52 2 5 5
------------
527·21 8 7 0
```

If an answer correct to 3 significant figures only is sufficient, it is better to simplify the expression £462 × $(1·045)^3$ by 4-figure logarithms.

No.	Log.
1·045	0·0191
$(1·045)^3$	0·0573
462	2·6646
527·1	2·7219

It works out to £527·1: but the 1 is not trustworthy. On the whole, we cannot give the answer beyond the nearest £, say £527. To obtain the shillings and pence, we should have to use at least 7-figure tables. But for practical purposes, a result to the nearest £ may be enough.

§186. In business, compound interest is calculated by a table such as that on p. 374.

EXERCISE XVIII k

1. Find the amount and compound interest in the following cases. If the method of contracted multiplication is used, answers should be to the nearest penny: if logarithm tables or compound interest tables are used, the answers should be only to the degree of accuracy allowed by such tables (not more than 4 and 6 significant figures respectively, and the last figure subject to error).

	Principal £	*s.*	*d.*	Rate %	No. of years
(1)	120	0	0	4	2
(2)	115	0	0	2	3
(3)	105	4	0	5	3
(4)	108	10	0	5	4
(5)	230	0	0	4	2
(6)	415	8	6	3	3
(7)	715	10	6	4	4
(8)	10537	0	0	$2\frac{1}{2}$	2
(9)	10	0		$3\frac{1}{4}$	3
(10)	118	11	6	5·2	4
(11)	127	18	9	$4\frac{3}{4}$	5
(12)	31	2	3	4	3
(13)	2116	7	11	$3\frac{1}{2}$	2

2. Find from the compound interest table on p. 374 the compound interest on:

(1) £43. 7*s*. 4*d*. for 5 years at $3\frac{1}{2}$ per cent.

(2) £2479. 16*s*. 3*d*. for 10 years at 4 per cent.

(3) £15645 for 7 years at $2\frac{1}{2}$ per cent.

3. Using the table on p. 374 and contracted division, find what sum will amount at compound interest to

(1) £56. 4*s*. 8*d*. in 6 years at 5 per cent.

(2) £7843. 10*s*. 6*d*. in 4 years at $6\frac{1}{2}$ per cent.

(3) £246842 in 3 years at $2\frac{1}{2}$ per cent.

4. Compare the simple and the compound interest on £21. 10*s*. at the end of 3 years, at 4 per cent. per annum.

5. If the annual increase of a city is 2·5 per cent. and the present number of inhabitants is 2,624,000, what will be the population in two years' time? What was it a year ago?

6. *A* lends *B* £20,000 at 5 per cent. interest, which *B* omits to pay. Every five years an account is taken and the simple interest due is added to the principal. Find to the nearest £100 the amount of the debt after 30 years.

7. An import duty of £4 a ton is placed on the raw material of an article which passes through the hands of six manufacturers, each of whom expects to make 5 per cent. profit on his outlay. What is the rise in price to the consumer owing to the tax (to the nearest shilling)?

Table showing the amount of £1 at compound interest for various periods.

Year	2 per cent.	$2\frac{1}{2}$ per cent.	3 per cent.	$3\frac{1}{2}$ per cent.	4 per cent.
1	1·02000	1·02500	1·03000	1·03500	1·04000
2	1·04040	1·05063	1·06090	1·07123	1·08160
3	1·06121	1·07689	1·09273	1·10872	1·12486
4	1·08243	1·10381	1·12551	1·14752	1·16986
5	1·10408	1·13141	1·15927	1·18769	1·21665
6	1·12616	1·15969	1·19405	1·22926	1·26532
7	1·14869	1·18869	1·22987	1·27228	1·31593
8	1·17166	1·21840	1·26677	1·31681	1·36857
9	1·19509	1·24886	1·30477	1·36290	1·42331
10	1·21899	1·28009	1·34392	1·41060	1·48024
11	1·24337	1·31209	1·38423	1·45997	1·53945
12	1·26824	1·34489	1·42576	1·51107	1·60103
13	1·29361	1·37851	1·46853	1·56396	1·66507
14	1·31948	1·41297	1·51259	1·61870	1·73168
15	1·34587	1·44830	1·55797	1·67535	1·80094
16	1·37279	1·48451	1·60471	1·73399	1·87298

Year	$4\frac{1}{2}$ per cent.	5 per cent.	$5\frac{1}{2}$ per cent.	6 per cent.	7 per cent.
1	1·04500	1·05000	1·05500	1·06000	1·07000
2	1·09203	1·10250	1·11303	1·12360	1·14490
3	1·14117	1·15763	1·17424	1·19102	1·22504
4	1·19252	1·21551	1·23882	1·26248	1·31080
5	1·24618	1·27628	1·30696	1·33823	1·40255
6	1·30226	1·34010	1·37884	1·41852	1·50073
7	1·36086	1·40710	1·45468	1·50363	1·60578
8	1·42210	1·47746	1·53469	1·59385	1·71819
9	1·48610	1·55133	1·61909	1·68948	1·83846
10	1·55297	1·62889	1·70814	1·79085	1·96715
11	1·62285	1·71034	1·80209	1·89830	2·10485
12	1·69588	1·79586	1·90121	2·01220	2·25219
13	1·77220	1·88565	2·00577	2·13293	2·40985
14	1·85194	1·97993	2·11609	2·26090	2·57853
15	1·93528	2·07893	2·23248	2·39656	2·75903
16	2·02237	2·18287	2·35526	2·54035	2·95216

CHAPTER XIX

STOCKS AND SHARES

§187. Suppose that the Canadian Government, or the City of Glasgow, or the Southern Railway Company wants to borrow money, say £1,000,000: in other words to "raise a loan." The loan is divided into convenient units, perhaps of £100, perhaps of £5; and people willing to lend the money are invited to apply for one or more of these units. In some cases these parts are called **shares**: in other cases we should speak of £100 **stock**; generally fractions of shares are not obtainable, but stock to any odd amount in £ *s*. *d*. may change hands.

§188. If a person is willing to lend £100 say, or to **buy £100 stock**, the borrowing party (Government or city or public company) may undertake to pay him interest at an agreed rate*, say 4 per cent. per annum; and perhaps undertakes to pay back the £100 at some definite future date. The lender receives from the borrower a **certificate**, inscribed on which he finds these undertakings. If the loan is divided into units of £100, and the lender buys £300 stock, he would become the possessor of three certificates of £100 each (or perhaps of one certificate for £300).

As long as he retains these certificates, he can claim his agreed interest yearly (or half-yearly or quarterly, as the case may be). But he cannot get his money back from the borrower till the agreed date for the repayment, or **redemption** of the loan. In some cases no date of redemption is fixed (e.g. in most loans to the British Government, and in some loans to British railways).

* In the case of "ordinary shares" no definite rate of interest is promised: the lender entrusts his money to the borrowing company and hopes that the business may be successful enough to pay him a good interest; the lenders are in fact partners in the business.

§189. How then is a borrower to act if he wants his money back before the date of redemption?

He can sell the certificate to anyone willing to buy it. The buyer of the certificate, i.e. the new possessor of the stock or shares, now receives the interest, and the former lender has no further concern with the matter.

§190. What price will be paid for the certificate?

This depends on various circumstances. If the seller is very anxious to sell, he may have to be content with receiving less than the original £100 (or whatever may be the original unit). For instance, in time of war many people are anxious to have cash rather than pieces of paper, and tend to sell their "securities"; and for a similar reason there is a scarcity of buyers. This will bring down the price of all securities; a man may find that he cannot get more than £80 for a certificate marked £100. The stock is then said to be **at a discount of £20.**

§191. Observe the consequence to the buyer. He buys £100 of 4 % stock for £80. But the original borrower will still have to pay £4 a year interest to the owner of the certificate. The latter therefore has paid £80, and receives £4 a year interest. This works out to $\frac{100}{80} \times 4$ per cent. on his investment; i.e. to 5 %. Thus, by buying at a discount, an investor will be securing more than the nominal rate of interest. The **yield** on the investment is £5 per cent.

§192. Under other circumstances, securities may sell at a price above their original face value. Thus, some years ago the City of Auckland in New Zealand borrowed £182,300 in bonds of £100. To get the money, the city had to offer 6 % interest. As the city increased in population and people were increasingly confident as to its ability to pay its debts, they began to argue that 6 % was a very attractive rate of interest, and became desirous of buying these bonds. The price accordingly rose, and later the £100 bonds were changing hands at about £120; they stood **at a premium of £20.** At this price, they yielded $\frac{100}{120} \times 6$ % interest; i.e. 5 %; and this is about the rate at which the City of Auckland could then borrow. Notice that an investor buying at a premium obtains less than the nominal rate of interest on the money he invests.

§ 193. A man wishing to buy stock generally commissions a broker to buy for him; and the broker, when he has bought the stock, sends his client a contract note such as that shown below.

LONDON, 28 *August*, 1914.

Bought for John Doe, Esq., £400 3 % 1*st Mortgage Sterling Bonds Grand Trunk Pacific Railway Co.* @ 73

Stamp		292	0	0
	Stamps		1	0
R. Roe & Co.	*Fees*			
	Brokerage	1	0	0
Members of the Stock Exchange, London.		£293	1	0

Subject to the Rules and Regulations of the Stock Exchange.

It will be noticed that the buyer has to make extra payments in the shape of stamps (a government tax), fees (to the company for the trouble of transferring stock to a new name) and brokerage (to the broker for managing the purchase). These charges vary for different stocks and may be ignored in working the exercises that follow.

EXERCISE XIX a (oral)

1. If £100 stock costs £90, how much can be bought for (1) £45, (2) £180, (3) £30, (4) £9? If the stock is paying 6 per cent., what will be the annual income in each case? What does (1) £200 stock, (2) £150 stock, (3) £10 stock cost?

For how much could you sell (1) £50 stock, (2) £300 stock, (3) £10 stock?

2. A company's shares, nominal value £4, and paying 5 per cent., are worth £3. How many shares can be bought for (1) £15, (2) £150? What income accrues in each case?

What will 25 shares cost and what income will they yield?

For how much could you sell (1) 15, (2) 20, (3) 100 shares?

§ 194. *Example* 1. For £1000, what quantity of Japanese 4 % Sterling Loan can be bought, the price being $82\frac{1}{2}$?

£100 nominal costs £82·5,

i.e. the nominal value is $\frac{100}{82\cdot5}$ of the real value.

∴ £1000 buys £1000 × $\frac{100}{82\cdot5}$ nominal = $\underline{£1212}$ to the nearest £.

Example 2. What is the yield per cent. on North Eastern Railway Consols, when the interest for the year is 6 % and the price is $129\frac{1}{4}$?

£$129\frac{1}{4}$ buys £100 Stock and brings in £6 per annum.

$$\therefore \text{ £100 brings in £}\frac{100}{129\frac{1}{4}} \times 6 \text{ per annum}$$

$$= \text{£}4\cdot642$$

$$= \text{£}4.\ 12s.\ 10d.$$

The yield per cent. is therefore $\underline{\text{£}4.\ 12s.\ 10d.}$

Example 3. At what price will $2\frac{1}{2}$ % Consols yield a return of $3\frac{1}{2}$ %?

If £$3\frac{1}{2}$ is the interest on £100 invested,

$$\text{£}2\tfrac{1}{2} \text{ is the interest on £}100 \times \frac{2\frac{1}{2}}{3\frac{1}{2}} = \frac{\text{£}500}{7} = \text{£}71\tfrac{3}{7}.$$

If I can buy £100 Consols for £$71\frac{3}{7}$, I receive $3\frac{1}{2}$ % on my investment. The price of Consols must therefore be $\underline{71\frac{3}{7}}$.

Example 4. What annual income is secured by investing £3000 in Birmingham Small Arms Preference Shares (5 %) when the £5 shares stand at $5\frac{7}{16}$?

The number of £5 shares that can be bought for £3000 at $5\frac{7}{16}$ is $\frac{3000}{5\frac{7}{16}}$.

Each £5 share pays 5*s*. a year interest.

∴ the annual income received is $\frac{3000}{5\frac{7}{16}} \times 5$ shillings, or

$$\text{£}\frac{3000}{4 \times 5\frac{7}{16}} = \text{£}\frac{3000}{20 + \frac{7}{4}} = \text{£}\frac{12000}{87} = \underline{\text{£}138} \text{ approx.}$$

The Exercise XIX b below is retained in its original form. Exercise XIX c has been added as being more up to date.

	Name of Security	Price on Jan. 29, 1914	%
A.	British Bonds (Consols)	76	$2\frac{1}{2}$
B.	Birmingham Corporation Stock	94	$3\frac{1}{2}$
C.	Belfast Harbour Stock	79	$3\frac{1}{2}$
D.	Dominion of Canada Bonds	92	$3\frac{1}{2}$
E.	Johannesburg Municipal Loan	92	4
F.	Argentine Government	103	5
G.	Russian Government	$123\frac{1}{2}$	5
H.	German Imperial	78	3
I.	Midland Railway debentures	$66\frac{1}{2}$	$2\frac{1}{2}$
J.	Canadian Pacific debentures	97	4
K.	Guinness Brewery preference	$146\frac{1}{2}$	6
L.	Harrod's Stores preference £5	$5\frac{3}{4}$	5
M.	Imperial Tobacco Co. preference £1 ..	$1\frac{5}{16}$	$5\frac{1}{2}$
N.	Hudson's Bay Co. preference £5 ..	$5\frac{11}{16}$	5
O.	Armstrong Whitworth preference £5 ..	$4\frac{3}{8}$	4
P.	Shell Transport preference £10	$10\frac{1}{2}$	5

EXERCISE XIX b

(*Where no prices are quoted, use those given in the preceding list.*)

1. What is the cost of (i) £3000 Consols, (ii) £450 Russians, (iii) £350 Harrod's preference, (iv) £730 Shell preference?

2. What amount is realised by the sale of (i) £820 Belfast Harbour Stock, (ii) £215 Hudson's Bay preference?

3. What amount of the following can be bought for the sums given, viz. (i) of British Consols for £3000, (ii) of Russian Government loan for £450, (iii) of Harrod's preference for £350, (iv) of Shell preference for £730, (v) of Hudson's Bay preference for £215, (vi) of Armstrong Whitworth preference for £750?

4. Calculate the yield per cent. of each of the named securities at the prices quoted.

5. At what price will (i) Consols yield $3\frac{3}{4}$ %, (ii) Birmingham Corporation Stock yield 4 %, (iii) Johannesburg Municipal loan yield $4\frac{1}{2}$ %, (iv) Argentine Government loan yield $4\frac{1}{2}$ %, (v) German Imperial loan yield $3\frac{3}{4}$ %, (vi) Midland Railway debentures yield $4\frac{1}{4}$ %, (vii) Guinness preference yield $4\frac{3}{4}$ %, (viii) Imperial Tobacco preference yield 5 %, (ix) Armstrong Whitworth preference yield $5\frac{1}{4}$ %?

6. What income is obtained from each of the investments mentioned in Exs. 1 and 2 above?

7. What income is obtained from each of the investments mentioned in Ex. 3 above?

8. What sum must be invested in each of the above securities to produce an income of £400 a year?

9. The shares of a certain company, nominal value £1, can be bought for 17*s.* 10*d.* What will be the cost of 1000 shares, and what income should I derive from them if the shares are paying $2\frac{3}{4}$ per cent.?

10. A man wishes to raise £1000, in order to invest it in stock paying 6 per cent., at 104. What would be the change in his income if he procured the money by

(1) selling out of Consols at 70,

or (2) accepting a loan of £1000 at 3 per cent.?

11. Find the change of income on selling out £20,000 stock at 90 paying 2 per cent., and buying railway stock at 108 paying 3 per cent.

12. A man buys $2\frac{1}{2}$ % Consols on Nov. 4, 1914, at $68\frac{5}{8}$. The income he receives is classed as "unearned" and taxed at 1/7 in the £. How much per cent. does he finally receive on his investment (to the nearest penny)?

EXERCISE XIX c

1. Find the cost of £3300 Consols at 55.

2. What amount of Consols can be purchased for £3300?

3. Calculate the income from £3300 Consols paying $2\frac{1}{2}$ %.

4. What is the income from £3300 invested in $2\frac{1}{2}$ % Consols at 55?

5. What is the yield per cent. (to the nearest shilling) on $2\frac{1}{2}$ % Consols at 55?

6. At what price would $2\frac{1}{2}$ % Consols yield £3. 15*s*. 0*d*. per cent.?

7. What sum must be invested in $2\frac{1}{2}$ % Consols at 55 to obtain an income of £400?

8. Find the market value of £450 Southern Railway Stock at 85.

9. How much will be realised by selling £615 Stock at 111?

10. If the price of £100 Stock is 110, how much can be bought for £575. 17*s*. 0*d*.?

11. If £1220 is invested in Levers 7 % Debentures at 106, what is the nominal value (to the nearest £1) of the holding?

12. How many £1 shares at 22/6 can be bought for £360?

13. What is the price of a 6 % Stock if the yield is £3. 15*s*. 0*d*. per cent.?

14. An investment in Stock at 77 yields £4. 10*s*. 11*d*. per cent. What is the nominal rate of interest?

15. If the yield on a 5 % Stock is £4. 7*s*. 6*d*., what must be the market value of a £20 share (to the nearest shilling)?

16. If interest at the rate of £5. 2*s*. 6*d*. per cent. is obtained from an investment in a 6 % Stock, at what price (to the nearest penny) does a £1 share stand?

17. At what price (to the nearest penny) will a £5 share in $4\frac{1}{2}$ % Stock yield £6. 5*s*. 0*d*. per cent.?

18. What is the yield per cent. if a £1 share paying 15 % is quoted at 96/-?

19. If Lyons 8 % Pref. Shares (£1) are quoted at 28/3, how much (to the nearest £1) must be invested in them to produce £100 a year?

20. What is the difference in income derived by investing £1700 in (1) a $3\frac{1}{2}$ % Stock at 75, and (2) a 7 % Stock (£1 share) at 21/6?

Make use of the prices given below (June 1925) in dealing with the following questions.

$2\frac{1}{2}$ % Consols	57	8 % Dunlop Debentures	$109\frac{1}{2}$
4 % Funding Loan	88	Lever 7 % Pref. (£1)	21/6
$3\frac{1}{2}$ % London County	75	Argentine Gov. 4 %	72
5 % Egyptian Loan	70	Lyons 8 % Pref. (£1)	28/3
4 % L.M.S. Debentures	81	Imp. Tobacco $22\frac{1}{2}$ % (£1)	96/-

21. If £500 Consols are sold and the proceeds invested in L.M.S. Debs., what nominal amount of the latter can be purchased and what is the income before and after the transaction?

22. How many Imperial Tobacco Shares can be bought by selling 515 Levers 7 %? Calculate the change in income.

23. If 759 Imperial Tobacco Shares are sold and the money invested in Funding Loan, how much of the latter can be bought, and what is the change of income?

24. Capital is transferred from Egyptian Loan to London County Stock. How much of the first must be sold to purchase £1050 of the latter and what is the change in the yield?

25. What is the change of income if 1200 Dunlops are sold and Argentines bought with the proceeds? What is the nominal value of the purchase?

26. How much capital must be invested in Lyons Pref. in order to produce the same income as £800 Funding Loan?

27. 100 Imperials were bought at 96/- and sold later at 101*s.* 3*d.* What was the profit on the transaction?

28. £540. 13*s.* 9*d.* Argentine Stock is sold and the proceeds invested in Levers. How many shares can be purchased and what is the change in income?

29. 60 £5 shares in a Company paying 5 % were lately bought at £6. 5*s.* 0*d.* per share and are sold at £7. 10*s.* 0*d.* What is the increase in the value of the holding and the decrease in the yield per share?

30. What annual income will be obtained by investing £1700 in Funding Loan? If the money had been invested in Levers 7 % Pref., how many shares could be purchased and what income would be obtained?

31. A man has £2400 to invest. If he divides it equally between Consols and Imperials, what amount of each will he hold and what income will he receive?

32. £1000 Argentine Stock is bought at 72, and £500 more at 75. Find the total cost and the yield on the whole investment.

33. Calculate the yield on Consols and on Dunlop Debentures and, if possible, account for the difference. What is the difference in income if £1000 is invested in each?

34. What income will be obtained from £600 L.M.S. Debs. if Income Tax at the rate of 4/- in the £1 is deducted before receipt?

35. If Income Tax at 4/6 in the £1 is deducted from dividends on 850 Levers and £300 Funding Loan, what sum will the holder receive?

36. If a man who holds £400 Stock actually receives £24 after Income Tax at 4/- in the £1 has been deducted, what rate of interest was declared on the Stock?

37. What sum must have been set aside by the directors, if a Company distributes $15\frac{1}{2}$ % "Income Tax paid at 4/6" on every £100 Stock? What fraction of the £15. 10*s*. 0*d*. received by the shareholder must be added to it to obtain the sum set aside?

38. If a shareholder receives a dividend of £6. 8*s*. 0*d*. per cent. "Income Tax at 4/- paid," how much was set aside for the purpose?

39. We bought 720 Lyons 8 % Pref. at 28/3 and sold them later at 29/-. What profit was made altogether if interest for the half-year less Income Tax at 4/- has been paid while we held the Shares?

40. After paying 6 % on £250,000 Debentures, $5\frac{1}{2}$ % on £100,000 Preference Shares, putting £50,000 to Reserves, etc., and carrying £3500 forward to the next annual account, what dividend can a Company pay on their £300,000 Ordinary Shares if the profits for the year were £59,000? If the Company distributed dividends on the Ordinary Shares "Income Tax at 5/- paid," what would a holder of 1600 shares receive, and at what rate would the dividend be declared?

REVISION PAPERS

On Chaps. I–XIX

Paper 55

1. Use tables to find the squares of 8·635, 924·3, ·002572.

2. Express as powers of 10

7; 24; 0·469; 0·006393.

3. Find the square root of 1522756.

4. What percentage of a gallon is a pint?

5. What are the values of (i) 3 per cent. of 25 miles, (ii) 14·25 per cent. of 400 kilometres?

6. If $\frac{2}{p} = \frac{1}{4} + \frac{1}{16} + \frac{1}{176}$ show that p is very nearly equal to the ratio of the circumference to the radius of a circle.

7. Find to within a penny the simple interest on £252. 12*s*. 7*d*. for three years at $2\frac{1}{4}$ %.

8. The perimeter of a field is $\frac{1}{3}$ of a mile: how many square feet are there in the inside face of a wall, 7′ 6″ high, enclosing it?

9. If a man pays 1/$4\frac{1}{2}$ in the £ income tax, what per cent. of his income does he pay?

10. The dimensions of a rectangular block are 27′ 9″ by 16′ 6″ by 7′ 3″. Use logs to find its volume in cubic feet.

11. A man motors 623·7 miles in 4·5 days, travelling 10·3 hours a day. How many hours a day must he travel to cover 3274 miles in 31·6 days?

Paper 56

1. Use tables to find the square roots of 74·61, 1·764, ·03856.

2. Express as powers of 10

0·2345, 172·8, 3·142, 0·00701, 70·01.

3. Extract the square root of 137641.

4. Find the value of $4\frac{1}{2}$ % of £1284 to the nearest penny.

5. What percentage of a mile is 1000 yards?

6. One inch is very nearly 2·54 cm. In saying that 1 inch = $2\frac{1}{2}$ cm., what percentage of error is there?

7. What is the area of the largest circle that could be cut out of a square piece of paper of side 8 in.?

8. A glass cylinder is about 7 cm. in diameter and 25 cm. high. It contains either 250, 500 or 1000 cubic centimetres. Find which it is with as little *working* as possible, but make the *method* quite clear.

9. Find the compound interest on £500 for 4 years at 4 per cent. to the nearest pound.

10. Find to the nearest ounce the weight of a rectangular sheet of crown glass 2 ft. 10 in. long, by 1 ft. 11 in. wide, and 0·085 in. thick. You may reckon that a cubic foot of crown glass weighs 156 lbs.

11. A block of metal whose dimensions are 15·76″ by 4·83″ by 4·32″ weighed 47·63 kilograms. What will be its weight if cut down to 15·32″ by 4·65″ by 3·14″?

Paper 57

1. Find, correct to two places, the square root of 478·3.

2. Use tables to find the squares of 8·263, 724·8, ·5632 and the square roots of 8·263, 724·8, ·5632.

3. Write down the numbers whose logarithms are $\bar{1}$·2345, 2·88, 4·0069, $\bar{4}$·761, 3.

4. Find the value of $3\frac{3}{4}$ per cent. of £423. 13*s*. 4*d*.

5. The value of the exports of the United Kingdom to Montenegro in 1909 was £561; in 1910 the value was £4700. Find the percentage increase.

6. A boy finds the weight of a cubic centimetre of mercury to be 13·12 gm. The actual weight is 13·6 gm. What is the error per cent.?

7. A new pencil is 14 cm. long and 8 mm. in diameter. Find its volume. Calculate also the area of a piece of paper which will just cover the curved surface.

8. If an investment of £2588 brings in an annual income of £76. 12*s*. 0*d*., how much per cent. is this?

9. Find the amount of £150 in three years at 4 per cent. compound interest.

10. Find the length of a square plot of ground the area of which is 0·2589 of an acre.

11. Find to within an inch the depth of a cubical tank which holds 10,000 gallons. (A gallon contains 277 cu. in.)

Paper 58

1. Write down from your tables the value of (i) $\sqrt{35{\cdot}21}$; (ii) $\sqrt{{\cdot}04271}$; (iii) $\sqrt{57890}$; (iv) 525^2.

2. Calculate (*a*) $10^{3{\cdot}2307}$; (*b*) $10^{{\cdot}1168}$; (*c*) $10^{\bar{2}{\cdot}1471}$.

3. Find the value of $3\frac{1}{2}$ % of £26. 11*s*. 4*d*. to the nearest penny.

4. A year's world production of beetroot sugar was 9,115,000 tons, and of cane sugar 7,007,000 tons. What percentage of the whole is beetroot sugar (to 3 sig. figs.)?

5. 400 grams of bronze consisting of 85 % of copper, 10 % of zinc and 5 % of tin are fused with 100 grams of bell metal consisting of 75 % of copper and 25 % of tin. Find the percentages of copper, zinc and tin in the mixture.

6. Find the square root of $81\frac{49}{100}$, and the square of $9\frac{7}{10}$.

[*See next page*

7. How far does the tip of the long hand of a clock move between 11.15 and 11.35 if its length is 1 ft. 2 in.?

8. Calculate how many cubic feet of gas are contained in a cylindrical gasholder, 112 feet high, whose diameter is 22 yards. If the curved surface is painted red, find the area painted.

9. A cubic decimetre of air weighs 1·293 grams, a cubic centimetre of water weighs 1 gram, and iron is 7·492 times as heavy as water. Find the edge of a cube of iron, which has the same weight as 1000 cubic feet of air.

10. Which is the more economical: to burn electric lamps costing $\frac{1}{2}$*d.* an hour, which last 1000 hours, or to burn those costing $\frac{5}{8}$*d.* an hour, which last 800 hours, but give 25 per cent. more light, the initial cost of a lamp being the same in each case?

11. A certain country imported mineral oils to the value of £6,123,889. These oils being worth 1*s.* 5*d.* per gallon, find (to five significant figures) how many gallons were imported. What revenue, to four significant figures, is raised by a tax of 3*d.* per gallon on these oils?

Paper 59

1. Find the value of (1) $37{\cdot}37 \times 3{\cdot}042$, (2) $\sqrt{0{\cdot}025849}$, (3) $\sqrt{0{\cdot}321^2 + 0{\cdot}0312^2}$.

2. Work out $\sqrt{0{\cdot}721}$ correct to 3 significant figures.

3. Find 5 per cent. of 16 tons.

4. What is the difference in area between a circle of radius 27·4 cm. and a square with side equal to the diameter of the circle? What percentage of the area of the square is the area of the circle?

5. A boy using his tables reads 4·792 instead of 4·972. What is his error per cent.?

6. The *Mauretania* is 750 ft. long and displaces 41,550 tons of water. The *Aquitania* is 865 ft. long and displaces 49,430 tons of water. Express *each* of these increases as a percentage correct to 3 significant figures.

7. Each person in the U.K. consumes on the average 111 eggs per year, namely 54 produced at home and 57 imported. What percentage is home produce?

8. In what time will £586 amount to £673. 18*s.* at $2\frac{1}{2}$ per cent. Simple Interest?

9. Find in kilograms the weight of a kilometre of wire whose diameter is 1·4 mm., if 1 c.c. of the wire weighs 8 gm.

10. The number of adult fowls in the U.K. is:—England 13,774,000; Wales 1,249,000; Scotland 2,429,000; Ireland 11,710,000. What percentage of the whole does each country possess (to 1 place of decimals)?

Paper 60

1. Find the value of (i) $(1{\cdot}714 + 1{\cdot}22) \times 6{\cdot}724$,

$$\text{(ii)}\ \sqrt{9\tfrac{49}{64}},\qquad \text{(iii)}\ \sqrt{3{\cdot}368^2 + 0{\cdot}2439^2}.$$

2. What is $3\frac{1}{2}$ per cent. of 3 tons 15 cwts.?

3. An author receives from his publisher a "royalty" of so much for every copy sold. But for every 25 copies sold he receives royalty on 24 only. By how much per cent. does this reduce his receipts?

4. The value of the exports from the U.K. were:

£460,700,000 in 1906,
£518,000,000 ,, 1907,
£456,700,000 ,, 1908.

Give (correct to 2 significant figures) the percentage increase in 1907, and the percentage decrease in 1908, in each case on the value of the preceding year's exports.

[*See next page*

5. Find to one significant figure the percentage error (+ or −) in the following measurements made by a class:

	(*a*)	(*b*)	(*c*)	(*d*)	(*e*)	(*f*)	(*g*)
Correct measurement	612	612	612	5·80	5·80	0·136	0·136 cm.
Measurement obtained	615	610	60·4	5·66	5·90	0·148	0·135 cm.

6. A pint of water weighs $1\frac{1}{4}$ lbs.; one kilogram = 2·206 lbs.; one cubic centimetre of water weighs 1 gram. Find to three significant figures the number of pints in a litre.

7. Find to the nearest penny the compound interest on £40. 17*s.* 1*d.* in 2 years at 4 per cent.

8. How many thousand gallons (to the nearest thousand) of water will be required to fill a water main 6 miles long and 1 foot in diameter? (1 cu. ft. = 6·24 galls.)

9. In 1914 England imported from Ireland 5,738,000 "great hundreds" of eggs, of value £2,044,000; in 1912, 6,313,000 "great hundreds," of value £2,926,000. Find the value per "great hundred" (i.e. 120) in each year (to within $\frac{1}{4}d.$).

Paper 61

1. Simplify $\frac{0{\cdot}72 \times 0{\cdot}008}{0{\cdot}72 - 0{\cdot}008}$. Give the result (*a*) correct to 3 decimal places, (*b*) correct to 3 significant figures.

2. Find by logarithms the value of 28·36 × ·006724.

3. Calculate without tables the value of $\sqrt{28{\cdot}03^2 + 0{\cdot}601^2}$.

Verify this from your tables, and find the percentage error, correct to one significant figure, of the result obtained from the tables.

4. Pure water is composed of two gases, oxygen and hydrogen, in the proportion by weight of 88·9 to 11·1. What weight of each is there in a cubic foot of water?

5. Calculate to 3 significant figures:

(i) 4·78 % of 469500,

(ii) The percentage that 72·96 is of 98696.

6. The correct result of a problem was 4·55 metres.

One person who attempted it obtained an answer which was too large by 2 per cent.; what was his answer?

Also, what was the error per cent. of another who obtained an answer 4·37 metres?

7. Gunpowder contains 74·8 % nitre, 10·3 % sulphur, 13·9 % charcoal, 1 % water. Find to the nearest pound the weight of each of these in a ton of gunpowder.

8. A person made £20. 0*s*. 6*d*. profit on £78. 2*s*. 6*d*. in $1\frac{1}{2}$ years; what was the rate per cent. per annum of simple interest?

9. The food bill at a school for a certain number of days was £2192. 8*s*. 10*d*. for 394 boys; the average per boy per day was 1*s*. 5·584*d*. How many days were there?

10. Three dozen tobacco tins are packed into and just fit a rectangular cardboard box. Each tin is cylindrical in shape, the diameter being 6·7 cm. and the depth 4·1 cm. The tins are arranged in layers of a dozen each, with the circular ends horizontal. Find, to the nearest cubic centimetre, the vacant space in the box.

Paper 62

1. Find correct to 3 significant figures the values of: (i) $316{\cdot}2^2$, (ii) $\cdot003167^2$, (iii) $\sqrt{8\frac{1}{9}}$, (iv) $\sqrt{45786}$, (v) $\sqrt[4]{678{\cdot}3}$, (vi) $\sqrt{28{\cdot}7^2 + 8{\cdot}68^2}$, (vii) $\sqrt[3]{0{\cdot}0135}$, (viii) $\sqrt[3]{4}$.

2. The area of a square is 139·808 square metres; find its side correct to the nearest millimetre.

3. An aviator, maintaining an average speed of 45 miles per hour, made ten complete circuits of a track in ten minutes. Assuming these to be circular, find the diameter of the track in yards.

[*See next page*

4. A closed cylindrical tin, capable of containing a given volume, requires least material for its construction when the height of the tin is equal to the diameter of its circular base. Find the height of such a tin if the volume contained is 1 cubic foot.

5. Find the number of gallons of water held by a tank 8 feet long, 4 feet wide, and 3 feet 6 inches deep, with a pipe 14 inches diameter running through it parallel to the longest sides. (1 cubic foot = 6·25 gallons.)

6. A certain quality of coal leaves 4·1 per cent. by weight, of ash. If a day's steaming yields 6 tons 4 cwts. of ash, how much coal has been burnt (to the nearest ton)?

7. A quantity of chloroform is found when analyzed to consist of 12 gm. of carbon, 1 gm. of hydrogen, and 107 gm. of chlorine. What percentage of each element does it contain?

8. It is estimated that in a certain airship 100 cu. ft. of gas can lift 7·33 lbs. Find the lifting force after 10 hours of an airship which contains 353,166 cu. ft. of gas originally, assuming that it loses 1 % of its volume of gas in 24 hours.

9. What would it cost to feed 414 schoolboys for 12 weeks 5 days at an average cost of 1*s.* 7·165*d.* per boy per day?

Paper 63

1. Calculate $\sqrt{894{\cdot}01}$.

2. Find the value of (1) $\sqrt{68{\cdot}79^2 - 15{\cdot}96^2}$,

(2) $24{\cdot}89 \times 0{\cdot}6847$, (3) $\dfrac{0{\cdot}00478}{2{\cdot}549}$.

3. A mill-sail is 6 yards long and goes round uniformly 9 times per minute. At what rate (in miles per hour) does its extremity move?

4. A sheet of paper is 25·7 cm. long and 20·25 cm. wide. If it is rolled to form a cylinder, with the two longer edges just meeting, find the volume of the space enclosed.

5. A man does two pieces of work in 6 hours 30 minutes. If one of these occupies him for 60 per cent. of this time, how long does it take him?

6. The number of seamen in the Navy in April 1902 was 88,700. During the following year this number increased 19 per cent.; what was it in April 1903?

Between April 1901 and April 1902 it had increased 4 per cent. What was it in April 1901?

7. Mocha coffee costs 1*s*. 8*d*. per lb. in beans, and 1*s*. $8\frac{1}{2}$*d*. per lb. when ground. How much per cent. is added to the price for grinding?

8. A return journey by train is equivalent to $1\frac{1}{2}$ single journeys. If the single journey be charged at the rate of 3*d*. a mile and an additional 5 per cent. on this amount for government duty, find the distance from A to B, a return ticket costing 7*s*. $10\frac{1}{2}$*d*.

9. Find the simple interest on £21. 17*s*. 5*d*. at $4\frac{1}{4}$ per cent. in $2\frac{1}{2}$ years to the nearest penny.

10. The cost of feeding a school for a term was £2838. 7*s*. 10*d*., and the number of daily rations during the term was 35679 (a daily ration is one boy's food for one day). Find to $\frac{1}{10}$*d*. the average cost per boy per day.

Paper 64

1. Find the value of (1) $\sqrt{720{\cdot}4^2 + 308{\cdot}5^2}$, (2) $\sqrt[3]{0{\cdot}2456}$.

2. How often must one go round a square field of ten acres to run one mile?

3. A running track is 440 yards long (measured along the inner edge). It consists of 2 straight pieces of 2 semicircles each 110 yards long. The track is 15 feet wide. Find the cost of covering it with cinders at 1/6 a sq. yd.

4. A cylindrical beaker 10 cm. in height inside just holds a litre of water. Find the inside diameter correct to the nearest millimetre.

[*See next page*

5. To obtain full marks for a certain measurement, the error was not to be greater than 1 % of the correct answer. If the correct answer were 9·87 cm., within what limits must the measured answer be, to obtain full marks? Give the answer to 3 significant figures.

6. A candidate fails in his examination by 2 % of the required minimum; he obtains 1470 marks. What is the required minimum? If this minimum is 40 % of the total, what is the total?

7. Milk has a percentage composition by weight as follows: water 87·00, fatty solids 3·90, other solids 9·10. If to this be added water in the proportion of 25 parts by weight of water to every 100 parts of milk, what is the percentage composition of the mixture?

8. The distance of the star Castor from the earth is approximately $9{\cdot}59 \times 10^{13}$ miles. The velocity of light is 186,330 miles per second. Find the number of years taken by light from Castor to reach us.

9. An increase of the tea tax from 5*d.* to 8*d.* per lb. was expected to reduce the consumption by 5 %, but to bring in £3,200,000 more revenue. What was the revenue from this source before the increase?

Paper 65

1. Find the value of: (1) $(28{\cdot}36)^3$, (2) $\sqrt{4{\cdot}689} - \sqrt{0{\cdot}4689}$, (3) $\sqrt{345^2 - 34{\cdot}5^2}$.

2. What is the length of the side of a square to the nearest mm. when its area is 47·54 sq. cm.?

3. A cylindrical vessel is 7·5 cm. high and has a diameter 3·5 cm.—both inside measurements. 50 gm. of water are poured in. Find (*a*) what space remains unfilled, (*b*) the depth of the water.

4. Find the area included between two concentric circles whose diameters are 17·8 and 21·4 inches.

What is the diameter of a circle whose area is equal to that which you have just found?

5. Find the weight of a cast-iron pipe 3″ bore, 8 feet long, metal $\frac{1}{2}$″ thick. [1 cu. in. of cast iron weighs 0·26 lb.]

6. A man sells a house for £1100, and thereby gains 3 per cent. What did he pay for the house?

7. Find the percentage error made in reckoning a pound $\frac{5}{11}$ of a kilogram, given that an ounce is 28·35 gm.

8. The weight of gold that a ton of ore yields increases from 0·004 % to 0·0044 %. How many more grains per ton does the ore yield? [1 lb. = 7000 grains.]

9. In how many years will £472. 10*s*. amount to £600 at $4\frac{3}{4}$ % simple interest?

Paper 66

1. Calculate $\sqrt{11}$ correct to 2 decimal places.

2. Find the value of

(1) $\sqrt{12{\cdot}3^2 + 17{\cdot}08^2}$, (2) $\sqrt{0{\cdot}37^2 + 0{\cdot}53^2}$, (3) $\dfrac{234{\cdot}6 \times 38{\cdot}9}{47{\cdot}58}$.

3. A chain-testing machine has a full power of 380 tons, an amount which is 90 % more than the Admiralty proof strain. Find the Admiralty proof strain.

4. The population of a town is 11,200. What will it be in two years if it increases by $7\frac{1}{2}$ per cent. each year?

5. Gun-metal contains by weight 88 per cent. of copper, 10 per cent. of tin, and 3 per cent. of zinc. Find the cost of the metal in 5 tons of the alloy when copper costs £58. 2*s*. 6*d*. per ton, tin £127. 10*s*., and zinc £24. 7*s*. 6*d*.

6. Find to the nearest penny the Simple Interest for 6 months on £350 at $7\frac{1}{4}$ per cent.

Deducting from this interest 1*s*. 2*d*. in the £ income-tax, what sum will be finally received?

7. A cylindrical glass, 7 cm. in diameter, is filled with water up to 3 cm. from the top. How many 2-cm. cubes of metal must be dropped in to make the water overflow?

[*See next page*

8. A glass tube has an internal diameter of ·1 cm. and an external diameter of ·4 cm. If the tube were fused and re-made into a solid cylindrical rod of the same length, what would its new diameter be?

9. Water is poured into a tank 24 ft. 9 in. long, 9 ft. 4 in. broad, and 7 ft. 6 in. deep, at the rate of 12 galls. a second; find the rate (in inches per minute) at which the water rises in the tank. How long will it take to fill the tank? (One gallon contains 277 cubic inches.)

Paper 67

1. Find the values of: (i) $(4325)^2$, (ii) $(\cdot 00173)^2$, (iii) $\sqrt{298560}$, (iv) $\sqrt{0{\cdot}104^2 + 0{\cdot}094^2}$, (v) $\sqrt[3]{0{\cdot}1468}$.

2. How many yards of fencing are required to enclose a square field whose area is 48841 sq. yards?

3. If 1 cubic foot of water contains $6\frac{1}{4}$ gallons, find how many inches the surface of the water in a cylindrical tank (4′ in diameter) will be lowered when 22 gallons are drawn off.

4. A motor track consists of two straight parallel sides, each 286 yards long, and two semi-circular ends. What must be the distance between the sides, if it is just half a mile round the inside of the track?

5. Population of England and Wales.

	Population	Increase per cent.
1891	—	11·65 between 1881 and 1891.
1901	32,527,843	12·17 between 1891 and 1901.
1911	36,075,269	—

Find (to 3 significant figures) the population in 1891 and the increase per cent. for the period 1901 to 1911.

6. In an English county the deaths in 1899 were 51,298. This was a death-rate of 3·25 per cent. What was the total population?

7. £140 is divided amongst *A*, *B*, and *C*, so that their shares are in the ratio of 5 : 7 : 4. *A* and *B* then pay *C* $1\frac{1}{2}$ per cent. of their share. What does each of the three finally get?

8. At what rate per cent. will the interest on £530 amount in 5 years to £92. 15*s.* 0*d.*?

9. The diameters of the earth and Venus are 7920 and 7660 miles respectively. If the earth is represented by a billiard ball $2\frac{1}{16}''$ in diameter, find correct to 2 significant figures the diameter of a billiard ball that would represent Venus on the same scale.

Paper 68

1. Calculate $\sqrt{548{\cdot}9}$ correct to 3 significant figures.

2. Find the value of

(i) $\sqrt{47{\cdot}84^2 + 33{\cdot}68^2}$, (ii) $(0{\cdot}978)^2 - (0{\cdot}3245)^2$,

(iii) $(0{\cdot}0295)^3$.

3. An article which costs 7*s.* 6*d.* in London sells for 4 dollars in New York. Find, to the nearest penny, the profit made by a man who buys 1000 of these articles in London and sells them in New York after paying 5*d.* per article for freight, etc. and a customs duty of 20 per cent. on the New York selling price. [£1 = 4 dollars 85 cents.]

4. What is the weight of a piece of machinery, which would weigh 2 tons 8 cwts. 2 qrs., if its weight were reduced by $3\frac{1}{2}$ %?

5. A gallon contains 277 cu. in. If this is taken as 0·16 cu. ft., what is the error per cent. correct to 1 significant figure?

6. Analysis shows that 4·75 gm. of sugar contain 1·994 gm. of carbon, ·3056 gm. of hydrogen, and the rest oxygen. What percentage of oxygen does sugar contain?

7. The coal imported into Russia in the first 4 months of 1912 amounted to 43,491,000 "poods," and was valued at 5,819,000 roubles. Taking a rouble as 2*s.* $1\frac{1}{3}$*d.* and a pood as 36 lbs., estimate in English money, to the nearest sixpence, the value of the coal per ton.

8. Find the compound interest on fifteen million pounds at 5 per cent. for 4 years (correct to within £10,000).

9. If telegraph wire weighs 450 lbs. per mile and 1 cubic foot of the metal of which it is made weighs 484 lbs., show that its cross-section is roughly 1/40 square inch.

Paper 69

1. Write down the values of: $\sqrt{64}$, $\sqrt[3]{27}$, 3^4, 10^5, $(\cdot 2)^2$, $\sqrt{\cdot 16}$, $(1\cdot 2)^2$, $\sqrt[3]{\cdot 001}$.

2. Find the values of: (*a*) $\sqrt{\cdot 03456}$, (*b*) $(\cdot 005678)^2$, (*c*) $\sqrt{12\cdot 73^2 + 2\cdot 56^2}$, (*d*) $\dfrac{31\cdot 46 \times (\cdot 0215)^2}{(24\cdot 87)^3 \times 124\cdot 3}$.

3. What rate of gain is given by a selling price of £2. 11*s*. 4*d*., and a cost price of £2. 5*s*. 10*d*.?

4. A model weighing 6 cwts. is made in two pieces, one of which weighs 65 per cent. of the whole. What is the weight of this piece to the nearest pound?

5. A boy gives up the number 0·123 as the result of an experiment and is told that his answer is too great by 30 per cent. What is the correct result?

6. Find to a square metre the area of a path 1·5 metres wide round a circular pond, whose diameter is 50 metres.

7. A cylinder will exactly fit into a box, whose internal dimensions are: length 9 cm., breadth 4 cm., height 4 cm. What is the volume of the cylinder? What is the difference between its volume and the volume of the box?

8. The food bill at a school was £1225. 6*s*. 4*d*. for 82 days; and this worked out at an average of 2*s*. 0½*d*. per boy per day. How many boys were there?

9. The following table shows, for the date 31st March, 1911, the numbers of pensioners respectively receiving the various pensions payable under the Old Age Pension Act:

Amount of Pension per Week	Number of Pensioners
Five shillings	533,507
Four shillings	16,033
Three shillings	15,165
Two shillings	7,193
One shilling	3,891

What was the total sum that was being distributed per week in pensions at the date mentioned? What percentage, correct to the nearest tenth, of this sum was being paid in pensions of less than 5*s*. per week?

Paper 70

1. Find the value of (1) $67{\cdot}24^2 - 28{\cdot}36^2$,

(2) $\sqrt{27{\cdot}38^2 + 0{\cdot}9207^2}$.

2. Caban Côch Reservoir has a total capacity of 8,000,000,000 gallons. The total area that drains into the reservoir is 45,562 acres. How many inches of rain must fall to fill the reservoir, if it were empty?

3. A man spends 76 per cent. of his income and saves £240 every year, what is his income?

4. An alloy is made by melting together 125 lbs. of copper and 31·25 lbs. of tin. Find the percentage of each metal in the mixture.

5. A strike raises wages from 3*s.* 6*d.* to 4*s.* a day. It lasts six weeks. Demand for the goods falls 10 per cent., and in consequence working hours (and wages) are diminished in the same proportion. In how many weeks will a man make up what he lost in the strike?

6. Find to within a shilling the cost of putting a fence, which costs 3*s.* 2*d.* per yard, round a square field of 5 acres.

7. Find, in yards, the diameter of a circular running track measuring 8 laps to the mile.

If the track is 10 yards wide, calculate the number of square feet of wood required to build a wood paling, 3′ high, outside the track.

8. Water flows at the rate of 4·78 ft. per second through a cylindrical pipe of $10\frac{1}{2}$ inches diameter. What is the supply in gallons per minute? [1 cu. ft. = 6·24 gallons.]

9. A person leaves his estate, valued at £14,500, to his widow and three children, so that, after estate duty has been paid, the widow shall receive double as much as a child. If estate duty is 5 per cent., find what each receives.

10. What is the cost of £1400 $2\frac{1}{2}$ % Consols at $68\frac{5}{8}$? What is the yield per cent. at this price? How does it compare with the yield of 4 % War Loan at 96?

Paper 71

1. Find the value of: (i) $\sqrt{12{\cdot}7^2 + 9{\cdot}08^2}$,

(ii) $\sqrt{0{\cdot}032^2 - 0{\cdot}0298^2}$, (iii) $\frac{28{\cdot}36^3 \times 0{\cdot}6724}{10000}$.

2. Find the weight (in lbs.) of the metal in a pipe 144 feet long, whose outer diameter is 7·2 inches and whose inner diameter is 6·8 inches, if the metal is 8 times as heavy as water. (1 cu. ft. of water weighs 62·4 lbs.)

3. *A* borrowed sixpence off *B* a week ago, and promised to pay him back ninepence to-day. What rate of interest per cent. per annum is this?

4. What will be the cost to nearest penny of gravel for a path 3′ wide round a circular pond, 50′ diameter; the gravel being 3″ deep and costing 10 shillings a cubic yard?

5. A square field contains $22\frac{1}{2}$ acres: how long will a man, walking 3 miles an hour, take to walk round it?

6. If the population of a county is now 10 millions and the births every year are one in twenty, and the deaths are 3 % of the population at the beginning of the year, find what the population will be at the end of two years.

7. In a certain Irish constituency there were 10,000 electors, 51 per cent. being Unionists and 49 per cent. Nationalists. Two candidates went to the poll; 94 per cent. of the Unionists voted and 98 per cent. of the Nationalists. Which candidate was elected, and by how many votes?

8. The basin of the Severn is 4350 sq. miles in area, and the yearly rainfall is 29 inches. Half of the rain that falls is again evaporated, and the rest flows into the sea. Find to the nearest million gallons how much water flows into the sea per hour (a cubic foot = 6·24 gallons).

9. A man buys a house for £3400. At what rent should the house be let so that, after deducting £12 yearly from the gross rent for repairs, the man may obtain 5 per cent. on his capital?

10. What quantity of L.N.W.R. ordinary stock can be bought for £3000, the price being $113\frac{3}{4}$?

Paper 72

1. Calculate $\sqrt{127 \cdot 706}$ correct to 1 decimal place.

2. Find the value of $(3 \cdot 671)^3 \div 0 \cdot 613$.

3. Find the weight of a copper tube 5 metres long, external diameter 7·5 cm., internal diameter 6·5 cm., having given that 1 cu. cm. of copper weighs 8·6 grams.

4. The parliamentary borough of Norwich contains 7,656 acres. It was estimated that during a certain 20 hours in August, 1912, 5,600,000 tons of rain fell on this area.

Taking the weight of 1 cu. ft. of water as 62·4 lbs., find (*a*) the volume of water which fell, in cu. ft.; (*b*) the number of inches of rainfall during this period.

5. *A* owns $\frac{1}{3}$ of a piece of land, *B* owns $\frac{1}{4}$ of it, *C* owns $\frac{1}{2}$ of *A*'s and *B*'s shares together, whilst the remainder belongs to *D*. If the property is sold for £450 and 4 per cent. of this is deducted as expenses of the sale, what does each receive?

6. 10 cwts. of lead are cast into a hollow pipe whose external diameter is 1 inch and internal diameter is $\frac{3}{4}$ inch. If 1 cubic foot of lead weighs 712 lbs., what length of pipe is made (i) if no portion of the lead is wasted, (ii) if 2 per cent. is wasted in casting?

7. Out of 380 tons of coal, 28 % was wasted; 72 % of what was left was sold at 25*s*. per ton, and the rest was given away. What money was received for the portion sold?

8. Vinegar costs 1*s*. 3½*d*. per gallon retail and £1. 14*s*. 6*d*. per 100 gallons wholesale. What is gained per cent. by buying wholesale and selling retail?

9. The asphalt in a courtyard is in the form of a rectangle and measures 44 yds. by 28 yds. Assuming that a cubic foot of water weighs 1000 oz., find the weight in tons, correct to the nearest $\frac{1}{10}$ ton, of the rain which fell during a rainfall of 0·75″.

If this water were collected into a cylindrical tank, whose cross-section was a circle of radius 3 ft., how high would the tank have to be in order to contain all the water?

10. Which of the three following Russian loans yield the best interest: the 5 % at 92, the $4\frac{1}{2}$ % at 84, or the 4 % at 73?

Paper 73

1. A tank 2 ft. 6 in. long, 1 ft. 6 in. broad, and 1 ft. 3 in. high is filled to the brim with water. In this is completely immersed a cylinder, 1 ft. diameter and 2 ft. 3 in. long. How much water will overflow and how much will be left in the tank? (Answer to the nearest cubic inch.)

2. Find the value of:

(i) $\sqrt{10{\cdot}07^2 + 0{\cdot}9191^2}$, (ii) $487{\cdot}6 \times 3{\cdot}498 \times (0{\cdot}0476)^2$.

3. A square field contains 10 acres; find the cost of fencing it with patent wire fencing at $8\frac{1}{2}d.$ per foot.

4. A crane, with a horizontal tie-rod 14 ft. long, raises a packing case from a vessel alongside a quay through a vertical distance of 25 feet, then swings it through an angle of 180°, and finally lowers it 5 ft. on to the quay. Find the total distance the packing case has been moved through.

5. A copper lode yielded 65 per cent. of copper ore; the copper ore yielded 45 per cent. of pure copper. Find the weight of pure copper extracted from $3\frac{3}{4}$ tons of the lode.

6. If bottles of soda-water are bought at 9*d.* a dozen and retailed at 2*d.* each, what percentage of the receipts is profit?

7. The populations of three towns are 10,941, 12,311, 25,992. If 684 policemen are to be distributed among them in proportion to the population, how many policemen will be sent to each town?

8. A sum of money has increased to £57. 3*s.* 6*d.* in 5 years at $2\frac{1}{2}$ % simple interest. Find the sum.

9. What income is obtained by investing £437. 10*s*. in Canada 4 % at $96\frac{1}{8}$ (answer to the nearest shilling)?

10. *A* and *B* enter into partnership, *A* providing £4000 and *B* £6000 capital. They make a profit of £450 per annum. How ought they to divide it? What percentage does each receive on his investment?

Paper 74

1. Find the value of: (i) $2{\cdot}785^2 + 6{\cdot}874^2$,
(ii) $\sqrt{0{\cdot}4657^2 + 0{\cdot}2843^2}$, (iii) $\sqrt{0{\cdot}00465} \div 0{\cdot}346^3$.
Give a rough check in each case.

2. The hole on a golf green is $4\frac{1}{2}$ inches in diameter. If the area of the hole were doubled, what would the diameter be?

3. What is the cubic content of a wall, 1 foot thick and 8 feet high, whose exterior boundary encloses a rectangular plot of ground of 5000 square yards, the length being double the breadth?

4. Hematite (iron ore) contains 70 per cent. of iron. If 2 per cent. of the iron is wasted in the process of extraction, how much hematite would be required to produce 4 tons 18 cwts. of iron?

5. If the manufacturer makes a profit of 10 per cent. on an article, and the shopkeeper a profit of 20 per cent., what is the cost of manufacturing an article which is sold in a shop for 5*s*. 6*d*.?

6. A sum of £2437 is to be divided among three communities according to population. If their populations are 2431, 3057, and 683, find to the nearest pound the share of each community.

7. A hollow cylinder has inside diameter 8 cm. and height 25 cm. How many litres of water will it hold? If, when it is partly filled with water, a steel cylinder 12 cm. long and 1·8 cm. in diameter is lowered into it, how much will the water rise?

[*See next page*

8. The diameter of a rivet may be calculated from the formula $d = 1{\cdot}2\sqrt{t}$, where d is the diameter of the rivet and t the thickness of the plates, both in inches. Find the diameter of the rivets when the plates are $\frac{3}{8}''$ thick.

9. A man wants to invest £2200 so as to obtain an income of £100 a year. If he chooses a 4 % stock, at what price must he buy?

10. A legacy of £1000 is left to be divided between A and B, so that each shall receive the same amount when legacy duty has been paid. If A is liable to pay no legacy duty, and B is liable to legacy duty at the rate of 10 per cent. on the whole amount left to him, how much will be paid in duty and how much will A receive (to the nearest £)?

Paper 75

1. Find $\sqrt{292{\cdot}0681}$ correct to 1 place of decimals.

2. Find the value of: (i) $\sqrt{3{\cdot}027^2 + 2{\cdot}001^2}$,
(ii) $\sqrt[4]{31{\cdot}42}$, (iii) $94{\cdot}67 \times 38{\cdot}52 + 62{\cdot}8 \times 91{\cdot}6$.

3. A body of men is drawn up in a hollow square 4 deep, with a front of 40 men: how many men are there in the square?

4. The iron cylinder of a garden roller is 32 inches long, $\frac{3}{4}$ inch thick, and has an inner diameter of 22 inches. Find its weight, if a cubic foot of iron weighs 468 lbs.

5. A sold a horse to B, gaining $7\frac{1}{2}$ per cent. on what it cost him. B sold it to C for £70. 19*s*., gaining 10 per cent. on what it cost him. What did A pay for the horse?

6. 5 % of the chickens on a farm are killed by rats, and 5 % of the remainder die during the winter. If 361 are still left, how many were there originally?

7. A cylindrical measuring-glass is graduated for marking cubic centimetres. If the diameter of the glass is 2·5 cm., how far apart will the graduations be?

8. The average annual rainfall at a certain place is 30 inches; but in 1913 it was found that 24 inches had fallen up to the end of September. The average number of inches that fall in October, November, December respectively is 4·43, 3·47, 3·43. What must be the actual rainfall for each of these months in 1913 in order that these proportions may be preserved and at the same time the total for the year may be 30 inches? (Answers to 3 significant figures.)

9. A company borrows the sum of £1,658,775 on the understanding that at the end of each year a portion of the principal is to be paid off, with interest at 4 per cent. per annum on the amount standing unpaid during that year. Prove that the debt can be cleared off in four years by an annual payment of £456,976.

10. What must be the rate of dividend on Great Western Deferred Stock standing at $44\frac{1}{2}$ in order that an investment in this stock may yield 6 %?

Paper 76

1. Find the value of $\frac{1}{6{\cdot}535}(3{\cdot}605^3 + 2{\cdot}93^3)$.

2. Find the value of $\sqrt{213^2 + 17{\cdot}2^2 - 32{\cdot}5^2}$ correct to the nearest whole number.

3. The edge of a cube is measured with a faulty ruler and found to be 6·22 cm. long. If the real length is 6·32 cm., what is the percentage error correct to one place of decimals?

If the volume of the cube is calculated from the wrong measurement, what is the absolute error to nearest c.c.?

Express this as a percentage of the actual volume correct to one place of decimals.

4. Three per cent. of the trees in a wood are destroyed in a storm, twelve per cent. of the remainder are cut down, and 6404 then remain. How many were there originally?

5. The receipts of a company amounted in one year to £354,000, and the expenses to £251,000. In the next year receipts fell off by $3\frac{1}{2}$ per cent., and expenses increased $\frac{1}{2}$ per cent. Find by what percentage of itself the profit in the first year exceeded the profit in the succeeding year.

[*See next page*

6. A body of men is drawn up in a hollow square four deep with 85 men in a side of the square. Show that they can also be drawn up in a solid square, and find the number of men there will then be in a side.

7. A water-ballast roller is ·75 m. wide and ·63 m. in diameter: if it is made of iron 5 mm. thick, how much water can be put into it? What will it weigh (1) when empty, (2) when full? (Iron weighs 7·2 times as much as water.)

8. A cubic foot is cut up into cubic inches; how many more inch blocks are needed in order that the whole may be arranged in one layer forming a solid square?

9. Find, to two significant figures, the number of gallons delivered per hour by a pump with a six-foot stroke, and making 15 strokes per minute, the internal diameter of the barrel being $4\frac{5}{8}$ in.

10. What is the least sum (containing no odd shillings or pence) which must be invested in $2\frac{1}{2}$ per cent. Consols at $69\frac{1}{2}$ so as to bring in an income which, after payment of income-tax at 1*s*. 2*d*. in the £, amounts to at least £100 per annum?

11. How many sovereigns could be made out of a bar of gold 33·4 inches long, 5·2 inches broad, and 3·7 inches thick, supposing all the material is used up and that the average diameter of a sovereign is 0·84 inch and the thickness 0·05 inch?

MISCELLANEOUS EXERCISES

On Chaps. I–XIX

1. A wheel of 6 ft. diameter, making 50 revolutions per minute, is connected by a band to a wheel of 1 ft. 6 in. diameter. How many revolutions will the latter wheel make in a minute?

2. A wheel of 28″ diameter is driven by belting from a wheel of 10″ diameter. At what speed must the small wheel rotate in order that the larger may make 40 revolutions per minute?

3. The engine of a motor-bicycle revolves $3\frac{1}{2}$ times for every time the back wheel revolves, i.e. it is said to have a gear of $3\frac{1}{2}$ to 1. How many revolutions per minute does the engine make, if the bicycle is going 26 miles per hour? The diameter of the back wheel is 26″.

4. The cog wheels of a bicycle have 24 and 15 cogs, the latter being connected with the axle of the driving wheel; the chain has 51 links. The diameter of the driving wheel is 28 inches. At a certain instant certain teeth of the cog wheels fit into certain links of the chain. How far will the bicycle have travelled before the same teeth fit again into the same links?

5. Three cylinders have their diameters in the ratio of 2 : 3 : 5, whilst their heights are in the ratio of 10 : 4 : 1. What is the ratio of their weights, provided that they are made of the same material?

6. Iron is 10 times as heavy as wood; an iron cylinder is 3 times as high as a wooden one, but the diameter of the wooden one is half as large again as that of the iron one; in what ratio are their weights?

7. Calculate, to within 10 c.c., the volume of a solid body whose plan is a rectangle 15 cm. by 9·5 cm., and whose elevation consists of a square with a semicircle described on the upper side. Side of square = 9·5 cm.

8. The speed of a steamboat is 12 knots, and for each knot the engine makes 30 revolutions per minute. The stroke of the piston is 14 inches and its diameter 16 inches. If the cylinder is filled with steam twice in each revolution, find, to the nearest cubic foot, the number of cubic feet of steam used per minute.

9. A steady rain was falling, and a rainwater tub was partly full. The average internal section of the tub just above the water was a circle of 2 feet diameter; and the water was pouring in at the rate of a half-pint in 10 seconds. Find in what time the water in the tub rose one inch.

10. A window blind 3 ft. 8 in. long is made of stuff $\frac{1}{20}$ in. thick. It is tightly rolled up on a cylindrical roller 1 in. in diameter. Find the average radius of the complete roll, and hence show that the blind is wrapped round the roller about $9\frac{1}{2}$ times.

11. Tape is wound on to an empty reel from a full one. At the start the full reel is 6 inches in diameter, and the empty reel is 2 inches in diameter; the tape is $\frac{1}{50}$ inch thick. If the empty reel is revolved at the rate of 25 revolutions a minute, find the radius of each wheel at the end of two minutes, neglecting the variation in length of the straight portion of the tape.

12. A penny is 3·1 cm. in diameter and 1·75 mm. thick. It is made of bronze, weighing 8·71 gm. per c.c. and containing 4 per cent. of tin by weight. What is the weight of tin in a penny, correct to $\frac{1}{100}$ of a gm.?

13. By selling an article for 5 guineas I gain 5 per cent. (*a*) What was the cost price? (*b*) What should I have sold the article for in order to gain 10 per cent.?

14. Sugar sold at 3*d*. per pound gives a profit of 10 %. Find the cost price per cwt., and the price at which 1 cwt. must be sold to gain 15 %.

15. Water in freezing expands in volume by 9 per cent.; by how much per cent. does ice contract on melting?

16. A dealer buys a car from the manufacturer for £600. (*a*) If he sells it at a gain of 40 per cent. what does he receive for it? (*b*) At what price must he mark it in his shop so that he may be able to allow a purchaser a reduction of 20 per cent. for cash, and yet make a profit for himself of 40 per cent.?

17. If the weight of an engine is diminished by 7 % it will then weigh 83 tons 15 cwts. What would it have weighed if its weight had been increased 5 %?

18. Two destroyers are contracted for at the same speed; one steams on her trials at 22·8 knots, and so falls short of the contract speed by 4 %; the other runs at 5 % in excess of the contract speed. Find the contract speed, and the speed of the second destroyer.

19. The inhabitants of a certain town speak either French or German; 76·4 per cent. speak French and 80·6 per cent. German. What percentage speak both languages?

20. Find the percentage error (to the nearest whole number) in the following rule: "To convert miles per hour to feet per second multiply by $1\frac{1}{2}$."

21. A dishonest tradesman marks his goods at an advance of 5 per cent. on the cost price, but uses a fraudulent balance, whose beam is horizontal when the weight in one scale is $\frac{1}{15}$ more than the weight in the other. What is his actual gain per cent.?

22. The nominal weight of a truck or coal is 8 tons 10 cwts., the actual weight is 8 tons 18 cwts. If 10 per cent. profit be made by selling the coal by the truck at 22*s*. per ton nominal weight, what rate of profit will be made by retailing the coal at 22*s*. 8*d*. per ton actual weight?

23. A tradesman sold half his stock of a certain article at a profit of 50 per cent. He disposed of half the remainder by allowing 3*d*. in the shilling off the original selling price and took £5 for what was left. If he made altogether a profit of $15\frac{5}{8}$ %, what sum did he pay for his stock?

24. By selling a house at £2576 a man gains 12 per cent. on his original outlay. How much per cent. would he have gained had the house cost him £100 less?

25. 100 lbs. of an alloy of brass and tin contain 75 per cent. of brass. How much brass must be added that there may be 80 per cent. of brass in the mixture?

26. A man bought an estate at 18 per cent. below its real value and sold it at 10 per cent. above its real value. If his profit was £1358, how much did he pay for the estate?

27. A dealer lost £1 by selling 2 cows at the same price, having gained 20 per cent. on one and lost 20 per cent. on the other; what was the selling price?

28. An article which cost 3*d.* per lb. to manufacture was sold at 7*d.* per packet. The cost of manufacture has risen 5 per cent.; the article is still sold at 7*d.* per packet, but each packet now contains only 15 oz. instead of 1 lb. What percentage profit does the manufacturer make on his outlay?

29. A merchant buys 2560 quarters of corn, $\frac{1}{8}$ of which he sells at a gain of 2 per cent., $\frac{1}{4}$ at a gain of 3 per cent., $\frac{3}{8}$ at a gain of 4 per cent., and the remainder at a gain of 5 per cent. If he had sold the whole at a gain of 4 per cent. he would have obtained £17. 12*s.* more than he did; what was the cost per quarter?

30. The purchaser of a farm lets it at a rent which, after reduction of 15 per cent. for repairs and necessary outgoings, will return 5 per cent. on the purchase price. What is the return in a year when the owner allows the tenant 25 per cent. reduction of rent?

31. Two casks, of capacities 36 gallons and 72 gallons, are filled with mixtures of wine and water; 80 per cent. by volume of the liquid in the first cask, and 60 per cent. of that in the second being wine. On two different occasions 18 gallons of liquid are drawn from each cask, and each cask then filled up with the liquid drawn from the other. What is the final percentage of wine in each cask?

32. Two adjacent sides of a rectangle are measured and are found to contain 1·534 metres and 0·662 metre respectively; find its area to 3 significant figures.

If in each measurement there may be an error of 2 millimetres, what is the greatest error in the area of the rectangle as calculated from the above measurements?

33. Find (to 1 significant figure) the percentage error between the measured value 4·63 inches of the edge of a cube, and the true value 4·75 inches.

What will be the resulting percentage error in calculating the volume of the cube?

34. The correct measurements of certain rectangles are:

Length (cm.)	10	12·5	30	60	150
Breadth (cm.)	7·5	6	2·5	1·25	·5

If I am liable to make an error of 1 mm. in measuring the length or breadth, by how much per cent. may the areas I obtain be (1) greater than, (2) less than, the actual areas?

35. If there is a possible error of 1 % in my measurements of length, show that (1) the area of a rectangle deduced from them may be as much as 2 % in error, (2) the volume of a cuboid may be as much as 3 % in error.

36. The three edges of a brick are measured with a faulty foot-rule, which makes everything appear 10 per cent. too small. What will be the percentage of error in the calculated volume?

37. A tea merchant finds that he saves appreciably in the cost of packing material if he makes the 1 lb. packets cubical rather than cuboidal. What percentage is saved when a cubical form is adopted in place of a cuboid whose edges are in the ratio 4 : 3 : 2?

38. Find to 1 significant figure the ratio of the densities of sun and earth, given that the diameters are 866,000 and 7920 miles and the ratio of the masses 300,000 : 1.

39. If we take a globe of one inch diameter to represent the earth, then we must have a globe of 9 feet in diameter at a distance of 323 yards to represent the sun on the same scale. Taking the diameter of the earth as 7920 miles, find the diameter and distance of the sun, to 2 significant figures. Also find the ratio (i) of the surfaces, (ii) of the volumes of sun and earth.

40. In a game of 24,000 points *A* gives *B* 9000 points start. When *B*'s score is 22,600, find what *A*'s should be that he may be level on the handicap.

41. An engine transmits power to a mill by means of belting: the driving-wheel is of 4′ radius and makes 15 complete revolutions in 2 minutes 12 seconds: the driven wheel is of 3′ radius: find (i) the angle turned through by the driving-wheel in 1 second, (ii) the number of revolutions made by the driven wheel in 1 minute, assuming that there is no slipping in the belting.

42. A brick is 12 inches long, $10\frac{2}{3}$ inches wide and 4 inches thick. Find its volume and surface.

What would be the length of the edge of a *cubical* brick having (1) the same volume, (2) the same total surface?

43. A merchant estimates that at the end of every month the value of his unsold goods is only 99 per cent. of their value at the beginning of the month. He calculates in this way that the value of part of his stock is only half what he gave for it. How long has he had the goods in his possession?

44. A metal plate, $\frac{3}{8}$ in. thick and weighing 15 lbs. for every square foot of area, is 5 ft. long by 2 ft. broad. How many holes, 1 inch in diameter, must be punched out to reduce the weight by at least 5 per cent.?

45. In one metre there are 39·37 inches. In measuring a metre rule in inches a boy gets the result 39·53 in. What percentage error has he made? If he afterwards used his own result to find the number of cubic inches in a sphere 21 cm. in diameter [$V = \frac{4}{3}\pi r^3$], what percentage error would he have in his answer?

46. Three persons enter into partnership, subscribing £800, £1000, £2000 capital respectively. If they make a profit of 20 % per annum, how much does each man derive from the business?

47. Two partners, A and B, provide £7500 and £2500 respectively of the capital of a business. Of the profits B first receives 20 per cent. as manager; interest is then paid on the capital at the rate of 4 per cent. per annum, the remaining profits being equally divided. In a year in which A receives £562, what does B receive?

48. Supposing the Channel tunnel were constructed at a cost of thirty millions of money, that one hundred trains per day passed each way, and that each train carried two hundred tons of goods; what must be the freight per ton that the goods traffic alone should produce sufficient to pay 4 per cent. per annum on the outlay, it being assumed that there are annually three hundred and sixty days, on the average, during which work might be carried on?

49. The average size of the *Standard* is 41 in. by 52 in., and the total number of copies issued during the month of February was 5,016,024. If the average has been the same since the beginning of the year, find to the nearest sq. mile what area of ground could be covered by the total issue of the newspaper up to March 4 inclusive.

Supposing the gross cost of production to be $1\frac{1}{2}d.$ a copy, and the newspaper to be sold to newsagents at $\frac{4}{5}d.$, what must be the daily income from advertisements to make it pay the proprietors 2 per cent. on their gross outlay?

50. A bankrupt is owed £470. 9*s*. 3*d*., this debt being good; and he has the following bad debts, namely: £283. 4*s*. 6*d*., £70. 6*s*. 8*d*., £110. 18*s*. 3*d*., for which he receives respectively 6, 7, and 10 shillings in the pound. His own liabilities are £4600. How much can he pay in the £?

51. The outstanding assets of an estate are being realised, and consist of the following: sums due to the estate, which will be paid in full, £4039. 13*s*. 6*d*.; a sum of £1145. 11*s*. 4*d*. due, on which 13*s*. 8*d*. in the £ will be paid; a sum of £972. 6*s*. 2*d*. due, on which 7*s*. 10*d*. in the £ will be paid; a sum of £650. 18*s*. 3*d*. due, on which 5*s*. $3\frac{1}{2}d.$ in the £ will be paid. Find the total cash value of these four assets; and find to the nearest integer how many shillings in the £ are being paid on the assets taken as a whole.

52. A sum of £57,683 is to be raised from three parishes in the proportion of 3 to 4 to 2. The sum to be raised in each parish is obtained by levying a rate of a certain amount per £1 of rateable value. The rateable values of the parishes are respectively £65,472, £84,271, and £55,641. Find the rate which must be imposed in each parish. Give the result to the nearest farthing.

53. On December 31st, 1908, the capital of the London and North-Western Railway consisted of £39,008,373 Debenture Stock on which 3 per cent. interest was paid, £15,100,406 Guaranteed and £23,080,620 Preference Stock on each of which 4 per cent. interest was paid, and £42,887,723 Ordinary Stock on which $5\frac{3}{4}$ per cent. interest was paid. Find the whole amount that was paid as interest, and what average rate per cent. of interest (to the nearest quarter of 1 per cent.) could have been paid on the whole capital with that amount.

54. What income is obtained by investing £1245 in a $3\frac{1}{2}$ per cent. stock at 90? and what will be the loss of capital if the stock is sold out at 87?

55. A Canadian invests £1000 in 4 per cent. mortgage bonds at 94. Find to the nearest cent the interest he will receive if £1 is equivalent to \$4·86. How much per cent., to within a penny, does he obtain on his investment?

56. I bought 104 shares in a company at £$93\frac{1}{3}$ each. Afterwards I sold 60 of them for £100 each. At what price must I sell the shares that remain so as neither to gain nor lose by the whole transaction?

57. On October 9th, 1914, the price of $2\frac{1}{2}$ per cent. Consols was $68\frac{1}{2}$. Neglecting brokerage, etc., how much must a man invest at this price to produce an income of £200 a year? Income tax was charged on income from investments at the rate of 1/2 in the £. How much must be invested to produce a *net* income of £200 a year after payment of tax? (Answer to the nearest £; logarithm tables may be used.)

58. A man has £9000 invested partly in a 3 per cent. stock at $93\frac{1}{2}$ and partly in a $4\frac{1}{2}$ per cent. stock at 102, in such a manner that he gets 4 per cent. for his money. How much has he invested in each stock?

59. A person who derives equal parts of his income from three investments, yielding respectively 3, 4 and 5 per cent. interest, consolidates them into a single investment, yielding 4 per cent.: find by how much per cent. his income is increased or diminished, and by how much the average rate of interest on his capital.

60. By paying to a certain Life Insurance Company a sum of £100, a man of 65 can obtain a yearly income till he dies of £10. 7*s.* 4*d.* How much must he pay the company in order to get a yearly income of £250?

61. A builder borrows £1261 from the bank to be paid back with compound interest at the rate of 5 per cent. per annum by the end of 3 years in 3 equal yearly instalments. Show that each instalment is £463. 1*s.*

62. *A* lends *B* a certain sum; at the same time he insures *B*'s life for £737. 12*s.* 6*d.*, paying annual premiums of £20. At the end of three years and just before the fourth premium is to be paid *B* dies, having never repaid anything. What must *A* have lent *B* in order that he may have just enough to recoup himself together with 5 per cent. per annum compound interest on the sum lent and on the premiums?

63. The capital of a bank is £50,000. Its current accounts, on which it pays no interest, amount on an average to £400,000, and its deposit accounts, on which it pays $2\frac{1}{2}$ per cent. interest, to £100,000. Keeping about £25,000 in hand, it invests £175,000 at 4 per cent. and £350,000 at 6 per cent. In a year it incurs losses amounting to £2000, its salaries and expenses amount to £5000, and it carries £2500 to a reserve fund. At what rate per cent. per annum can it pay dividends?

64. The total National Debt, funded and unfunded, for the financial year 1911–12, was £718,406,428. It consisted mainly of $2\frac{1}{2}$ per cent. Consols, and the average rate of interest paid was 2·65. The sum of £24,500,000 was set apart for the service of the debt; it was applied in the first instance in the payment of the interest for the year; the remainder was applied at the end of the year to the extinction of debt. Find, to the nearest £1, the sum available for the latter purpose.

65. Calculate the total amount produced by a County Rate of 2*s*. 3½*d*. in the pound from a town which is assessed as follows:

(*a*) Net annual value of Agricultural Land £4252.

(*b*) Net annual value of buildings and other hereditaments not being Agricultural Land, £160,164.

The County Rate is calculated on the full value of item (*b*) added to one-half of item (*a*).

66. A builder contracted to build a house in twelve weeks, and agreed to pay £5 a day or portion of a day that the work remained incomplete after the end of that period. To fulfil his contract he engaged 15 men to work 10 hours a day. At the end of three weeks 10 of the men struck, and remained away from their work for a fortnight; after their return the number of working hours was reduced to nine. What was due from the builder for his breach of contract?

67. A builder takes a contract from which he expects to make a profit of £500, but when the work is half finished, the cost of materials being raised 20 per cent. and wages being raised 15 per cent., he calculates that he will lose £550. He therefore finishes the work with inferior materials, only 4/5 as good as those first used, and gains £350. What was the contract price of the work?

68. A rectangular cistern, without a lid, measures 6 ft. 10 in. by 4 ft. and is 4 ft. 3 in. deep. It rests with one of the short edges on the ground and the other is propped up on supports so that it is 1 ft. 6 in. above the ground. How much water will it hold? (If you cannot do this entirely by calculation, draw to scale a side-view and make any measurements you need.)

69. One of the Birmingham reservoirs in the Welsh mountains has a total capacity of 8×10^9 gallons. The top-water area is 500 acres. What is the average depth of the water? (One gal. contains 277 cub. in.)

70. A stretch of 10 miles of a river of an average width of 120 yards rose 5 ft. 7 in. in consequence of a heavy rainfall. How many million gallons of additional water in this stretch of the river did this rise represent? If the area on which the rain fell was 30 square miles per mile length of the river and the rainfall was 5·2 inches, find what percentage of the total rainfall was represented by the additional water in this stretch of the river.

ANSWERS

PART I

EXERCISE I a

PAGE								
2.	**1.**	1368.	**2.**	1962.	**3.**	1516.	**4.**	1283.
	5.	1515.	**6.**	14326.	**7.**	4635.	**8.**	9822.
	9.	5287.	**10.**	11904.	**11.**	14561.	**12.**	14564.
	13.	16403.	**14.**	22204.	**15.**	21756.	**16.**	162995.
3.	**17.**	105842.	**18.**	90609.	**19.**	137214.	**20.**	188353.
	21.	18611.	**22.**	26020.	**23.**	19092.	**24.**	24063.
	25.	173004.	**26.**	127205.	**27.**	144846.	**28.**	130235.
	29.	117873.	**30.**	70744.	**31.**	150886.	**32.**	85225.
	33.	197543.	**34.**	166036.	**35.**	366260.	**36.**	239447.
	37.	4297399.	**38.**	4266304.	**39.**	3962739.		

4.	**40.**	4165860.	**41.**	5611597.	**42.**	4768321.
	43.	5965762.	**44.**	6781482.	**45.**	7693996.
	46.	6167061.	**47.**	5000000.	**48.**	4403655.

EXERCISE I b

5.	**1.**	16.	**2.**	20.	**3.**	27.	**4.**	25.	**5.**	27.
	6.	37.	**7.**	39.	**8.**	51.	**9.**	126.	**10.**	175.
	11.	211.	**12.**	156.	**13.**	120.	**14.**	248.	**15.**	252.
	16.	176.	**17.**	1099.	**18.**	1193.	**19.**	1199.	**20.**	1350.
	21.	1275.	**22.**	1310.	**23.**	1793.	**24.**	2647.	**25.**	822.
	26.	294.	**27.**	3131.	**28.**	1084.	**29.**	1526.	**30.**	3262.

EXERCISE I c

1. *A*, 345; *B*, 245.

6. **2.** 635. **3.** 45221615. **4.** 45, 60, 20.

5. (1) 345. (2) 345. (3) 117. (4) 115. (5) 106. (6) 306. (7) 306. (8) 322. (9) 254. (10) 104. (11) 70. (12) 120.

7. (13) 62. (14) 64. (15) 194. (16) 636.

EXERCISE I d

PAGE								
8.	**1.**	32.	**2.**	42.	**3.**	134.	**4.**	454.
	5.	552.	**6.**	506.	**7.**	307.	**8.**	105.
	9.	109.	**10.**	82.	**11.**	189.	**12.**	176.
	13.	1219.	**14.**	1332.	**15.**	2003.	**16.**	912.
	17.	2509.	**18.**	3331.	**19.**	889.	**20.**	5564.
	21.	1622.	**22.**	4258.	**23.**	1269.	**24.**	3086.
	25.	81749.	**26.**	14368.	**27.**	22097.	**28.**	11281.
9.	**29.**	9289.	**30.**	291882.	**31.**	139848.		
	32.	4089878.	**33.**	408812.	**34.**	1719686.		
	35.	1434352.	**36.**	5128889.				

EXERCISE I e

	1.	7835.	**2.**	721.	**3.**	65.	**4.**	58.	**5.**	1111.
	6.	27.	**7.**	12.						
10.	**8.**	9, 14, 10.	**9.**	£6.	**10.**	70, 8, 31.				
	11.	5.	**12.**	13, 24, 46.	**13.**	Always 1089.				

EXERCISE I f

11.	**1.**	6.	**2.**	6.	**3.**	6.	**4.**	3.	**5.**	12.
	6.	3.	**7.**	3.	**8.**	2.	**9.**	9.	**10.**	3.
	11.	11.	**12.**	12.	**13.**	56.	**14.**	115.	**15.**	16.
	16.	52.	**17.**	34.	**18.**	30.	**19.**	18.	**20.**	8.

EXERCISE I g

12.	**1.**	6.	**2.**	3.	**3.**	12.	**4.**	6.	**5.**	6.
	6.	12.	**7.**	24.	**8.**	24.	**9.**	5.	**10.**	11.
	11.	5.	**12.**	5.	**13.**	11.	**14.**	13.	**15.**	5.
	16.	13.	**17.**	13.	**18.**	13.	**19.**	5.	**20.**	5.
	21.	20.	**22.**	4.	**23.**	11.	**24.**	12.	**25.**	23.
	26.	21.	**27.**	94.	**28.**	41.	**29.**	18.	**30.**	1.

EXERCISE I h

14.	**1.**	108876.	**2.**	101556.	**3.**	87699.
	4.	80717.	**5.**	708522.	**6.**	826361.
	7.	648225.	**8.**	2424870.	**9.**	5499542.
	10.	3387664.	**11.**	2487372.	**12.**	2224908.
	13.	5420425.	**14.**	585170.	**15.**	3921354.
	16.	3802953.	**17.**	1917630.	**18.**	3729494.
	19.	2131344.	**20.**	7154132.	**21.**	35625506.
	22.	25392510.	**23.**	39323781.	**24.**	37420377.
	25.	37928226.	**26.**	61702176.	**27.**	23473782.
	28.	40713075.	**29.**	43962324.	**30.**	65591465.
	31.	82271.	**32.**	795984.	**33.**	1861194.
	34.	91907810.	**35.**	41414282.	**36.**	203718735.
	37.	338767625.	**38.**	432185628.	**39.**	408097990.
	40.	309176136.				

PAGE

EXERCISE I i

14. **1.** 252.

15. **2.** 53280. **3.** 112455. **4.** 2240, 31536000.
5. 91512000 mi. **6.** £170650.
7. £54937500 + £22284450 + £70543700 = £147765650.
8. 1800, 93600. **9.** 1248668 lbs.

16. **10.** £3988305. **11.** 165939480. **12.** 640965.

EXERCISE I j

1. 529. **2.** 2197. **3.** 14641.
4. 20736. **5.** 59049. **6.** 3375.
7. 279841. **8.** 1124864. **9.** 10218313.
10. 234256. **11.** 360. **12.** 4725.
13. 6125. **14.** 21168. **15.** 107811.
16. 42875. **17.** 20808. **18.** 98496.
19. 15876000. **20.** 93170000.

EXERCISE I k

17. **1.** 18940, 1; 11364, 1; 8117, 2; 5165, 6; 4735, 1.
2. 693, 27; 69, 237; 6, 6537. **3.** 842, 80; 84, 300; 8, 4700.
4. 1070, 63; 107, 63; 10, 8463. **5.** 212, 8; 97, 23; 11, 191.
6. 301, 2; 234, 4; 36, 40. **7.** 335, 9; 164, 6; 77, 91.
8. 18, 30; 10, 6; 5, 76. **9.** 684, 6; 62, 348; 7, 2642.
10. 1063, 47; 560, 54; 46, 292. **11.** 807, 29; 262, 52; 10, 1176.
12. 93148, 33; 44583, 110; 83, 57622.

EXERCISE I l

18. **1.** 52, 25. **2.** 284, 125. **3.** 158, 386.
4. 37, 286. **5.** 48, 439. **6.** 83, 539.
7. 161, 31. **8.** 79, 766. **9.** 184, 141.
10. 95, 455. **11.** 79, 1755. **12.** 30, 3361.
13. 343, 569. **14.** 103, 5005. **15.** 110, 1516.
16. 62, 3942. **17.** 118, 1886. **18.** 110, 2106.
19. 267, 1361. **20.** 30, 3318.

EXERCISE I m

19. **1.** 104, 26, 13. **2.** £11. **3.** 4, 8. **4.** 4, 9.
5. 16, 160. **6.** 6, 7 in. **7.** 10, 12 in. **8.** 4, 160.

EXERCISE I n

20. **1.** 68. **2.** 156. **3.** 32. **4.** 132. **5.** 33.
6. 5. **7.** 116. **8.** 292. **9.** 148. **10.** 11.
11. 10 **12.** 16. **13.** 16. **14.** 10. **15.** 8.
16. 8. **17.** 16. **18.** 16. **19.** 8. **20.** 4.
21. 30. **22.** 12. **23.** 2. **24.** 81. **25.** 41.
26. 55. **27.** 25. **28.** 55. **29.** 24. **30.** 24.
31. 446. **32.** 1598. **33.** 10. **34.** 780.

21. **35.** 36. **36.** 720. **37.** 47. **38.** 1371. **39.** 2704.
40. 2080. **41.** 2. **42.** 6. **43.** 37. **44.** 31.
45. 17.

EXERCISE I c

PAGE

22. **1.** 216. **2.** 974. **3.** 32. **4.** 2592.
5. 2493. **6.** 2000. **7.** 3528. **8.** 17.

23. **9.** 29. **10.** 6204. **11.** 834. **12.** 37.
13. 120 mi. **14.** 188.

EXERCISE II b

27. **1.** 499. **2.** 3016. **3.** 5166. **4.** 3798.
5. 10549. **6.** 24309. **7.** 16960. **8.** 8389.

28. **9.** 50795. **10.** 57583. **11.** £9. 16*s*. 5*d*.
12. £5. 11*s*. 9¾*d*. **13.** £145. 16*s*. 8*d*. **14.** £36. 16*s*. 8*d*.
15. £13. 13*s*. 4¼*d*. **16.** £76. 18*s*. 5¼*d*. **17.** Table:—

1 hour	1 day	6 days	30 days	78 days	310 days
	£ *s*. *d*.	£ *s*. *d*.	£ *s*. *d*.	£ *s*. *d*.	£ *s*. *d*.
½*d*.	4	2 0	10 0	1 6 0	5 3 4
2*d*.	1 4	8 0	2 0 0	5 4 0	20 13 4
2½*d*.	1 8	10 0	2 10 0	6 10 0	25 16 8
3*d*.	2 0	12 0	3 0 0	7 16 0	31 0 0
etc.	etc.	etc.	etc.	etc.	etc.

EXERCISE II c

1. 2256. **2.** 403. **3.** 41488. **4.** 31151.
5. 14160. **6.** 1685. **7.** 3888. **8.** 1733.
9. 413. **10.** 12860. **11.** 16 cwts. 54 lbs. 7 oz.
12. 10 cwts. 80 lbs. **13.** 12 cwts. 56 lbs.
14. 1 ton 6 cwts. 44 lbs. **15.** 6 tons 11 cwts. 63 lbs.
16. 2 tons 15 cwts. 90 lbs. **17.** 13 tons 6 cwts. 83 lbs.
18. 2 tons 3 cwts. 86 lbs. 7 oz. **19.** 6 tons 13 cwts. 4 lbs.
20. 4 tons 9 cwts. 32 lbs.

EXERCISE II d

29. **1.** 86400. **2.** 219600. **3.** 62045. **4.** 259367.
5. 12900. **6.** 946800. **7.** 88995. **8.** 299640.
9. 1841460. **10.** 428429.
11. 8 d. 12 h. 9 m. 25 s. **12.** 5 d. 4 h. 29 m.
13. 2 d. 20 h. 25 m. 57 s. **14.** 574 w. 5 h.
15. 25 w. 2 d. 11 h. 47 m. **16.** 5 w. 2 d. 14 h. 42 m. 12 s.
17. 212 w. 3 d. 4 h. **18.** 8 d. 14 h. 34 m. 53 s.
19. 85 w. 11 h. **20.** 55 w. 6 d. 6 h. 10 m.

EXERCISE II e

PAGE

29. **1.** 276. **2.** 447. **3.** 4057.
4. 2597. **5.** 89280. **6.** 517.
7. 128232. **8.** 69504. **9.** 65640.
10. 164520. **11.** 104 y. 1′ 9″. **12.** 1 mi. 883 y.
13. 555 y. 1′ 8″. **14.** 1 mi. 1087 y. 2′. **15.** 3 mi. 1053 y. 1′.
16. 165 y. 3″. **17.** 3 mi. 1269 y. **18.** 5 mi. 1200 y.
19. 29 mi. 413 y. 1′. **20.** 548 y. 2′ 5″.

EXERCISE II f

30. **1.** 31. **2.** 30. **3.** 41. **4.** 93. **5.** 96. **6.** 198.
7. 13 b. 1 g. 3 q. **8.** 26 b. 7 g. 1 p.
9. 42 b. 1 g. 2 q. 1 p. **10.** 34 b. 3 g.
11. 156 b. 2 g. **12.** 54 b. 5 g. 2 q.

EXERCISE II h

31. **1.** £17. 0*s.* 4*d.* **2.** £36. 5*s.* 9*d.* **3.** £18. 14*s.* 11*d.*
4. £32. 15*s.* 11½*d.* **5.** £10. 18*s.* 1*d.* **6.** £62. 0*s.* 7½*d.*
7. £6. 14*s.* 5½*d.* **8.** £2. 10*s.* 11½*d.* **9.** £1424. 14*s.* 0*d.*
10. £2. 10*s.* 3*d.* **11.** £11. 11*s.* 0*d.* **12.** £35. 1*s.* 3*d.*
13. £28. 15*s.* 5*d.* **14.** £43. 10*s.* 4½*d.* **15.** £18. 3*s.* 2½*d.*
16. £22. 15*s.* 1*d.* **17.** £961. 5*s.* 0*d.* **18.** £14. 8*s.* 3¼*d.*
19. £424. 16*s.* 8*d.* **20.** £108. 18*s.* 6*d.* **21.** £587. 15*s.* 7*d.*
22. £134. 17*s.* 7*d.* **23.** £701. 14*s.* 8*d.* **24.** £51. 15*s.* 11*d.*

32. **25.** £44. 10*s.* 9*d.* **26.** £288. 19*s.* 11*d.* **27.** £135. 6*s.* 8*d.*
28. £697. 14*s.* 2*d.* **29.** £1446. 9*s.* 6*d.* **30.** £549. 19*s.* 0*d.*

EXERCISE II i

1. 5 yds. 0′ 2″. **2.** 12 yds. 1′ 2″. **3.** 14 yds. 1′ 0″.
4. 73 yds. 0′ 7″. **5.** 9 mi. 320 yds. **6.** 5 mi. 115 yds.
7. 15 yds. 1′ 2″. **8.** 17 yds. 1′ 1″. **9.** 6 mi. 544 yds.
10. 1 mi. 1471 yds. 0′ 6″. **11.** 1 mi. 1411 yds.
12. 464 yds. 0′ 3″.

EXERCISE II j

33. **1.** 10 lbs. 8 oz. **2.** 7 st. 12 lbs.
3. 2 tons 7 cwts. 0 qr. 0 lb. **4.** 17 lbs. 11 oz.
5. 40 st. 9 lbs. **6.** 4 tons 9 cwts. 6 st. 9 lbs.
7. 14 lbs. 2 oz. **8.** 24 st. 4 lbs.
9. 5 tons 2 cwts. 0 qr. 26 lbs. **10.** 15 lbs. 5 oz.
11. 55 st. 12 lbs. **12.** 9 tons 0 cwts. 1 st. 0 lb.
13. 18 cwts. 2 qrs. **14.** 2 tons 17 cwts. 0 qr. 20 lbs.
15. 3 tons 12 cwts. 2 qrs. **16.** 4 tons 8 cwts. 0 qr. 24 lbs.
17. 2 tons 15 cwts. 2 qrs. 4 lbs.

EXERCISE II k

PAGE

34. **1.** 7 h. 12 m. 7 s. **2.** 5 h. 28 m. 52 s.
3. 4 h. 23 m. 35 s. **4.** 8 d. 7 h. 59 m.
5. 5 d. 3 h. 47 m. **6.** 5 d. 15 h. 37 m. 35 s.

EXERCISE II l

1. 4 g. 2 q. 0 p. **2.** 4 g. 2 q. 0 p. **3.** 8 g. 0 q. 1 p.
4. 5 b. 7 g. 2 q. 1 p. **5.** 8 b. 5 g. 0 q. 0 p.

EXERCISE II m

35. **1.** £1. 1*s*. 5*d*. **2.** £2. 3*s*. 11*d*. **3.** £30. 12*s*. 6*d*.
4. £2. 17*s*. 7$\frac{3}{4}$*d*. **5.** £1. 17*s*. 9$\frac{1}{2}$*d*. **6.** £18. 1*s*. 6$\frac{1}{4}$*d*.
7. 13*s*. 11*d*. **8.** £2. 18*s*. 7*d*. **9.** £3. 3*s*. 2$\frac{1}{2}$*d*.
10. £8. 15*s*. 7*d*. **11.** £12. 16*s*. 11$\frac{3}{4}$*d*. **12.** £6. 16*s*. 9$\frac{3}{4}$*d*.

EXERCISE II n

1. 7 yds. 1′ 11″. **2.** 1 yd. 2′ 2″. **3.** 1 yd. 2′ 10″.
4. 1190 yds. **5.** 1726 yds. **6.** 1 mi. 1388 yds.

EXERCISE II o

1. 7 st. 8 lbs. **2.** 1 st. 12 lbs.
36. **3.** 1 cwt. 59 lbs. **4.** 3 cwts. 3 qrs.
5. 1 ton 9 cwts. 110 lbs. **6.** 1 ton 7 cwts. 6 st. 12 lbs.

EXERCISE II p

1. 47 m. 18 s. **2.** 3 h. 49 m. 31 s. **3.** 3 d. 18 h. 50 m.
4. 10 h. 20 m. **5.** 2 w. 4 d. 21 h. **6.** 4 d. 6 h.

EXERCISE II q

37. **1.** £2. 2*s*. 0*d*., £2. 12*s*. 6*d*., £8. 15*s*. 0*d*., £17. 10*s*. 0*d*.
2. £26. 10*s*. 6*d*., £44. 4*s*. 2*d*., £106. 2*s*. 0*d*., £358. 1*s*. 9*d*., £318. 6*s*.
3. £75. 10*s*. 2*d*., £528. 11*s*. 2*d*., £849. 9*s*. 4$\frac{1}{2}$*d*., £1359. 3*s*. 0*d*.
4. £316. 15*s*. 0*d*. **5.** £11391. 1*s*. 3*d*. **6.** £10171. 2*s*. 0*d*.
7. £30142. 17*s*. 3*d*. **8.** £14412. 19*s*. 10*d*. **9.** £6700. 2*s*. 4*d*.
10. £29630. 0*s*. 6*d*.
11. 15 tons, 45 tons, 75 tons, 86 tons 5 cwts., 461 tons 5 cwts.
38. **12.** 28 lbs. 14 oz., 1 qr. 16 lbs. 10 oz., 3 qrs. 7 lbs. 14 oz., 19 qrs. 27 lbs. 2 oz.
13. 11 c. 5 st. 8 lbs., 2 t. 18 c. 3 st. 12 lbs., 3 t. 1 c. 7 st. 13 lbs., 8 t. 18 c. 7 st. 9 lbs., 14 t. 15 c. 7 st. 5 lbs.
14. 142 t. 10 cwts. 80 lbs. **15.** 25 t. 14 cwts. 2 qrs. 16 lbs.
16. 5 yds. 2′ 2″, 34 yds. 1′ 0″, 65 yds. 2′ 5″.
17. 513 mi. 1120 yds., 616 mi. 640 yds., 652 mi. 560 yds.
18. 79 d. 0 h. 40 m. 3 s. **19.** 4 years 344 d. 11 h. 20 m.
20. 106 d. 15 h. 11 m. 1 s.

EXERCISE II r

PAGE

40. **1.** £1. 15*s*. 0*d*., 11*s*. 3*d*., 5*s*. 3*d*., 5*s*., 2*s*. 4*d*.
2. £20. 16*s*. 8*d*., £4. 3*s*. 4*d*., 8*s*. 4*d*., £3. 6*s*. 8*d*., 13*s*. 4*d*.
3. £1. 10*s*. 0*d*., 18*s*., 16*s*., 15*s*., 10*s*.
4. 18*s*. 4*d*., 9*s*. 2*d*., 8*s*. 4*d*., 1*s*. 10*d*., 2*s*. 1*d*.
5. £3. 1*s*. 3*d*. **6.** £2. 3*s*. 5*d*. **7.** £4. 11*s*. 2*d*.
8. £2. 13*s*. 5*d*. **9.** 1 yd. 1 ft. 1 in. **10.** 10 yds. 1 ft. 1 in.
11. 2 yds. 1 ft. 11 in. **12.** 5 yds. 2 ft. 7 in. **13.** 3 lbs. 14 oz.
14. 1 cwt. 2 qrs. 2 lbs. **15.** 3 tons 3 cwts. 74 lbs.
16. 10 cwts. 1 qr. 14 lbs. **17.** 5 cwts. 1 qr. 11 lbs. 3 oz.
18. 13 cwts. 14 lbs. **19.** 1 ton 4 cwts. 1 qr. 14 lbs.
20. 1 ton 14 cwts. 3 qrs.

EXERCISE II s

41. **1.** 12. **2.** 15. **3.** 23. **4.** 32.
5. 33, 1*s*. **6.** 11, 2*s*. 8*d*. **7.** 14, 2*s*. 1*d*. **8.** 222, 1*s*.
9. 27, 6*s*. 3*d*. **10.** 27. **11.** 16, 10*s*. 3*d*.
12. 11, 10*s*. 1*d*. **13.** 11, 6 in.

42. **14.** 79, 2 in. **15.** 257, 1 in. **16.** 21, 9 in.
17. 423, 9 in. **18.** 17, 1 ft. 11 in. **19.** 10, 80 lbs.
20. 7, 27 lbs. **21.** 118, 1 lb. 14 oz. **22.** 29, 121 lbs.
23. 78, 15 lbs. 12 oz. **24.** 6, 10 lbs. 2 oz. **25.** 36, 1 lb. 4 oz.

EXERCISE II t

1. £20. 9*s*. 9½*d*.

43. **2.** £56. 1*s*. 5*d*. **3.** £322. 6*s*. 1*d*. **4.** 1 ton 2 cwts. 2 qrs.
5. £20. 16*s*. 8*d*. **6.** 5 mi. 1520 yds. **7.** 1*s*. 4*d*.
8. 8 lbs. 12 oz. **9.** 72 days. **10.** £28. 0*s*. 8*d*.
11. 17*s*. 6*d*.

44. **12.** £22. 11*s*. 8*d*. **13.** 4*s*. 6*d*. **14.** 14*s*. 9*d*. **15.** 5.
16. 17 galls. 2 qts. **17.** 3*s*. 10*d*. **18.** 6*d*. **19.** 7½*d*.

45. **20.** 240. **21.** 25. **22.** 3 mi. 1320 yds.

EXERCISE II v

46. **1.** £180. **2.** £104. 12*s*. 6*d*. **3.** £122. 10*s*. 0*d*.
4. £234. 10*s*. 0*d*. **5.** £38. 10*s*. 6*d*. **6.** £57. 15*s*. 0*d*.
7. £1405. 16*s*. 0*d*. **8.** £184. 17*s*. 11¾*d*.
9. £152. 13*s*. 11½*d*. **10.** £617. 8*s*. 7½*d*.
11. 114 tons 2 cwts. 2 qrs. **12.** 236 t. 15 c. 3 q. 21 lbs.
13. 305 t. 15 c. 2 q. 9 lbs. **14.** 914 t. 3 c. 0 q. 16 lbs.
15. 938 t. 1 c. 3 q. **16.** 241 yds. 2 ft.
17. 1 mi. 952 yds. 2 ft. 8 in. **18.** 903 yds. 1 ft. 8½ in.

EXERCISE II w

PAGE

47. **1.** £23. 17*s*. 9*d*. **2.** £17. 17*s*. $5\frac{1}{4}$*d*. **3.** £10. 1*s*. $0\frac{1}{2}$*d*.
4. £5. 12*s*. $1\frac{1}{2}$*d*. **5.** £53. 15*s*. 9*d*. **6.** £4. 17*s*. 6*d*.
7. £7. 19*s*. $7\frac{3}{4}$*d*. **8.** £9. 11*s*. $8\frac{1}{2}$*d*. **9.** £22. 6*s*. $10\frac{1}{2}$*d*.
10. £4. 6*s*. $4\frac{3}{4}$*d*. **11.** £60. 9*s*. 7*d*. **12.** £242. 3*s*. 8*d*.
13. £645. 18*s*. $3\frac{1}{2}$*d*.

EXERCISE III b

49. **1.** 64. **2.** £9. **3.** 1*s*. $10\frac{1}{2}$*d*. **4.** 8, 7.
5. £22. 10*s*. 0*d*. **6.** 450. **7.** 500.

50. **8.** 1*s*. 8*d*. **9.** 3 lbs. **10.** 336. **11.** 36 lbs.
12. 84. **13.** 1200 lbs. **14.** 3840, 2400.

EXERCISE III d

51. **1.** 18 days. **2.** 10 days. **3.** 15, 1, 300, 12 days.

52. **4.** 16 hrs. **5.** 16 mins., 12 mins.

EXERCISE III e

1. 468. **2.** 4 hrs. **3.** £27. 10*s*. 0*d*. **4.** 10 days.
5. 5*s*. $7\frac{1}{2}$*d*.

53. **6.** 576 miles. **7.** 12 mins. **8.** £4. 7*s*. 6*d*.
9. 20 mi. **10.** 16. **11.** 40, 12.
12. 108. **13.** £16, 15 weeks. **14.** 30.

EXERCISE IV a

56. **1.** 21 sq. ft. **2.** 60 sq. in. **3.** 52 sq. yds.
4. 6 sq. mi. **5.** 24000 sq. yds. **6.** 29900 sq. yds.
7. 391 sq. in. **8.** 986 sq. in. **9.** 91 sq. ft.
10. 989 sq. ft. **11.** £48. **12.** £48.

57. **13.** £373. 13*s*. 0*d*. **14.** £15. **15.** £50. 1*s*. 0*d*.
16. £92. 19*s*. 0*d*. **17.** 4 sq. in., 128 sq. in. **18.** 10 ft.
19. 23 ft. **20.** 210 yds. **21.** 30 yds.
22. 37 in. **23.** 60 ft. **24.** 70 ft. **25.** 90.

58. **26.** 229 sq. ft. **27.** 2 ft. square = 4 sq. ft.

EXERCISE IV b

1. 576. **2.** 8208. **3.** 2. **4.** 5. **5.** 45.
6. 5184. **7.** 15. **8.** 2. **9.** 7. **10.** 9.
11. 13. **12.** 17. **13.** 384. **14.** 1760 sq. ft.

59. **15.** 15000 ft.^{2}, 2400 ft.^{2} **16.** 1026. **17.** 90, 10.
18. 1152, 8. **19.** £5. 5*s*. 0*d*. **20.** £2. 5*s*. 0*d*.
21. £3. 15*s*. 0*d*. **22.** 8. **23.** 48.
24. 200 ft., 480 ft. **25.** 3000 ft. **26.** 50.

EXERCISE IV c

PAGE

60. **1.** 5. **2.** 60.

61. **3.** 273. **4.** 39843. **5.** 6400. **6.** 25.
7. 60. **8.** 19360. **9.** 108900. **10.** 1521.
11. 69120. **12.** 330 yds. **13.** 220 yds.

EXERCISE IV d

1. 336 sq. ft.

62. **2.** 936 – 600. **3.** 282 sq. ft. **4.** 125 sq. ft.
5. 100 sq. in. **6.** 96, 84; 156, 84 sq. in.
7. 18000, 12800 sq. yds. **8.** 31 ft., 93 sq. ft.

63. **9.** 6″ all round; 35 sq. ft.
10. (1) 60. (2) 15. (3) 21. (4) shaded 24; unshaded 72.
(5) s. 36; u. 60. (6) s. 96; u. 64.
(7) s. 120; u. 40. (8) s. 122; u. 98. **11.** 65 lbs.

64. **12.** 2934 sq. in., 1512 sq. in. **13.** 9*s*.

EXERCISE IV e

66. **1.** 142. **2.** 568. **3.** 862. **4.** 3448.
5. 1728. **6.** 432. **7.** 13. **8.** 4.
9. 53. **10.** 88. **11.** 40, 64, 126 sq. in., 3 sq. ft.
12. 160, 256, 504 sq. in., 12 sq. ft.
13. 1008, 840, 950 sq. ft. **14.** $2h(l+b)$ sq. ft.

EXERCISE IV f

69. **1.** (i) 72. (ii) 432. (iii) 1296. (iv) 1080. (v) 9360. (vi) 5103.
2. (i) 120 cu. ft. (ii) 1617 cu. in. (iii) 12 cu. in. (iv) 720 cu. in.
3. $V=216$ cu. ft.; $S=228$ sq. ft.; $V=160$ cu. ft.; $S=232$ sq. ft.
4. $V=512$ cu. ft. in each case.
$S=$(i) 320, (ii) 320, (iii) 416, (iv) 528 sq. ft.
5. 48. **6.** 2880, $14\times10\times13=1820$, 1060 cu. in.

70. **7.** (i) 2816, 1504. (ii) 2652, 624. (iii) 954, 558. (iv) 4095, 2817.
8. (i) 1265, 685. (ii) 12936, 2904. (iii) 2160, 1744.
9. (i) 15 cu. in. (ii) 18 cu. ft. (iii) 38376 cu. in.
10. (i) 120 cu. in. (ii) 144 cu. ft. (iii) 307008 cu. in.

71. **11.** Fig. 1. 62 sq. in. Fig. 2. 48 sq. ft.
12. (i) 3228 sq. in. (ii) 3060 sq. in.

EXERCISE IV g

1. 8, 27, 64. **2.** 8, 27, 64, 216.
3. 8, 64, 216, 1728. **4.** 4. **5.** 20.

EXERCISE V a

PAGE

73. **1.** 2^5. **2.** $2^2 . 3 . 5$. **3.** 2^6.
4. 71. **5.** $5 . 17$. **6.** $2^2 . 23$.
7. $2^3 . 3 . 7$. **8.** $2^6 . 3$. **9.** 211.
10. $2^5 . 7$. **11.** 251. **12.** 2^8.
13. $3^2 . 5 . 7$. **14.** 7^3. **15.** $2^3 . 3^2 . 5$.
16. $2^2 . 11^2$. **17.** 571. **18.** $3^2 . 7 . 11$.
19. $3^3 . 7^3$. **20.** $2^3 . 3^3 . 7^3$. **21.** $3^2 . 7 . 11 . 41$.
22. $3^2 . 5^2 . 7^2$. **23.** $2^4 . 3^2 . 7 . 89$. **24.** $3 . 37$.
25. $11 . 101$. **26.** 11^4. **27.** $3 . 7 . 11 . 13 . 37$.
28. $2^9 . 7^2$.

EXERCISE V c

75. **1.** 2. **2.** 18. **3.** 12. **4.** 1. **5.** 1.
76. **6.** 24. **7.** 5. **8.** 16. **9.** 240. **10.** 49.
11. 11. **12.** 185. **13.** 2. **14.** 14. **15.** 8.
16. 8. **17.** 25. **18.** 24. **19.** 9. **20.** 37.

EXERCISE V d

1. 23. **2.** 24. **3.** 41. **4.** 87. **5.** 33.
6. 113. **7.** 59. **8.** 127. **9.** 239. **10.** 323.
11. 571. **12.** 874.

EXERCISE V f

78. **1.** $3^3 . 5^2$. **2.** $2^3 . 5 . 7 . 13$. **3.** $2 . 3^2 . 5 . 7 . 13$.
4. $2^2 . 3^2 . 5^2 . 7^2$. **5.** $2^3 . 3^3 . 7^3$. **6.** 120.
7. 240. **8.** 240. **9.** 84.
10. 1152. **11.** 180. **12.** 4608.
13. 123321. **14.** 686. **15.** 1500.
16. $2^4 . 3^2 . 5 . 7$. **17.** $2 . 3^2 . 7^2$. **18.** $2^2 . 3 . 7 . 11$.
19. $3^4 . 7^3$. **20.** $3 . 5 . 11 . 37 . 101$. **21.** $2^6 . 3^2 . 5 . 7$.
22. $2^4 . 3^4 . 11$. **23.** 2^8. **24.** $2^2 . 3^2 . 7^2 . 19 . 127$.
25. $2 . 3 . 5 . 7^2 . 11 . 101$. **26.** $2^2 . 3 . 5 . 7^2 . 13$.

EXERCISE V g

79. **1.** 16. **2.** 16. **3.** 16. **4.** 36. **5.** 39. **6.** 3.
7. 24. **8.** 96. **9.** 1260 in. = 35 yds.
80. **10.** After 6, 12, 18 secs.; 8.
11. 6 steps; never. **12.** 72 days; never.

EXERCISE V h

81. **1.** 16, 27, 256, 14, 33. **2.** 98. **3.** 30. **4.** 72. **5.** 99.
6. 175. **7.** 300. **8.** 12. **9.** 32. **10.** 30.
11. 42. **12.** 84. **13.** 105. **14.** 132. **15.** 55.
16. 350. **17.** 108. **18.** 1152. **19.** 153. **20.** 216.
21. 1200. **22.** 15. **23.** 160. **24.** 18. **25.** 77.
26. 640. **27.** 56. **28.** 945. **29.** 19600. **30.** 429.

EXERCISE V i

PAGE

82. **1.** 2, 9, 5, 6, 18, 98. **2.** 60. **3.** 77.
4. 66. **5.** 150. **6.** 11858. **7.** 3600.
8. 12. **9.** 14. **10.** 60. **11.** 50.
12. 15. **13.** 42. **14.** 24. **15.** 18.
16. 8. **17.** 45. **18.** 330. **19.** 72.

REVISION PAPERS

Paper 1

83. **1.** 286. **2.** 1, 3, 1. **3.** 383. **4.** 1826.
5. 16. **6.** £5698. 15*s*. 0*d*. **7.** £2. 5*s*. 0*d*.

Paper 2

84. **1.** 316566465. **2.** 32, 14. **3.** £1. 3*s*. 2*d*. **4.** 3*d*.
5. 145717, 1564511, 9625, 559845 pence.
6. 14 wks. 2 days; 36 wks. 5 days. **7.** £211. 7*s*. 11*d*.

Paper 3

1. 53094. **2.** 11, 11, 21, 103. **3.** 2 tons 2 cwts. 3 qrs. 12 lbs.
85. **4.** 13*s*. 0½*d*. **5.** £1. 9*s*. 8½*d*. **6.** 2*s*. 11*d*. **7.** £1259. 10*s*. 10*d*.

Paper 4

2. £10. 2*s*. 4½*d*. **3.** £42. 1*s*. 9*d*. **4.** 8*d*. **5.** £43. 8*s*. 6¼*d*.
86. **6.** 98. **7.** £721. 7*s*. 1*d*.

Paper 5

1. 12307965. **2.** 72; 63; 0. **3.** 9632.
4. £2. 13*s*. 10*d*. **5.** 210 tons 12 cwts. 1 qr. 24 lbs.
6. £7. 10*s*. 0*d*.; £5. 4*s*. 0*d*. **7.** £411. 4*s*. 0*d*.

Paper 6

87. **1.** 1232766. **2.** £1. 3*s*. 5*d*.; 1*s*. 7*d*. **3.** 231.
4. £17. 18*s*. 10½*d*. **5.** 5. **6.** 5*d*. **7.** £448. 17*s*. 6*d*.

Paper 7

1. 293715964. **2.** 2; 6. **3.** 2 mi. 80 yds.
4. £26. 2*s*. 11½*d*. **5.** £159. 17*s*. 4*d*. **6.** £48. 17*s*. 1*d*.

Paper 8

88. **1.** 43596125; 348420231. **2.** £1. 19*s*. 0*d*.
3. 138 yds. 2 ft. 8 in. **4.** £1. 6*s*. 8*d*.; 4*d*.
5. 3 mi. 1030 yds. 2 ft. 7 in. **6.** £25867. 12*s*. 6*d*.
7. £4. 15*s*. 0*d*.; £1. 5*s*. 0*d*.

Paper 9

PAGE

89. **1.** 1174575. **2.** 3, 5, 27, 29, 30.
3. 3960. **4.** lose 1*d.*; 0; 0; gain 6*s.* 8*d.*
5. $2\frac{1}{2}$*d.* **6.** £107. 6*s.* 8*d.* **7.** 3 mi. 1320 yds.

Paper 10

90. **1.** 23. **2.** 274; 183. **3.** 4*s.* $10\frac{1}{2}$*d.*
4. $2\frac{3}{4}$*d.*; $1\frac{1}{2}$*d.*; 1*s.*; $6\frac{1}{4}$*d.*; $8\frac{1}{4}$*d.* **5.** £20. 18*s.* 0*d.*
6. 9545. **7.** 1; 4800.

Paper 11

91. **1.** 50; 79. **2.** 106. **3.** $2\frac{1}{2}$ tons. **4.** 114.
5. £27. 11*s.* 3*d.* **6.** 40. **7.** 36 ft. **8.** 35.

Paper 12

92. **1.** 1 ton 6 cwts. 2 qrs. 19 lbs. 4 oz.
2. 9 tons 7 cwts. 16 lbs. **3.** 261. **4.** £4. 7*s.* 6*d.*
5. Japan 317, B.I. 379. **6.** 403 yds. 1 ft.
7. 2431; 593 cu. in.
93. **8.** 120.

Paper 13

1. 4 cwts. 1 qr. 9 lbs. **2.** 76. **3.** 88. **4.** £1. 16*s.* $6\frac{3}{4}$*d.*
5. 546. **6.** 1225. **7.** 2^{10}, $2^3 . 3^3 . 7^3$; 32, 42.

Paper 14

94. **1.** 5*s.* 10*d.* **2.** £156. 11*s.* 0*d.* **4.** 2800 sq. ft.
5. 144 cu. in. **6.** 40, 16. **7.** 42, 6.

Paper 15

1. £34277. 2*s.* 0*d.* **2.** £43749.
95. **3.** 6216. **4.** £30. 19*s.* 4*d.* **5.** 240, 188, 15000.
6. 10 lbs. 5 oz. **7.** 21, 4410.

Paper 16

1. 29′, 1450′. **2.** £26. 4*s.* 0*d.*
96. **3.** £102. **4.** 72. **5.** 240 sq. ft.
6. 77; 15015. **7.** 48; 45.

Paper 17

PAGE

96. **1.** £80. 18*s.* 2*d.* **2.** £37. 6*s.* 7*d.* **3.** £720.
4. 10. **5.** £210. 7*s.* 6*d.*; 29′ 8″.

97. **6.** 1 ton 19 cwts. **7.** 720.

Paper 18

1. 4*s.* 7*d.*; 3*s.* 6*d.*; 18*s.* 8*d.*; £16.
2. £471. 6*s.* 1*d.* **3.** £93. 13*s.* 4*d.* **4.** 430.
5. 975. **6.** 4, 19200. **7.** 66, 22.

Paper 19

98. **1.** 13*s.* **2.** £307. 3*s.* $4\frac{1}{2}d.$ **3.** 472, 332 sq. in.
4. £15. 1*s.* 6*d.* **5.** 1728, 48 tons 4 cwts.
6. 360 cu. in., 150 lbs. **7.** 720.

Paper 20

99. **1.** £357. 6*s.* 0*d.* **2.** £68. 19*s.* $1\frac{1}{2}d.$ **3.** £21. **4.** 11696.
5. 15 cwts. $7\frac{1}{2}$ lbs. **6.** 99, 110, 180. **7.** 108, 4.

MISCELLANEOUS EXERCISES

100. **1.** 65536, 4294967296. **2.** 12, 32; 16, 256; 5 and 6.
3. 666. **4.** 39. **5.** 2080. **6.** 12, 12, 6, 12 mins.

101. **7.** £1. **8.** £2. 8*s.* 0*d.* **9.** 9*d.*
10. 1, 2, 3, 4, 6, 12; 1080, 1440, 1800.
12. five 3 lb. jars. 6*s.* $10\frac{1}{2}d.$ **13.** 200.

102. **14.** 0, 8*s.* 8*d.*, 17*s.* 4*d.*, £1. 6*s.* 0*d.*, £1. 14*s.* 8*d.*, £2. 3*s.* 4*d.*, £2. 12*s.*
15. £3. 11*s.* 10*d.* **16.** 41. **17.** 457×38.
18. 15 secs. **19.** £13. 6*s.* 0*d.* **20.** 5, 11, 6.

103. **21.** 21*s.* **22.** 80, 64.
23. 256, 512, 8, 16. Unbroken in 1st and 4th. **24.** 7 ft.
25. (1) Ends 2′ 10″ by 2′ ; bottom 2′ 10″ by 3′ 10″.
or „ 2′ 10″ by 1′ 11″; „ 2′ 10″ by 4′.
(2) „ 2′ 10″ by 1′ 11″; „ 3′ by 4′.
(3) „ 3′ by 2′ ; „ 2′ 10″ by 3′ 10″.
(4) „ 3′ by 1′ 11″ ; „ 3′ by 4′.
or „ 3′ by 2′ ; „ 3′ by 3′ 10″.
In every case 38 sq. ft. 28 sq. in. are required.

104. **26.** noon July 21st. **27.** 2*s.* 11*d.* **28.** £12014433. 17*s.* 2*d.*
29. £13. **30.** £8. 8*s.* 0*d.*
31. (i) 2*d.*; 21*s.* + 2*s.* 6*d.*, 15*s.* + 8*s.* 6*d.* (ii) $5\frac{1}{2}d.$: 3 at 21*s.* + 15*s.*, 4 at 15*s.* + 18*s.*

105. **32.** Complete at 4.15. 17 or 18. **33.** 30 miles.
34. (i) Feb., Mar., Nov. (ii) Jan., Ap., Jul.

PART II

EXERCISE VI b

PAGE

108. **1.** 5*s*. **2.** 5*s*. 10*d*. **3.** 8*s*. 2*d*. **4.** 18*s*. 4*d*.
5. £1. 2*s*. 2*d*. **6.** 16*s*. 6*d*. **7.** 6 ft. **8.** 12 yds. 1′ 11″.
9. 600 yds. **10.** 6 st. 7 lbs. 9 oz.

109. **11.** 25 lbs. 5 oz. **12.** 8 tons 4 cwts. 2 qrs.
13. 9 galls. 1 qt. 1 pt. **14.** 7 galls. 3 qts. 1 pt.
15. $\frac{31}{200}$. **16.** $\frac{43}{720}$. **17.** $\frac{59}{1200}$. **18.** $\frac{61}{120}$.
19. $\frac{11}{36}$. **20.** $\frac{17}{30}$. **21.** $\frac{23}{28}$. **22.** $\frac{27}{80}$.
23. $\frac{51}{70}$. **24.** $\frac{53}{5280}$.

EXERCISE VI e

112. **1.** $\frac{3}{10}$. **2.** $\frac{19}{25}$. **3.** $\frac{2}{5}$. **4.** $\frac{5}{9}$. **5.** $\frac{2}{3}$.
6. $\frac{9}{20}$. **7.** $\frac{2}{3}$. **8.** $\frac{3}{20}$. **9.** $\frac{7}{8}$. **10.** $\frac{2}{7}$.
11. $\frac{22}{101}$. **12.** $\frac{2}{3}$. **13.** $\frac{91}{128}$. **14.** $\frac{11}{25}$. **15.** $\frac{9}{20}$.
16. $\frac{3}{7}$. **17.** $\frac{2}{5}$. **18.** $\frac{8}{9}$. **19.** $\frac{5}{21}$. **20.** $\frac{31}{47}$.
21. $\frac{27}{35}$. **22.** $\frac{6}{13}$. **23.** $\frac{13}{14}$. **24.** $\frac{13}{23}$. **25.** $\frac{49}{55}$.
26. $\frac{3}{20}$. **27.** $\frac{7}{12}$. **28.** $\frac{3}{13}$.

EXERCISE VI f

113 **1.** $\frac{5}{6}$. **2.** $\frac{3}{4}$. **3.** $\frac{11}{15}$. **4.** $\frac{4}{9}$.
5. $\frac{4}{9}$. **6.** $\frac{10}{13}$. **7.** $\frac{5}{24}$. **8.** $\frac{17}{21}$.
9. $\frac{31}{64}$. **10.** $\frac{11}{36}$. **11.** $\frac{13}{50}$. **12.** $\frac{22}{7}$.
13. $\frac{3}{4}, \frac{2}{3}, \frac{1}{2}$. **14.** $\frac{4}{5}, \frac{3}{4}, \frac{2}{3}$. **15.** $\frac{2}{5}, \frac{1}{3}, \frac{3}{10}$.
16. $\frac{7}{10}, \frac{2}{3}, \frac{7}{12}$. **17.** $\frac{5}{6}, \frac{7}{9}, \frac{3}{4}$. **18.** $\frac{7}{12}, \frac{9}{16}, \frac{5}{9}$.

EXERCISE VI h

115. **1.** $\frac{93}{7}$. **2.** $\frac{331}{13}$. **3.** $\frac{454}{11}$. **4.** $\frac{655}{9}$.
5. $\frac{327}{17}$. **6.** $\frac{223}{18}$. **7.** $\frac{317}{14}$. **8.** $\frac{197}{15}$.

116. **9.** $\frac{139}{8}$. **10.** $\frac{339}{23}$. **11.** $\frac{665}{24}$. **12.** $\frac{822}{43}$.
13. $\frac{304}{13}$. **14.** $\frac{251}{19}$. **15.** $\frac{791}{34}$. **16.** $\frac{1323}{32}$.
17. $\frac{627}{25}$. **18.** $\frac{534}{31}$. **19.** $\frac{2053}{64}$. **20.** $\frac{3854}{75}$.
21. $23\frac{17}{41}$. **22.** $13\frac{16}{17}$. **23.** $14\frac{31}{43}$. **24.** $20\frac{5}{36}$.
25. $10\frac{23}{112}$. **26.** $48\frac{25}{53}$. **27.** $17\frac{73}{101}$. **28.** $38\frac{78}{97}$.
29. $4\frac{125}{128}$. **30.** $16\frac{41}{77}$. **31.** $55\frac{1}{202}$. **32.** $30\frac{45}{143}$.
33. $83\frac{54}{89}$. **34.** $58\frac{144}{157}$. **35.** $94\frac{21}{47}$. **36.** $36\frac{59}{95}$.

EXERCISE VI l

122. **1.** $\frac{3}{4}$. **2.** $\frac{7}{10}$. **3.** $\frac{9}{14}$. **4.** $1\frac{1}{10}$. **5.** $\frac{11}{42}$.
6. $\frac{7}{18}$. **7.** $\frac{17}{30}$. **8.** $\frac{41}{60}$. **9.** $\frac{37}{100}$. **10.** $\frac{44}{105}$.
11. $5\frac{5}{6}$. **12.** $9\frac{3}{4}$. **13.** $7\frac{5}{6}$. **14.** $5\frac{7}{8}$. **15.** $9\frac{5}{14}$.
16. $5\frac{13}{15}$. **17.** $4\frac{5}{9}$. **18.** $5\frac{19}{42}$. **19.** $4\frac{1}{4}$. **20.** $4\frac{1}{10}$.
21. $2\frac{1}{4}$. **22.** $4\frac{7}{36}$. **23.** $7\frac{19}{24}$. **24.** $6\frac{19}{66}$. **25.** $7\frac{11}{24}$.
26. $5\frac{7}{20}$. **27.** $15\frac{11}{20}$. **28.** $11\frac{7}{8}$. **29.** 6. **30.** $8\frac{1}{7}$.
31. $5\frac{5}{18}$. **32.** $2\frac{2}{9}$. **33.** $5\frac{23}{36}$. **34.** $4\frac{3}{4}$. **35.** $4\frac{1}{30}$.
36. $3\frac{1}{7}$. **37.** $5\frac{9}{10}$. **38.** $\frac{2}{97}$.

PAGE

EXERCISE VI m

123. 1. $\frac{1}{2}$. 2. $\frac{2}{3}$. 3. $\frac{1}{4}$. 4. $\frac{1}{2}$. 5. $\frac{1}{6}$. 6. $\frac{4}{21}$.

124. 7. $\frac{1}{20}$. 8. $\frac{1}{4}$. 9. $\frac{1}{10}$. 10. $\frac{1}{6}$. 11. $\frac{7}{12}$.
12. $\frac{1}{24}$. 13. $\frac{5}{36}$. 14. $\frac{1}{6}$. 15. $\frac{1}{6}$. 16. $\frac{1}{18}$.
17. $1\frac{5}{16}$. 18. $1\frac{1}{4}$. 19. $3\frac{1}{2}$. 20. $2\frac{1}{8}$. 21. $3\frac{1}{6}$.
22. $2\frac{1}{2}$. 23. $5\frac{1}{8}$. 24. $3\frac{1}{14}$. 25. $\frac{3}{4}$. 26. $\frac{5}{6}$.
27. $\frac{7}{8}$. 28. $\frac{11}{12}$. 29. $2\frac{11}{12}$. 30. $1\frac{7}{8}$. 31. $3\frac{7}{8}$.
32. $2\frac{1}{2}$. 33. $4\frac{3}{8}$. 34. $\frac{3}{4}$. 35. $\frac{4}{7}$. 36. $\frac{1}{5}$.
37. $\frac{3}{8}$. 38. $\frac{4}{11}$. 39. $\frac{7}{12}$. 40. $1\frac{5}{8}$. 41. $3\frac{1}{7}$.
42. $9\frac{5}{9}$. 43. $2\frac{1}{24}$. 44. $1\frac{17}{24}$. 45. $1\frac{4}{5}$. 46. $1\frac{19}{20}$.
47. $3\frac{1}{10}$. 48. $1\frac{11}{12}$. 49. $4\frac{1}{42}$. 50. $1\frac{34}{35}$. 51. $4\frac{5}{14}$.
52. $1\frac{49}{60}$. 53. $1\frac{5}{14}$. 54. $1\frac{163}{252}$.

EXERCISE VI n

125. 1. $5\frac{1}{3}$. 2. $6\frac{1}{2}$. 3. $3\frac{1}{10}$. 4. $5\frac{5}{8}$. 5. $3\frac{1}{3}$.
6. $4\frac{1}{24}$. 7. $3\frac{2}{3}$. 8. $7\frac{29}{36}$. 9. $\frac{2}{3}$. 10. $10\frac{1}{4}$.
11. $2\frac{101}{144}$. 12. $1\frac{149}{300}$. 13. $11\frac{61}{294}$. 14. $2\frac{21}{32}$. 15. $\frac{3}{8}$.
16. 1. 17. $3\frac{2}{5}$. 18. $4\frac{1}{7}$. 19. $4\frac{17}{36}$. 20. $5\frac{11}{60}$.
21. $1\frac{1}{2}$. 22. $\frac{1}{10}$. 23. $\frac{11}{12}$. 24. $\frac{5}{12}$. 25. $\frac{1}{8}$.
26. $\frac{5}{6}$.

EXERCISE VI o

126. 1. $\frac{13}{16}''$. 2. $16\frac{5}{6}$ y. 3. $\frac{1}{20}$. 4. $\frac{7}{8}$, $\frac{1}{8}$. 5. $1\frac{19}{30}$.
6. $\frac{13}{60}$. 7. $\frac{1}{3}$, 30/-, 20/-. 8. $\frac{7}{10}$. 9. $\frac{1}{50}$.
10. $\frac{7}{40}$; 10, 32, 24, 14 miles. 11. $\frac{1}{3}$, $\frac{1}{4}$, $\frac{7}{12}$.

127. 12. $\frac{73}{184}$. 13. $1\frac{1}{3}$ ft. 14. $\frac{1}{156}$ ft. $=\frac{1}{13}''$.

EXERCISE VI q

128. 1. 6. 2. $1\frac{1}{4}$. 3. $10\frac{1}{2}$. 4. 4. 5. 9.
6. 3. 7. 15. 8. 4. 9. 2. 10. $\frac{1}{2}$.
11. $\frac{5}{6}$. 12. $1\frac{1}{3}$. 13. $\frac{2}{3}$. 14. $\frac{3}{4}$. 15. $\frac{2}{3}$.
16. $4\frac{1}{2}$. 17. $7\frac{1}{2}$ 18. $4\frac{1}{2}$.

129. 19. $7\frac{1}{2}$. 20. $7\frac{1}{2}$. 21. $3\frac{1}{3}$. 22. $1\frac{1}{3}$. 23. $3\frac{3}{4}$.
24. $10\frac{5}{8}$. 25. 10. 26. 22. 27. $6\frac{2}{3}$. 28. $8\frac{1}{4}$.
29. $19\frac{1}{6}$. 30. $9\frac{1}{2}$. 31. $12\frac{1}{2}$. 32. $15\frac{1}{3}$. 33. $31\frac{1}{2}$.
34. $9\frac{1}{3}$. 35. $27\frac{1}{3}$. 36. $147\frac{1}{2}$.

EXERCISE VI t

132. 1. $\frac{6}{11}$. 2. $\frac{10}{13}$. 3. $\frac{1}{8}$. 4. $\frac{4}{15}$.
5. $\frac{2}{3}$, $\frac{1}{10}$, $\frac{5}{14}$, 1. 6. $\frac{3}{10}$, $\frac{5}{24}$, 1, 2. 7. $\frac{1}{3}$, $\frac{5}{24}$, 1, $\frac{4}{9}$.
8. $\frac{4}{35}$, 1, 2, 6. 9. 1, 2, $1\frac{1}{2}$, 6. 10. $3\frac{3}{11}$, $\frac{4}{15}$, $2\frac{1}{5}$, $1\frac{1}{4}$.
11. $2\frac{1}{4}$. 12. $1\frac{7}{9}$. 13. $1\frac{9}{16}$. 14. $6\frac{1}{4}$. 15. $3\frac{1}{16}$.
16. $\frac{1}{24}$. 17. $\frac{1}{4}$. 18. $1\frac{1}{2}$. 19. $\frac{1}{12}$.

133. 20. 1. 21. $15\frac{5}{8}$. 22. $37\frac{1}{27}$. 23. $\frac{1}{16}$.
24. $\frac{1}{10000}$. 25. $\frac{1}{32}$. 26. 8. 27. $\frac{5}{8}$.
28. 4. 29. $8\frac{2}{5}$. 30. $\frac{7}{16}$. 31. 3.

EXERCISE VI w

PAGE

135. **1.** $1\frac{3}{25}$. **2.** $\frac{25}{28}$. **3.** $\frac{15}{16}$. **4.** $1\frac{1}{15}$. **5.** $\frac{2}{5}$.
6. $2\frac{1}{2}$. **7.** $\frac{2}{3}$. **8.** $1\frac{1}{2}$. **9.** $\frac{4}{15}$. **10.** $3\frac{3}{4}$.
11. $3\frac{3}{4}$. **12.** $3\frac{3}{4}$. **13.** $1\frac{1}{2}$. **14.** $\frac{2}{3}$. **15.** $7\frac{1}{2}$.
16. $\frac{2}{15}$. **17.** 3. **18.** $1\frac{2}{3}$. **19.** $1\frac{1}{7}$. **20.** $3\frac{1}{3}$.
21. $\frac{2}{5}$. **22.** $\frac{4}{15}$. **23.** $1\frac{8}{17}$. **24.** $\frac{14}{27}$. **25.** $\frac{8}{35}$.
26. $\frac{30}{49}$. **27.** $5\frac{4}{7}$.

EXERCISE VI x

136. **1.** $\frac{1}{3}$. **2.** $\frac{5}{7}$. **3.** $\frac{1}{17}$. **4.** $\frac{14}{15}$. **5.** $1\frac{1}{14}$.
6. $\frac{1}{2}$. **7.** 2. **8.** 3. **9.** $\frac{1}{3}$. **10.** $\frac{6}{7}$.
11. $\frac{3}{4}$. **12.** $\frac{4}{21}$. **13.** 2. **14.** $\frac{3}{4}$. **15.** $16\frac{1}{4}$.
16. $2\frac{2}{3}$. **17.** $\frac{4}{5}$. **18.** $1\frac{1}{4}$. **19.** $1\frac{25}{27}$. **20.** $1\frac{1}{5}$.
21. $2\frac{1}{2}$. **22.** $1\frac{3}{4}$. **23.** $1\frac{5}{7}$. **24.** $1\frac{4}{9}$. **25.** $\frac{2}{7}$.
26. $\frac{4}{13}$. **27.** $1\frac{1}{5}$. **28.** $1\frac{2}{7}$. **29.** $\frac{4}{17}$. **30.** $\frac{25}{26}$.

EXERCISE VI y

137. **1.** 13. **2.** 16. **3.** 13. **4.** $5\frac{1}{2}$. **5.** 4.
6. $5\frac{1}{2}$. **7.** $7\frac{1}{2}$. **8.** $7\frac{1}{2}$. **9.** $\frac{3}{10}$. **10.** $1\frac{1}{5}$.
11. $\frac{7}{12}$. **12.** $\frac{5}{24}$. **13.** $\frac{7}{12}$. **14.** $\frac{7}{12}$. **15.** $1\frac{5}{6}$.
16. $3\frac{1}{3}$. **17.** $1\frac{5}{6}$. **18.** $\frac{2}{3}$. **19.** $\frac{2}{3}$. **20.** $\frac{3}{8}$.
21. 6. **22.** $1\frac{31}{120}$. **23.** $1\frac{61}{120}$. **24.** $\frac{17}{24}$. **25.** $\frac{17}{24}$.
26. $1\frac{31}{120}$. **27.** $1\frac{31}{120}$. **28.** $1\frac{31}{120}$. **29.** $6\frac{1}{2}$. **30.** $6\frac{1}{2}$.
31. $\frac{3}{4}$. **32.** $\frac{39}{49}$. **33.** $3\frac{19}{26}$. **34.** $\frac{3}{13}$.

EXERCISE VI z

138. **1.** $\frac{3}{4}$. **2.** $\frac{184}{1149}$. **3.** $88\frac{8}{9}$ tons.
4. $2\frac{1}{2}$, $6\frac{1}{4}$, 30, $4\frac{3}{8}$, $7\frac{3}{16}$, $62\frac{1}{2}$, 250 lbs. **5.** 6/3.

139. **6.** $36\frac{4}{11}$, $109\frac{1}{11}$, 1920, 2640. **7.** 16, 21. **8.** 5/-.
9. 93 mi. **10.** $8\frac{4}{19}$ cwts. **11.** 6*d.*, 4/-, £1.
12. £1, 1*s.*, £2. 10*s.*, £2. 5*s.*, £14. 15*s.*
13. $\frac{1}{5}$, $\frac{4}{5}$, 8, 2, $2\frac{2}{3}$, $\frac{1}{4}$; $11\frac{1}{5}$. **14.** £15. **15.** $15\frac{3}{4}$ mi.
16. $5\frac{1}{3}$ days. **17.** $\frac{4}{11}$. **18.** $\frac{1}{57}$, $\frac{8}{57}$, 7 days and 1 hr.

140. **19.** $\frac{17}{32}$, 7 mill. **20.** 42′. **21.** £180, £168.
22. £3. 15*s.* **23.** $\frac{8}{15}$, $\frac{4}{5}$. **24.** $\frac{1}{6}$. **25.** 5*s.* 10*d.*
26. 5. **27.** 28. **28.** 1*s.*, 11*s.*

141. **29.** $\frac{7}{25}$, $\frac{18}{25}$, 25,000, 2500, 4500. **30.** £4. 10*s.*
31. £21. 12*s.* **32.** $\frac{5}{8}$. **33.** £3658. 13*s.* 4*d.*
34. $\frac{271}{1000}$, $\frac{729}{1000}$. **35.** $33\frac{3}{22}$ ft. **36.** 4000 m.
37. 18, 1 ft. **38.** 20, 1 in. **39.** $\frac{2}{3}$. **40.** $\frac{19}{24}$.

142. **41.** (*a*) $\frac{43}{60}$, (*b*) $\frac{43}{84}$, (*c*) 0, (*d*) 0, (*e*) $\frac{1}{2}$, (*f*) $\frac{125}{132}$.
42. Greater. **48.** $\frac{5}{13}$.

EXERCISE VI aa

PAGE

143. **1.** $\frac{14}{15}$. **2.** $1\frac{1}{6}$. **3.** $1\frac{1}{8}$. **4.** $4\frac{1}{5}$.
5. $9\frac{2}{7}$. **6.** $4\frac{1}{5}$.

144. **7.** $1\frac{13}{15}$. **8.** $5\frac{13}{15}$. **9.** $1\frac{1}{9}$. **10.** $\frac{3}{5}$.
11. $\frac{7}{12}$. **12.** 1. **13.** $9\frac{29}{90}$. **14.** $6\frac{1}{9}$.
15. $\frac{11}{130}$. **16.** $1\frac{6}{13}$. **17.** $\frac{95}{1474}$. **18.** 4.
19. $\frac{58}{429}$. **20.** $1\frac{103}{129}$. **21.** $2\frac{2}{3}$. **22.** $\frac{22}{105}$.
23. $\frac{3}{5}$. **24.** $\frac{49}{66}$. **25.** $2\frac{101}{272}$. **26.** $\frac{3}{4}$.
27. $2\frac{11}{36}$. **28.** 4. **29.** 1. **30.** $\frac{11}{24}$.

145. **31.** $\frac{406}{449}$. **32.** $1\frac{69}{70}$. **33.** $9\frac{3}{7}$. **34.** $3\frac{151}{250}$.
35. $1\frac{89}{207}$. **36.** $2\frac{5}{8}$. **37.** 3. **38.** $1\frac{395}{1204}$.
39. 28. **40.** $3\frac{17}{21}$. **41.** $\frac{1}{7}$. **42.** $\frac{1}{9}$.

EXERCISE VII a

148. **1.** 10, $\frac{2}{3}$, $\frac{2}{5}$. **2.** 126 fr. **3.** £39, 7.
4. 6 days. **5.** 72 y., 8*s*. **6.** 16, 24.
7. £3. 10*s*. 0*d*., 15. **8.** £1. 6*s*. 8*d*., 11. **9.** 14 d.
10. 315, 12*s*. 6*d*. **11.** £121, 19.

149. **12.** 20, 12, 15. **13.** 120, 36 days. **14.** 16*s*. 8*d*., 17.
15. £2. 3*s*. 4*d*., 13. **16.** 20. **17.** 54 miles, 40 mins.
18. 246, 9*s*. 7*d*. **19.** 8 hr., 2 h. 24 m., 1 h. 36 m.
20. £7. 9*s*. 10*d*., 15. **21.** £47. 10*s*. 0*d*. **22.** 72*d*.
23. £1. 4*s*. 2*d*., $4\frac{1}{2}$. **24.** £2. 7*s*. 10*d*., 9.
25. 400, $266\frac{2}{3}$ days.

150. **26.** $6\frac{1}{4}$ m., 77 min. **27.** $3\frac{3}{4}$ h., 35 mi./hr.
28. 40, £1. 1*s*. $10\frac{1}{2}$*d*. **29.** £13. 14*s*. 5*d*.
30. 48, 30 days. **31.** £5. 12*s*. 6*d*.
32. 66. **33.** $15\frac{5}{8}$ miles, $10\frac{2}{5}$ in. **34.** 72.

EXERCISE VII b

151. **1.** 2, 30*d*., £5. 6*s*. 8*d*., 3. **2.** £6, 4, $12\frac{4}{5}$.
3. 260 days altogether. **4.** 36, 18. **5.** 24 wks.
6. $4\frac{2}{3}$. **7.** 6*s*. 8*d*. **8.** 8. **9.** £23. 15*s*. 0*d*.

152. **10.** £6. 18*s*. 8*d*. **11.** B. **12.** 2 h. 59 m. 59 s.
13. 6.15$\frac{3}{7}$ A.M. **14.** 6 h. 2 m. 16 s.
15. 1 Aug. 5 A.M. **16.** 4 h. 7 m. 10 s. **17.** Fri. 7 A.M.
18. 7 Ap. 1 P.M. 8 h. 59 m. $40\frac{4}{5}$ s.
19. 15 Mar. 7 A.M. $4\frac{9}{77}$ mins. before noon.

EXERCISE VIII d

161. **1.** 334·832. **2.** 55·183. **3.** 38·4365. **4.** 85·6191.
5. 21·3328. **6.** 650·5766. **7.** 50·6673. **8.** 8·0684.
9. 65·5547. **10.** 97·22125. **11.** 24·8993. **12.** 10·423.
13. 285·233. **14.** 251·9009. **15.** 1·58785.

EXERCISE VIII e

PAGE

161. **1.** 0·823. **2.** 1·359. **3.** 1·633. **4.** 0·2246.
5. 36·353. **6.** 1·978. **7.** 4·2719. **8.** 88·898.
9. 0·38854. **10.** 0·299937. **11.** 0·99038. **12.** 1·7635.
13. 0·85664. **14.** 0·008446. **15.** 0·0130861. **16.** 1·93815.

162. **17.** 1·534316. **18.** 3·238351. **19.** 6·237416.
20. 7·321726. **21.** 0·999889. **22.** 6·348527.
23. 6·237316. **24.** 8·287927. **25.** 13·380216.

EXERCISE VIII g

1. 32·742. **2.** 318·886. **3.** 1·4766. **4.** 2·74661.
5. 2·8561. **6.** 66·0795. **7.** 2·4695. **8.** 3·91637.
9. 1·2361. **10.** 65·0919.

EXERCISE VIII i

163. **1.** 5·1. **2.** 403. **3.** 5·775. **4.** 0·5904.
5. 0·426. **6.** 155·4. **7.** 7·502. **8.** 0·1458.
9. 24·76. **10.** 0·6454. **11.** 35·16. **12.** 0·5412.

EXERCISE VIII j

164. **1.** 0·1196. **2.** 0·03441. **3.** 23·22. **4.** 35·1.
5. 0·001768. **6.** 49·58. **7.** 6·63. **8.** 38·61.
9. 180. **10.** 298·8. **11.** 0·1848. **12.** 0·0476.

EXERCISE VIII l (See Exercise VIII j, p. 164)

166. **1.** 188. **2.** 3250. **3.** 282. **4.** 0·288. **5.** 0·264.
6. 3·64. **7.** 203. **8.** 0·0117. **9.** 0·000284.
10. 36·21. **11.** 0·4824. **12.** 197·892. **13.** 417·6.
14. 0·01271. **15.** 2·14731. **16.** 1·08603. **17.** 0·78694.
18. 4·4457. **19.** 0·3621528. **20.** 54·74794. **21.** 4·186763.
22. 1209·5168.

EXERCISE VIII n

168. **1.** 241·75. **2.** 6·45. **3.** 0·068. **4.** 0·0154.
5. 1·35. **6.** 0·025375. **7.** 0·80375. **8.** 61·4.
9. 1·61. **10.** 0·122. **11.** 176·5. **12.** 0·58.
13. 0·0048875. **14.** 0·081. **15.** 0·082. **16.** 0·0051.

EXERCISE VIII o

170. **1.** $0{\cdot}\dot{3}$. **2.** $0{\cdot}1\dot{6}$. **3.** $0{\cdot}\dot{1}4285\dot{7}$. **4.** $0{\cdot}\dot{1}$. **5.** $0{\cdot}\dot{0}\dot{9}$. **6.** $0{\cdot}\dot{6}$.
7. $0{\cdot}\dot{2}8571\dot{4}$. **8.** $0{\cdot}\dot{2}$. **9.** $0{\cdot}\dot{1}\dot{8}$. **10.** $0{\cdot}\dot{4}2857\dot{1}$. **11.** $0{\cdot}\dot{2}\dot{7}$.
12. $0{\cdot}\dot{5}7142\dot{8}$. **13.** $0{\cdot}\dot{4}$. **14.** $0{\cdot}\dot{3}\dot{6}$. **15.** $0{\cdot}8\dot{3}$. **16.** $0{\cdot}\dot{7}1428\dot{5}$.
17. $0{\cdot}\dot{5}$. **18.** $0{\cdot}\dot{4}\dot{5}$. **19.** $0{\cdot}\dot{8}5714\dot{2}$. **20.** $0{\cdot}\dot{5}\dot{4}$.
21. $0{\cdot}\dot{7}$. **22.** $0{\cdot}\dot{6}\dot{3}$. **23.** $0{\cdot}\dot{8}$. **24.** $0{\cdot}\dot{7}\dot{2}$. **25.** $0{\cdot}\dot{8}\dot{1}$.
26. 4·333. **27.** 3·25. **28.** 1·857. **29.** 1·625.

PAGE

EXERCISE VIII o (continued)

170. **30.** 1·222. **31.** 9·667. **32.** 2·556. **33.** 2·091.
34. 8·2. **35.** 7·429. **36.** 2·182. **37.** 8·875.
38. 3·917. **39.** 3·222. **40.** 6·375. **41.** 11·455.
42. 40·833. **43.** 18·75. **44.** 12·083. **45.** 5·286.
46. 37·125. **47.** 14·727. **48.** 6·714. **49.** 21·583.
50. 39·909. **51.** 12·857. **52.** 22·111. **53.** 29·917.
54. 21·545. **55.** 38·889.

EXERCISE VIII p

171. **1.** 21. **2.** 13. **3.** 50·25. **4.** 104. **5.** 3·209. **6.** 0·104.
7. 0·012. **8.** 0·00248. **9.** 0·203. **10.** 5·4. **11.** 81·7.
12. 312. **13.** 88. **14.** 7·8. **15.** 0·00648. **16.** 0·005.
17. 0·08302. **18.** 0·672.

172. **19.** 0·746. **20.** 0·4. **21.** 1·6. **22.** 6·25. **23.** 0·0859375.
24. 0·032. **25.** 0·0032. **26.** 0·1953125.

EXERCISE VIII q

1. 240·6. **2.** 24163000. **3.** 40. **4.** 0·004.
5. 4900. **6.** 0·04. **7.** 0·541. **8.** 22·74.
9. 0·853. **10.** 3764. **11.** 3·212. **12.** 7·86.
13. 6230. **14.** 0·0865. **15.** 6·842. **16.** 0·4506.
17. 0·00525. **18.** 0·24. **19.** 0·1007. **20.** 10300.

173. **21.** 43·1. **22.** 0·082. **23.** 62. **24.** 0·12.
25. 290. **26.** 0·20. **27.** 0·010. **28.** 0·0012.

EXERCISE VIII s

174. **1.** $\frac{1}{2}$. **2.** $\frac{1}{4}$. **3.** $\frac{3}{4}$. **4.** $\frac{1}{16}$.
5. $\frac{5}{8}$. **6.** $\frac{7}{20}$. **7.** $\frac{7}{40}$. **8.** $\frac{3}{20}$.
9. $\frac{3}{40}$. **10.** $5\frac{13}{20}$. **11.** $9\frac{16}{25}$. **12.** $1\frac{7}{25}$.
13. $\frac{16}{625}$. **14.** $3\frac{63}{125}$. **15.** $10\frac{3}{8}$. **16.** $12\frac{51}{125}$.

EXERCISE VIII u

175. **1.** 0·2625. **2.** 0·079. **3.** 0·117. **4.** 0·246.
5. 0·579. **6.** 0·621. **7.** 0·7125. **8.** 0·779.
9. 0·883. **10.** 0·958. **11.** 0·917. **12.** 0·920.
13. 2·858. **14.** 1·860. **15.** 5·4125. **16.** 4·415.
17. 6·416. **18.** 7·417. **19.** 0·006. **20.** 0·206.
21. 0·2375. **22.** 0·3625. **23.** 0·756. **24.** 0·331.
25. 0·3375. **26.** 0·794. **27.** 0·156.

EXERCISE VIII v

1. 10*s.* **2.** 5*s.* **3.** 12*s.* **4.** 13*s.* **5.** 16*s.* **6.** 19*s.*
7. 6*d.* **8.** 2*s.* 6*d.* **9.** 2*s.* 10*d.* **10.** 14*s.* 6*d.*
11. 3*s.* 7*d.* **12.** 18*s.* 6*d.* **13.** 17*s.* 8*d.*
14. £2. 5*s.* 9*d.* **15.** £3. 2*s.* 10*d.* **16.** £17. 8*s.* 10*d.*
17. £18. 8*s.* 1*d.* **18.** £24. 0*s.* 1*d.* **19.** £17. 12*s.* 10*d.*
20. £17. 12*s.* 11*d.* **21.** £17. 12*s.* 11*d.* **22.** £17. 12*s.* 11*d.*
23. £17. 12*s.* 11*d.* **24.** £8. 5*s.* 11*d.*

EXERCISE VIII w

PAGE

177. **1.** 0·250. **2.** 0·750. **3.** 0·167. **4.** 0·200.
5. 2·917. **6.** 0·833. **7.** 0·258. **8.** 0·671.
9. 0·846. **10.** 0·871. **11.** 0·489. **12.** 0·270.
13. 2 h. 30 m. **14.** 2 h. 33 m. 36 s. **15.** 13 h. 34 m. 30 s.
16. 6 h. 14 m. 6 s. **17.** 12 h. 0 m. 7 s. **18.** 9 h. 8 m. 49 s.

178. **19.** 0·708. **20.** 0·354. **21.** 0·427. **22.** 0·531.
23. 1·172. **24.** 1·266. **25.** 3 yds. 1′ $1\frac{1}{2}$″.
26. 2′ 7″. **27.** 2′ 8″. **28.** 586 yds. 3″. **29.** 176 yds.
30. 17 yds. 1′ 10″. **31.** 0·150. **32.** 0·400. **33.** 0·0625.
34. 0·007. **35.** 0·045. **36.** 0·098. **37.** 0·187.
38. 0·595. **39.** 2 cwts. **40.** 46 cwts. 45 lbs.
41. 22 lbs. **42.** 2 lbs. **43.** 1 cwt. 16 lbs. **44.** 76 lbs.

EXERCISE VIII x

1. 0·19375. **2.** £13. 12*s*. 3*d*. **3.** £0·308.
4. £4. 7*s*. $1\frac{1}{2}$*d*. **5.** 1·105. **6.** £5. 12*s*. 2*d*.
7. 0·180. **8.** 13*s*. 5*d*. **9.** 0·09375.

179. **10.** 7 cwts. 15 lbs. **11.** £17·226, £17. 4*s*. $6\frac{1}{4}$*d*.
12. 7000. **13.** 16*s*. 10*d*. **14.** £34. 7*s*. 9*d*. **15.** £4. 16*s*. 11*d*.
16. £5. 1*s*. 2*d*. **17.** 0·246875, £1. 0*s*. 9*d*.
18. £17. 13*s*. 1*d*. **19.** 0·17.
20. £65233. 3*s*. 0*d*., £53223. 5*s*. 0*d*. **21.** £6. 3*s*. 4*d*.

180. **22.** (i) 2·4, (ii) 6·5, (iii) 9·1, (iv) 14·0, (v) 20·0, (vi) 36·6, (vii) 40·4.
23. (i) £16. 4*s*. 7*d*., (ii) £29. 8*s*. 11*d*., (iii) £189. 14*s*. 6*d*., (iv) £10353. 16*s*. 1*d*.

EXERCISE VIII y

1. 6·25. **2.** $\frac{1}{5}$. **3.** 218. **4.** 2388, 2387·95 lbs.
5. 60·3. **6.** 0·5, 0·46, 0·46, 0·43, 0·41. **7.** 8·15 h.

181. **8.** 51, 59, 50, 55, 57, 55, 54, 55, 57, 50, 51. **9.** 21·2.
10. 62·3, 50·2, 46·3, 59·6 mi./hr. **11.** 0·033 secs., 5·02 secs.
12. 2·016.

182. **13.** 0·23, 0·59, 0·457, 0·6689. **14.** 0·5126. **15.** 0·0998.
16. 3059 yds. **17.** 6·58, 3·58, 1·00, 0·88. **18.** 997.
19. 2·5, 7·1, 13·8. **20.** 0·464. **21.** 6*s*. **22.** 1*s*. $7\frac{1}{2}$*d*.
23. 6*s*. 8*d*. **24.** $16\frac{2}{3}$, 25. **25.** £5. 16*s*. $2\frac{1}{2}$*d*.

EXERCISE IX k

189. **1.** 1 Dm. 6 m. 2 dm. 4 cm. **2.** 4 Dm. 2 m. 5 dm.
3. 1 Km. 2 Hm. 8 Dm. 3 m. 6 dm. **4.** 18 Km. 3 Hm. 3 Dm.

190. **5.** 8·27 m. **6.** 7·33 m. **7.** 21·8 m. **8.** 15·159 Km.
9. 197561 mm. **10.** 420·724 m. **11.** 68·435 m.
12. 93·421 m. **13.** 11814 mm. **14.** 0·023155 Km.

EXERCISE IX k (continued)

PAGE

190. **15.** 9·2022 Kgm. **16.** 8282·5423 m. **17.** 0·9822 Km.
18. 9·437 Km. **19.** 0·08403 Km. **20.** 9·08 fr. **21.** 26·25 fr.
22. 2·62. **23.** 0·18. **24.** 3·255. **25.** 1·994.
26. 0·91. **27.** 0·12. **28.** 8·458. **29.** 0·549.
30. 3947. **31.** 0·936. **32.** 1·99257. **33.** 1·238.
34. 0·16. **35.** 0·568. **36.** 0·15. **37.** 4·95 fr.
38. M 8,95. **39.** 5·25 fr. **40.** 6·13 fr.

191. **41.** 1·884. **42.** 30 m. **43.** 5·011 Km. **44.** 2·5 cm.
45. 11 m. **46.** 7 cm. **47.** No. **48.** 4·5 mm. **49.** 0·22 cm.

EXERCISE IX l

192. **1.** 1·035 m. **2.** 2·03 m. **3.** 26·38... fr. **4.** 3240 fr.
5. 7·97... fr. **6.** 8·35... fr. **7.** 3082·16 fr. **8.** 96·93... fr.
9. 52 gm. **10.** 53·802 gm. **11.** 11·4009 Kgm. **12.** 5·876 gm.
13. 23·932 gm. **14.** 1133·05 gm. **15.** 1338·82 Kgm.
16. 145·7 gm. **17.** 60. **18.** $4·7304 \times 10^{14}$.

193. **19.** 0·48. **20.** 885 fr. **21.** 9·585 fr.
22. 167·25 fr. **23.** 17·25 fr. **24.** 1077·375 fr.

EXERCISE IX m

1. 10·8 cm. **2.** 12 c. **3.** 1·32 fr.
4. 2789 fr. **5.** 0·005 cm. **6.** 121·9 mins.

194. **7.** 7, 1 cm. **8.** 59, 4·7 cm. **9.** 76, 40 mm.
10. 11, 75 c. **11.** 188.

EXERCISE IX n

1. 0·008705. **2.** 2290. **3.** 6·25. **4.** 1 l. 25 cl.
5. 75·022 Km. **6.** 0·063907. **7.** 11·3 fr. **8.** 2·998.

195. **9.** 18 m. **10.** £1. 7*s.* 9*d.* **11.** 121 f. 52 c. **12.** 1·3 min.
13. 9*s.* 7½*d.* **14.** 30·375 Kgm. **15.** 222. **16.** 2867·49 fr.
17. 10 c. **18.** £3. 6*s.* 3*d.* **19.** 37 fr. 20 c. **20.** 13*s.* 6*d.*
21. 2 days. **22.** 787·08 dollars. **23.** 39·68 gm.
24. £5. 8*s.* 8*d.* **25.** 778 mm.

196. **26.** 11 lbs. 7 oz. **27.** 26, 16 mm. **28.** (i) $1\frac{3}{5}=1·6$, (ii) 4, (iii) $2\frac{1}{8}=2·125$, (iv) 400, (v) $91\frac{73}{197}=91·3\ldots$, (vi) 0·394, (vii) 2200, (viii) $\frac{5}{11}=·45\ldots$, (ix) $1018\frac{2}{11}=1018·2\ldots$, (x) $28\frac{9}{22}=28·4\ldots$, (xi) $7\frac{7}{8}=7·875$, (xii) $\frac{4}{7}=0·57\ldots$, (xiii) $2\frac{4}{7}=2·57\ldots$, (xiv) $9\frac{3}{5}=9·6$, (xv) 1·25, (xvi) $10\frac{5}{12}=10·4\ldots$, (xvii) 16*s.*, (xviii) 1 ton, (xix) 2 in., (xx) 6250, (xxi) French.
29. 3·0, 16·5 in. **30.** 29·92. **31.** 2·20.
32. 12 st. = 168 lbs. 76 Kgm. = 167·6 lbs.

197. **33.** P. to M., 0·0975*s.*, L. to W., 0·106*s.* **34.** 220. **35.** 582·5 y.
36. 65 c. per Km. = 10*d.* per mi. **37.** English greater by 1*s.* 11¼*d.*
38. £2. 16*s.* 8*d.*

EXERCISE X b

PAGE

200. **1.** 5 : 1. **2.** 1 : 4. **3.** 3 : 2. **4.** 8 : 11.
5. 0·66 : 1. **6.** 371 : 743. **7.** 22 : 7. **8.** $3^2 : 2^2$.
9. $12^2 : 7^2$. **10.** 5 : 7. **11.** $2^2 : 3^2$. **12.** $1 : 0{\cdot}5^2$.
13. 1·05. **14.** 2·4. **15.** 1·25. **16.** 1·07.
17. 0·96. **18.** 0·9775. **19.** 1·61. **20.** 1·09.
21. $\frac{7}{8}$. **22.** 1·8. **23.** 1·25. **24.** 1·625.
25. 0·05. **26.** 2·89. **27.** $2\frac{7}{9}$. **28.** 0·16.
29. 9. **30.** 26. **31.** 1·5. **32.** $1\frac{1}{3}$.

201. **33.** 1 : 21, 1 : 22. **34.** 1 : 12, 1 : 13. **35.** 3 : 5. **36.** 7 : 5.
37. 1 : 4, 1 : 4, 1 : 9. **38.** 1 : 8, 1 : 8. **39.** 1·03.
40. 1·62. **41.** 1 : 295. **42.** 2·13.

202. **43.** 20 : 19. **44.** 171 : 320. **45.** 4 : 1. **46.** 1·14 : 1.
47. 140 lbs. **48.** 16. **49.** 1·6. **50.** 16 : 21. **51.** $6\frac{1}{4}$*d.*

EXERCISE X c

204. **1.** 8·57″. **2.** 9·91″. **3.** 1·8 cm. **4.** 20 cm., 5·83 cm.
5. 12·5 Km., 0·0507 in. **6.** 3·925″, 75·4′.
7. 3·65″, 3 mi. 1505 yds. **8.** 400. **9.** 4·24, 2048.

205. **10.** 600 : 200 : 1; 900, 300, 1·5 sq. in. **11.** 0·74″, 4·84 sq. in.
12. 36·864. **13.** 281·25 sq. in. **14.** 23·75 sq. in. **15.** $\frac{1}{16}$ sq. in.
16. 4·27 mi.; 1·76 sq. mi. **17.** 9400. **18.** 8664.
19. 12 mi. to 1″; 384·48 sq. mi.

206. **20.** 115·2. **21.** $183\frac{1}{3}$ ac. **22.** 33600.
23. 126·72; 62·208 sq. in. **24.** £2,867,857, 31285714 cwts.
25. 9.33 A.M. **26.** 156. **27.** 37′ 4″. **28.** 1190, 850.

207. **29.** 15. **30.** No; 16·9, $9\frac{16}{17}$; $4\frac{12}{13}$, 12·8.

EXERCISE X d

1. 130. **2.** £445. 10*s.*, £267. 6*s.*, £133. 13*s.*, £44. 11*s.*
3. 300, 900, 1200 tons. **4.** £13. 10*s.*
5. £2000, £3000, £5000; $\frac{1}{5} = 0{\cdot}2$, $\frac{3}{10} = 0{\cdot}3$, $\frac{1}{2} = 0{\cdot}5$.

208. **6.** £15. 1*s.* 3*d.*, £10. 0*s.* 10*d.*, £8. 0*s.* 8*d.*
7. £2. 12*s.* 6*d.*, £1. 17*s.* 6*d.*, £1. 2*s.* 6*d.*
8. £10. 5*s.* 1*d.*, £35. 17*s.* $9\frac{1}{2}$*d.*, £97. 8*s.* $3\frac{1}{2}$*d.*
9. 6″, 8″, 6″. **10.** £21, £26. **11.** About 9 hrs.
12. First and second quarters, £192. 10*s.* 0*d.*; third quarter, £193 if 365 days taken as the year, or £193. 10*s.* 7*d.* if 91 days in a quarter; fourth quarter £193. 15*s.* 0*d.*

EXERCISE XI a

211. **1.** (i) $8\frac{3}{4}$ in.², (ii) $3\frac{15}{16}$ in.², (iii) $2\frac{59}{64}$ in.², (iv) $3\frac{45}{128}$ in.²,
(v) $10\frac{1}{8}$, (vi) $6\frac{1}{4}$, (vii) $5\frac{5}{9}$, (viii) $30\frac{1}{4}$ yds.²
2. (i) 178·75 cm.², (ii) 1·7875 dm.², (iii) 0·017875 m.²
3. 91 sq. in., 182 sq. in. **4.** (i) 100 cm.², (ii) 133 cm.², (iii) 194 cm.² **5.** 6·45. **6.** 27. **7.** 147·32 ac., 1584 houses.
8. 158·55 lbs. **9.** 18·2 cm. **10.** 60 ft.²

EXERCISE XI a (continued)

PAGE

212. **11.** $366\frac{2}{3}$ ft.^{2} **12.** 14400. **13.** (i) 61·92 m.^{2}, (ii) 962·5 ft.^{2}, (iii) $1096\frac{1}{2}$ ft.^{2} **14.** (i) 149·2 cm.^{2}, (ii) 110·6 cm.^{2}, (iii) 172·125 cm.^{2}, (iv) 49·25 cm.^{2} **15.** (i) 596·8 cm.^{2}, (ii) 442·4 cm.^{2}, (iii) 688·5 cm.^{2}, (iv) 197 cm.^{2} **16.** 2*s.* 4*d.*
17. £16, £3, £9. 3*s.* 4*d.*, 1440 fr., 1031·25 fr. **18.** 7*s.* 6*d.*
19. £7. 13*s.* $1\frac{1}{2}$*d.*, £6. 6*s.* 2*d.*, 471·31 fr. **20.** $\frac{2}{7}$*d.*, $\frac{10}{21}$*d.*, $5\frac{1}{4}$*d.*

213. **21.** 6250, 9375 m. **22.** £8. 13*s.* 4*d.*, £1. 13*s.* $9\frac{1}{3}$*d.*
23. £2. 6*s.* 8*d.* **24.** £4. 12*s.* $1\frac{1}{2}$*d.* **25.** £12. 4*s.* 9*d.*
26. £1. 9*s.* $7\frac{1}{2}$*d.* **27.** 6*s.* 2*d.* **28.** £10, £1. 14*s.* $1\frac{1}{2}$*d.*
29. £3. 15*s.* 4*d.* **30.** 6*s.* 6*d.*

214. **31.** £1. 6*s.* 3*d.* **32.** £851. **33.** 540. **34.** 38·4 in.^{2}

EXERCISE XI b

215. **1.** 512. **2.** (i) $\frac{1}{2}$ in.^{3}, (ii) 1 in.^{3}, (iii) 2 in.^{3}, (iv) 3 in.^{3}, (v) 1·32 dm.^{3}, (vi) 1·54 dm.^{3}, (vii) 3·024 cm.^{3}, (viii) 0·99876 cm.^{3}
3. (i) $97\frac{1}{2}$ in.^{3}, (ii) 65 in.^{3}, (iii) 75 in.^{3}, (iv) 37·5 in.^{3} **4.** $5\frac{1}{2}$.
5. 14*s.* 6*d.* **6.** 65120. **7.** 193 nearly.

216. **8.** 13090. **9.** 620 lbs. **10.** 12·5. **11.** $2058\frac{1}{3}$.
12. $2\frac{1}{3}''$. **13.** $831\frac{1}{4}$. **14.** 6″. **15.** 2′ 6″. **16.** 28*s.* $1\frac{1}{2}$*d.*, 6*s.* 5*d.*
17. 1′ 7″ about. **18.** 4500 yds.

217. **19.** $7\frac{2}{3}$ ft.^{2} **20.** £26. 2*s.* 6*d.* **21.** 480 lbs. **22.** $435\frac{3}{8}$ lbs.
23. 2724, $89\frac{67}{72}$. **24.** 489, 3·1 lbs. **25.** 36 ft.^{3}, $\frac{2}{3}$ in.
26. 5·3 in. **27.** 6 cm.

218. **28.** 24, 40. **29.** 1·28 in./min. **30.** 21 in.^{2}, $2\frac{13}{16}$ in.^{3}
31. 101 tons. **32.** 997.

EXERCISE XI c

219. **1.** 0·4 m.^{3}, 0·26 m.^{3} **2.** $22\frac{2}{9}$ cm.

220. **3.** 78000 Kgm. **4.** 1300 cm.^{2} **5.** 108. **6.** 196.
7. $7\frac{1}{3}$ ft.^{3}, $7\frac{5}{9}$ ft.^{3} **8.** 8662500. **9.** 48 in.^{3}
10. 500 c.c., $\frac{1}{2}$ l.

221. **11.** 65·52 m.^{3} **12.** 5·586. **13.** 3456. **14.** 107200.
15. 100000. **16.** $140\frac{1}{4}$. **17.** 400. **18.** 24000, 15 in. **19.** 575.

222. **20.** A, 60 ft.^{3}; B, 84 cm.^{2}; C, 57·46 in.^{3}; D, 100 ft.^{3}; E, 64 ft.^{3}; F, 240 in.^{3}; G, 252 ft.^{3}; H, 114 ft.^{3}. **21.** 22·7.
22. 120 ft.^{2}, 7920000. **23.** $44\frac{8}{9}$ tons.

223. **24.** 1000 ft.^{3}, 10. **25.** If 2′ is depth of tank, (1) 0·58 in., (2) 3·01 in.

EXERCISE XII a

224. **1.** (i) 30, (ii) 27·5, (iii) 20, (iv) 37, (v) 51. **2.** (i) 29·5, (ii) 49·2, (iii) 1·8. **3.** (i) 19·7, (ii) 1090·9, (iii) 45·5, (iv) 60·4, (v) 69·1, (vi) 31·1.

225. **4.** 32 nearly. **5.** 1500. **6.** 61·65, 90·4. **7.** 67.
8. 132. **9.** 4 d. 17 h. **10.** $26\frac{1}{4}$. **11.** 66.
12. 25. **13.** 24·4 secs. **14.** 672 ft.

EXERCISE XII b

PAGE

226. **1.** 663. **2.** 54, 63. **3.** H. 55, D. 37, G. 53, F. 36, B. 21, D. 20, F. 19, H. 20.

227. **4.** 5′ 11″. **5.** 11 st. 10 lbs. 2 oz., 11 st. 3 lbs. 2 oz. **6.** 18. **7.** 55·7, 56·9. **8.** $57\frac{6}{7}$, 75. **9.** 11 st. 12 lbs. 14 oz. **10.** 3 mins. 52 secs. **11.** Places: A, D, C, B, E. Marks: D, A, E, C, B.

EXERCISE XII c

231. **1.** 1 h. 36 m. **2.** 2.54 P.M., 2·7 mi. from N. **3.** 1.34 P.M., 57 mi. **4.** 3.35 P.M., $34\frac{1}{6}$ mi. from L. **5.** 3 h. 9 m. A.M. Wed. **6.** In 10 mins., 2 mi. **7.** About 9 A.M., $168\frac{3}{4}$.

232. **8.** 56. **9.** 99 mi., 11.30. **10.** 16 mins. altogether. **11.** 9.0, 9.20 A.M. **12.** 7·5 mi. **13.** 3 h., 4 h. 12 m. **14.** $12\frac{3}{11}$ secs. **15.** 10 mi./hr.

233. **16.** 2 A.M. Feb. 18. **17.** $143\frac{2}{11}$ secs., $13\frac{2}{121}$ secs. **18.** 45. **19.** $3\frac{3}{4}$ secs. **20.** 9. **21.** 22 mi./hr. **22.** 3 h. **23.** About $22\frac{1}{2}$ mi. from *A*, 5.56 P.M. **24.** 3 o'clock, 51 mi.

234. **25.** 4 and 3 mi./hr. **26.** (i) $4.54\frac{6}{11}$, (ii) $4.21\frac{9}{11}$, (iii) $4.5\frac{5}{11}$ and $4.38\frac{2}{11}$. **27.** $72\frac{1}{12}$ n. mi. **28.** 6 d. 6 h. **29.** 38·3 m. **30.** 1·5 mi./hr., 30 m. **31.** $4\frac{4}{9}$ h.

235. **32.** $\frac{20}{21}$ h. = 57 mins.

EXERCISE XII d

1. $2\frac{1}{12}$ d. **2.** $13\frac{1}{3}$ mins.

236. **3.** $13\frac{1}{3}$ mins. **4.** 36 mins. **5.** $78\frac{3}{4}$ mins. **6.** $39\frac{9}{14}$ mins. **7.** $14\frac{2}{5}$ d., 6 d. more. **8.** 8 d. more. **9.** 9 d. $4\frac{18}{19}$ h. **10.** 8 d. **11.** $\frac{1}{2}$ h.

237. **12.** 40 mins., 375, 12·5, 3·125 galls. **13.** $66\frac{2}{3}$ h.

REVISION PAPERS

Paper 21

238. **1.** $\frac{7}{90} > \frac{1}{13} > \frac{3}{40}$. **2.** 452886, 45·29. **3.** 0·123, 0·03005, 0·0024, $0\cdot\dot{1}4285\dot{7}$, $0\cdot\dot{6}$. **4.** 20, 2000, 0·0002. **5.** £5·3375. **6.** $43\frac{1}{3}$ Km. **7.** 9*d.* **8.** 39·37 in., $\frac{5}{8}$, 2·54, 2·2, $1\frac{3}{4}$. **9.** $73\frac{1}{3}$.

Paper 22

1. 76. **2.** 0·00003, 0·1503, 0·15033.

239. **3.** 3·6125. **4.** $\frac{5}{32}$, 6, $3\frac{1}{5}$, 1, 0·0009, 40, 0·025, 3. **5.** 1·325. **6.** 10. **7.** 1 cu. dm., 1 Kgm., 1 gm. **8.** £2. 1*s.* 8*d.* **9.** 13·1″.

Paper 23

PAGE

239. **1.** $\frac{41}{84} > \frac{11}{24} > \frac{49}{108}$. **2.** $\frac{5}{24}, \frac{7}{12}, \frac{1}{25}$. **3.** 2514·0768, 0·2448568. **4.** 3502000. **5.** 3·86875. **6.** 16*s*. $4\frac{1}{2}$*d*. **7.** $21\frac{3}{7}$ secs. **8.** $\frac{1}{800}$*d*. **9.** 1800.

Paper 24

240. **1.** 1, $11\frac{1}{2}$, $10\frac{3}{5}$. **2.** 0·5, 0·25, 0·125, 0·0625, 0·875, 0·025, 0·02, 0·14, 0·0125. **3.** 0·1501. **4.** 0·12, 0·3, 0·048, 0·09, 0·5, 0·3, 0·001. **5.** 0·04375. **6.** 1*s*. 9*d*. **7.** £1000. **8.** 36. **9.** 13824.

Paper 25

1. $\frac{8}{21} > \frac{8}{22} > \frac{5}{14} > \frac{1}{3} > \frac{2}{7} > \frac{21}{77}$. **2.** 3·35382245, 33·5382245. **3.** 344·44. **4.** 0·1614. **5.** 1·10592. **6.** $\frac{1}{12}$, 45, 10.

241. **7.** 26 d. **8.** 21, $\frac{1}{2}$ in. **9.** 4000, 4500 Kgm.

Paper 26

1. $\frac{7}{9}$. **2.** 13·45, 1·345, 1345000. **3.** $\frac{5}{14}, \frac{23}{84}, 1\frac{23}{40}, \frac{1}{8}$, 1·728. **4.** £8. 6*s*. 4*d*. **5.** 26·4. **6.** $\frac{16}{25}, \frac{3}{200}$, 0·015. **7.** £178. 10*s*. 0*d*., £85. **8.** £111. 12*s*. $4\frac{3}{4}$*d*. **9.** $5\frac{1}{3}$.

Paper 27

242. **1.** $\frac{43}{60}$. **2.** 22·92356. **3.** $1\frac{2}{15}, 2\frac{59}{65}$, 45, $1\frac{44}{65}$, 10. **4.** 0·74, 6081·901, 60819·01, 0·06081901. **5.** 16 cwts. 28 lbs. **6.** £450. **7.** 36. **8.** 405 in.^{3}, 333 in.^{2} **9.** 6·45, 16·39.

Paper 28

1. 13·8, 31·8. **2.** 5612·5. **3.** $7\frac{5}{6}, 11\frac{23}{24}$.

243. **4.** £3. 16*s*. 0*d*. **5.** 5*s*. **6.** 3 h. **7.** £3. 6*s*. 0*d*. **8.** 63. **9.** 1375.

Paper 29

1. $\frac{2}{19}$. **2.** 2·43. **3.** 0·142551. **4.** $1\frac{3}{10}$, 1 ton 6 cwts. **5.** 3, 1*s*. **6.** 9 cwts. 1 qr. 13 lbs. **7.** $236\frac{4}{9}$.

244. **8.** 32. **9.** 132.

Paper 30

1. 65. **2.** 0·04, 0·048, 0·00407. **3.** 44, $2\frac{34}{77}, 3\frac{31}{75}$. **4.** 0·2433. **5.** $52\frac{1}{2}$. **6.** 90. **7.** 164, 1 cm. **8.** 1224, 1224000.

Paper 31

1. 0·08016, 0·07984, 0·0000128, 500. **2.** 1·6176342. **3.** $6\frac{23}{60}, 2\frac{211}{240}$.

245. **4.** 48. **5.** 1 ton 14 cwts. 1 qr. $3\frac{1}{4}$ lbs. **6.** 18. **7.** 4798. **8.** Torpedo boat $1013\frac{1}{3}$, Motor $1026\frac{2}{3}$ yds./min. **9.** Between 113 and 114 of *B*.

Paper 32

PAGE

245. **1.** $\frac{25}{28}$, 1. **2.** $\frac{1}{3}$. **3.** 1·16, 2705·8, 0·02337, 108·88.
4. 4·43. **5.** 2·5, 2·5 tons. **6.** $111\frac{1}{9}$.

246. **7.** 2808. **8.** 0·00127. **9.** 6 hrs. 19 mins. nearly.

Paper 33

1. 0·2401828. **2.** 0·2. **3.** $2\frac{7}{30}$. **4.** 2.
5. $3\frac{11}{60}$, £3. 3*s*. 8*d*. **6.** £20. 15*s*. 3*d*. **7.** 364 Kgm.
8. $132\frac{1}{2}$ oz. **9.** 336.

Paper 34

247. **1.** 1·11. **2.** 0·282282. **3.** 1·42, 6293·75.
4. $\frac{59}{70}$, 1, 10, $\frac{7}{33}$. **5.** £13. 5*s*. $7\frac{1}{2}$*d*. **6.** £1. 4*s*. $4\frac{1}{2}$*d*.
7. 2 h. $37\frac{1}{7}$ m. **8.** 10*s*. 3*d*.

Paper 35

1. 0·134689. **2.** 0·0075, 0·12. **3.** $10\frac{1}{2}$, 2, 1·11645, 1·45.
4. $1\frac{19}{30}$, $1\frac{1}{14}$, $11\frac{17}{36}$.

248. **5.** £2. 14*s*. $9\frac{1}{2}$*d*. **6.** 65, 5·5. **7.** $115\frac{1}{5}$. **8.** 7615 sq. yds.

Paper 36

1. $\frac{13}{140}$. **2.** $\frac{5}{9}$. **3.** 0·0125, $\frac{1}{40}$. **4.** 148·696.
5. 0·7*d*. **6.** 21 weeks. **7.** 12 Kgm. **8.** 151·38.
9. $10\frac{1}{2}$*d*.

Paper 37

249. **1.** $\frac{3}{5}$. **2.** $\frac{1}{20}$. **3.** 35·532. **4.** 0·12, 0·0156, 3·31, 58·2, 25. **5.** 1 gall. per sec. **6.** 50·7 oz. **7.** £2. 14*s*, 11*d*.
8. 0·1. **9.** $24\frac{4}{5}$ secs.

Paper 38

1. $8\frac{1}{30}$. **2.** $\frac{53}{122}$. **3.** 0·0748. **4.** $39\frac{3}{8}$.

250. **5.** 23 yrs. $7\frac{1}{2}$ mths. **6.** 111. **7.** 88, 3.10 P.M. **8.** $5\frac{5}{9}$.
9. £12934. 12*s*. 3*d*.

Paper 39

1. $1\frac{37}{60}$. **2.** $2\frac{1}{3}$. **3.** 9·821, 100·74, 1·0327. **4.** 11·126.
5. £1. 6*s*. 3*d*. **6.** 13, 1·64 in. **7.** 47 in.

251. **8.** £55. 10*s*. **9.** 10·6*d*.

Paper 40

1. 0·21. **2.** $13\frac{59}{60}$. **3.** 4. **4.** 175, 565.
5. £1121. 2*s*. 8*d*. **7.** 100°. **8.** 8·4 mi./hr.
9. 4 mths. longer.

Paper 41

252. **1.** $\frac{11}{20}$. **2.** $\frac{44}{81}$. **3.** 150·1876, 150, 150·19. **4.** $286\frac{7}{8}$ lbs.
5. £3. 1*s*. 4*d*. **6.** 43, or 13 not out. **7.** 6*s*. 9*d*.
8. 9·54*d*., 25·15 fr. **9.** 143.

Paper 42

PAGE

252. **1.** $\frac{2}{3}$. **2.** $32\frac{8}{9}$.

253. **3.** 0·24, 0·1007. **4.** £68. 19*s*. $1\frac{1}{2}$*d*. **5.** 7350000, 406000 cm.^{2}, 7350. **6.** 16*s*. **7.** $1856\frac{1}{4}$ yds. **8.** 40. **9.** 1*s*., £2, $\frac{1}{100}$.

Paper 43

1. $24\frac{7}{40}$, $2\frac{89}{120}$, 1.

254. **2.** 1. **3.** 0·002. **4.** £15. 12*s*. 8*d*. **5.** 5. **6.** 18 ft.^{3}, 6 in. **7.** 1 : 3·67. **8.** 2211·84. **9.** $9\frac{3}{4}$*d*., $10\frac{3}{4}$*d*., $5\frac{1}{4}$*d*., 1*s*. 10*d*., 1*s*. 8*d*.

Paper 44

1. $\frac{1}{12}$. **2.** $\frac{1}{19}$. **3.** 534, 0·019825.

255. **4.** 486. **5.** 282·25 cm.^{2} **6.** 100 lbs. **7.** 2.31 P.M., 45. **8.** 1·46. **9.** $10\frac{1}{4}$*d*.

Paper 45

1. 1. **2.** $\frac{13}{35}$. **3.** 0·9, 90, 104, 6230. **4.** 18·4. **5.** $3\frac{1}{4}$ mi.

256. **6.** 11·34*d*. **7.** 2897, 7 mm. **8.** $\frac{18}{13}$ = 1·385. **9.** $32\frac{1}{2}$ mi./hr. **10.** Currants $58897\frac{1}{2}$, Prunes $1430\frac{1}{7}$, Raisins $34386\frac{2}{7}$, Tea $129561\frac{9}{14}$.

Paper 46

1. $48\frac{1}{8}$, $\frac{296}{1503}$. **2.** 0·58. **3.** 884. **4.** 2 : 1. **5.** £7. 13*s*. 5*d*. **6.** 21, 140 tons, 0·83 tons.

257. **7.** 642·6 lbs., £2. 9*s*. 0*d*. **8.** 2*s*. $3\frac{3}{4}$*d*.

Paper 47

1. $2\frac{5}{16}$ = 2·31. **2.** $\frac{84}{95}$. **3.** 261·06. **4.** 39, 0·094. **5.** £1. 15*s*. 6*d*. **6.** 3·942. **7.** £1. 2*s*. $9\frac{1}{4}$*d*. **8.** 78. **9.** 1·84, 0·2875, 0·1725 gm.

Paper 48

258. **1.** 0·111. **2.** 4·52. **3.** 3. **4.** $3\frac{1}{4}$, £3. 5*s*. **5.** 51 yds. **6.** 13*s*. 3*d*. **7.** 5660. **8.** £14. 5*s*. 9*d*. **9.** H. 39787, 27; N. 29028, 40.

Paper 49

259. **1.** $3\frac{1}{11}$. **2.** $\frac{1}{2}$. **3.** 600. **4.** £10. 18*s*. 9*d*. **5.** 5·067. **6.** £121. 15*s*. 8*d*., £152. 4*s*. 7*d*., £334. 18*s*. 1*d*. **7.** $\frac{1}{8}$ in. **8.** 1 lb. 13 oz., 3 lbs. 10 oz. **9.** £134000.

Paper 50

260. **1.** $3\frac{1}{2}$. **2.** $2\frac{2}{9}$. **3.** 15·4, 10·89. **4.** 0·365. **5.** 25. **6.** 35. **7.** £3. 4*s*. 11*d*. **8.** 12. **9.** 62·5.

Paper 51

PAGE

261. **1.** $\frac{1}{40}$. **2.** 226·11, 9·106, 0·0635, 0·0006535, 18·7, 3·2.
3. 284. **4.** £16. 4*s*., £2. 11*s*. 9*d*., 900 ft.^{2}, 4158 ft.^{3}
5. £70, £49, £21. **6.** 16. **7.** £4. 4*s*. $11\frac{1}{2}d$. **8.** £1200.

262. **9.** $26\frac{7}{8}$ miles.

Paper 52

1. 2. **2.** 3·1552, 0·4865535, 2·00. **3.** 0·0013.
4. $\frac{8}{45}$, $\frac{10}{45}$, $\frac{27}{45}$. **5.** French, 9*d*. **6.** *A*. **7.** 7.

263. **8.** 1·2. **9.** 4·2 mi./hr.

Paper 53

1. 0·04575. **2.** $2\frac{2}{5}$, 1. **3.** 0·0272, 0·9375.
4. £2100. **5.** 8*s*. 4*d*., £1. 10*s*., £6. 13*s*. 4*d*. **6.** 0·019. **7.** 7*s*.

264. **8.** $9\frac{1}{3}$ mi., 1 h. 52 m. **9.** $7.37\frac{1}{2}$ P.M. Thurs.

Paper 54

1. $\frac{13}{42}$. **2.** 0·530. **3.** 0·779. **4.** $10\frac{1}{2}d$.
5. 858. **6.** £5.

265. **7.** Men £19. 14*s*. 9*d*., Women £13. 3*s*. 2*d*., Children £6. 11*s*. 7*d*.
8. 21 mins. **9.** 5·5 secs., 5 P.M. June 9.

MISCELLANEOUS EXERCISES on Part II

266. **1.** 5 opposite to 4, 10 to 8, 15 to 12, etc. **3.** 365·2422.
4. 0·0003. **6.** 80. **7.** $13\frac{1}{5}$ secs. **8.** 6 mi./hr.

267. **9.** 910. **10.** 344. **11.** (i) £15. 11*s*. 0*d*., (ii) £34. 8*s*. 8*d*., (iii) £19. 13*s*. 10*d*., (iv) £7. 12*s*. 10*d*., (v) £24. 1*s*. 1*d*., (vi) £1. 6*s*. 1*d*., (vii) £1. 0*s*. 4*d*., (viii) 6*s*. 10*d*., Total £104. 0*s*. 8*d*.
12. 2·8 mi./hr. **13.** 5*s*. $4\frac{1}{4}d$. **14.** 242. **15.** £4. 3*s*. 4*d*.

268. **16.** American $£103\frac{1}{8}$. **17.** $\frac{6}{7}$, $\frac{5}{6}$, $\frac{4}{5}$, $\frac{3}{4}$, $\frac{5}{7}$, $\frac{2}{3}$, $\frac{3}{5}$, $\frac{4}{7}$, $\frac{1}{2}$, $\frac{3}{7}$, $\frac{2}{5}$, $\frac{1}{3}$, $\frac{2}{7}$, $\frac{1}{4}$, $\frac{1}{5}$, $\frac{1}{6}$, $\frac{1}{7}$. **18.** $\frac{4}{15}$, $\frac{2}{5}$, $\frac{1}{3}$. **19.** 15 mi.
20. 11 h. 2 m. 20 s. **21.** £1153. 5*s*. 8*d*. **22.** 273 tons.
23. £6. 13*s*. 4*d*.

269. **25.** Gold £10. 10*s*., Silver £12, Copper 6*s*. $6\frac{3}{4}d$.
26. 1156, 1105. **27.** Larger, Yes. **28.** $\frac{3}{4}$ in.
29. 21 yds. 2′ $2\frac{5}{6}$″. **30.** 5*s*. $1\frac{13}{16}d$. **31.** 2*s*. 4*d*.

270. **32.** (i) 14, 70·02 cm.^{2}, (ii) 15, 39·02 cm.^{2} **33.** 17*s*. 10*d*.
34. 1216, 3720 ft. **35.** 464, $8\frac{5}{18}$ in.^{2} **36.** 3·7 tons.
37. 16 ft., £1398946. **38.** 4·31 in.^{2}

271. **39.** £91. 13*s*. 4*d*. **40.** £1040, £700. **41.** 10.
42. 20 yds. **43.** The latter; *A* has 4 yds. less in which to overtake *B*. **44.** After 2 mins., 1 min. **45.** 310 yds.
46. $14\frac{1}{2}$ yds. **47.** *B* by $\frac{1}{3}$ sec., or about $2\frac{1}{2}$ yds.

272. **48.** *A* wins by $\frac{1}{88}$ min. in $44\frac{43}{88}$ mins. **50.** 9 miles.
51. 4·8 Km./hr. **52.** 4 : 5. **53.** £360. **54.** 8 mi. **55.** 55·46 ft.^{3}

273. **56.** 8 : 5. **57.** $£358\frac{1}{3}$, £400, $£366\frac{2}{3}$. **58.** $3\frac{2}{3}$ miles, $32\frac{2}{9}$ mins.
59. $33\frac{1}{3}$ mi./hr., $48\frac{1}{3}$ miles. **60.** £6. 5*s*. 0*d*., £6. 16*s*. 3*d*., £8. 10*s*. 0*d*. **61.** The same.

274. **62.** 27·52, 12·48. **63.** $1\frac{3}{4}d$., $\frac{1}{4}d$. **64.** 75 lbs. **65.** 12.

PART III

EXERCISE XIII b

PAGE

277. **1.** 51. **2.** 22 tons. **3.** 1452. **4.** 99.

278. **5.** 3360, 448, 672. **6.** 3·593, 0·798, 4·011, 0·055.
7. 65 lbs. **8.** 6953·25 lbs. **9.** 59 yds., 6·03 in.

EXERCISE XIII d

279. **1.** 422266. **2.** 589. **3.** £23. 2*s*., £15. 15*s*., £336, £483, £2268. **4.** 3654. **5.** 0·234 oz. **6.** £230. 17*s*., £159. 18*s*., £152. 2*s*., £118. 8*s*. 1*d*., £210. 9*s*. $7\frac{1}{4}$*d*., £124. 12*s*. $8\frac{1}{2}$*d*.
7. $\frac{1}{5}$*d*., 5. **8.** 30. **9.** 8. **10.** 80.

EXERCISE XIII f

280. **1.** $14\frac{6}{7}$. **2.** 15. **3.** 6, 7·5, 3·5, 3·3, 12·9, 32·7.

281. **4.** $+33\frac{1}{3}$, $-4·5$, $-3·8$, $+7·1$, $-7·5$, $+17·1$.
5. 39·2, 57·2, 3·7. **6.** 74·8, 10·3, 13·9, 0·982. **7.** 57·7.
8. 66·7. **9.** $76\frac{8}{17} = 76·5$ %. **10.** 8·5 nearly. **11.** 3.
12. 44. **13.** $42\frac{6}{7}$.

EXERCISE XIII h

283. **1.** £760. **2.** £625, £125. **3.** £300. **4.** 20000.
5. 1500. **6.** £40. **7.** £400. **8.** £50. **9.** £100, £60, £37. 10*s*., £83. 6*s*. 8*d*., £95. 4*s*. 9*d*., £100, £150, £105. 5*s*. 3*d*. **10.** 8 for 10*d*.

EXERCISE XIII i

284. **1.** 13 lbs. **2.** $6\frac{1}{4}$. **3.** 40. **4.** 32·9. **5.** £24. 10*s*.
6. $57\frac{1}{7}$, $62\frac{1}{2}$. **7.** 2·6. **8.** £6. 13*s*. 4*d*. **9.** 525.
10. 11·7. **11.** 23 nearly. **12.** 11·9, 24·0.

285. **13.** £21. 10*s*. **14.** 6 tons, 2·6 tons. **15.** 0·85, 81.
16. 30. **17.** 6*s*. 8*d*. **18.** $12\frac{1}{2}$ lbs. **19.** £57. 8*s*.
20. $3\frac{3}{4}$. **21.** No difference. **22.** 119.
23. £143. 19*s*. $7\frac{1}{2}$*d*., 28·8.

286. **24.** 63·5. **25.** Gain, 14·2. **26.** As. 9·9, 42·35; Ak. 39·3, 35·4. **27.** 20 tons. **28.** $2\frac{1}{2}$. **29.** Fall of 2 %.
30. 28. **31.** Tonnage: 3688. B. 3091, G. 666.

287. **32.** 26, 13, 51. **33.** +5 %, +7 %. **34.** 1·4, 28·4, 26·6.
35. $78\frac{3}{4}$ lbs., 8·1.

288. **36.** £454. 19*s*. 10*d*.

EXERCISE XIII j

PAGE

288. **1.** 22. **2.** $77\frac{7}{9}$. **3.** $14\frac{7}{12}$.

289. **5.** £3. 6*s*. 8*d*. **6.** 10 for 9*d*. **7.** £1100. **8.** 120. **9.** $\frac{1}{2}$% gain. **10.** 102 lbs., 76 lbs. nearly. **11.** 6·4. **12.** £462, $15\frac{1}{2}$. **13.** $14\frac{2}{7}$. **14.** $14\frac{2}{7}$ oz.

290. **15.** 60. **16.** $33\frac{1}{3}$. **17.** 30·5 about. **18.** £1 or 1*s*. 8*d*. per lb., 2*d*., 10. **19.** 3·4. **20.** 4. **21.** 18*s*. **22.** $\frac{1}{4}$% loss. **23.** 44. **24.** 72·8. **25.** 15·3, 28·3. **26.** 9·75, 14·26.

EXERCISE XIV b

294. **1.** ±0·5 cm., ±0·005, 0·5. **2.** ±0·005 gm., ±0·001, 0·1. **3.** ±0·005 in., ±0·01, 1. **4.** ±0·0005 l., ±0·02, 2. **5.** ±0·5 cm., ±0·0007, 0·07. **6.** ±0·5 mgm., ±0·01, 1. **7.** ±0·5 Km., ±0·0003, 0·03. **8.** ±0·5 gm., ±0·005, 0·5. **9.** ±0·5 f., ±0·0025, 0·25. **10.** ±0·5 f., ±0·0005, 0·05. **11.** ±6*d*., ±0·001, 0·1. **12.** $\pm\frac{1}{2}$ min., ±0·004, 0·4. **13.** $\pm\frac{1}{2}$ oz., ±0·006, 0·6. **14.** ±6 in., ±0·02, 2. **15.** 1. **16.** 1. **17.** 0·4. **18.** 0·3. **19.** 0·2. **20.** 0·1.

295. **21.** 2.
22. *A* 0·042, 1·3 %, 2, 2. *B* 0·078, 2·5 %, 3, $\bar{1}$. *C* 0·002, 0·1 %, 4, 3. *D* 0·142, 4·5 %, 1, 1. *M* = 3·116, 0·026, 0·83 %.
23. 0·28. **24.** 0·1 %. **25.** 0·3, 0·9, 0·2 %.

EXERCISE XIV c

298. **1.** (*a*) The nearest million. (*b*) The nearest mile. (*c*) Foot. (*d*) Inch. (*e*) Mm. or $\frac{1}{10}$ inch. (*f*) $\frac{1}{100}$ inch or $\frac{1}{100}$ cm. (*g*) lb. (*h*) 6×10^{21} tons. (*i*) *Parcels* nearest lb. *Letters Post* nearest oz. (*j*) 13·6, 13·596. (*k*) Nearest £100,000. (*l*) Penny. (*m*) Sixpence or shilling. (*n*) $\frac{1}{5}$ sec. (*o*) A million years? (*p*) $\frac{1}{10}$ knot. (*q*) Very accurate. 36578′ to nearest inch.
2. 4·950 m., 3 mm., 4·953 m., 4·947 m.

299. **3.** 2 c.c. **4.** 2 lbs., $\frac{1}{2}$ lb. **5.** 47·65 cm., $\frac{1}{10}$ mm. **6.** 147·65 − 99·95, 0·1 c.c. **7.** 2 Kgm., 10 mgm. **8.** 0·881 gm., ±0·005 gm. **9.** Circum. 4·062″, ±0·0005″.

300. **10.** 3·4. **11.** 23·83 gm., 0·03 gm. **12.** ±0·005, ±0·004; ±0·009; two. **13.** (i) ±0·01, ±0·01; ±0·03; one. (ii) ±0·001, ±0·0005; ±0·002; two. (iii) ±0·0006, ±0·0006; ±0·001; two. **14.** No. **15.** 0·9982 ± 0·0006 gm. **16.** 13·73 ± 0·07 gm. **17.** 0·82 ± 0·0005 gm. **18.** ±1·6 in.[3]

EXERCISE XIV d

PAGE

301. **1.** 192·6. **2.** 0·18. **3.** 21. **4.** 2·6. **5.** 7·6. **6.** 4·85.

EXERCISE XIV e

303. **1.** 13·47. **2.** 20·0. **3.** 0·32. **4.** 637. **5.** 4700. **6.** 1460. **7.** 0·023. **8.** 210·4. **9.** 14·08. **10.** 0·0572. **11.** 741. **12.** 883000. **13.** 66200000. **14.** 45500000. **15.** 100000.

EXERCISE XIV f

304. **1.** 115. **2.** 972. **3.** 109. **4.** 0·0120. **5.** 0·00862. **6.** 0·00631. **7.** 0·000000201. **8.** 0·318. **9.** 40·24. **10.** 0·00089. **11.** 100. **12.** 0·10. **13.** 0·395.

EXERCISE XIV g

1. 16 f. **2.** 7973 f. **3.** £648, £492, £422. **4.** £1139, £439, £274.

305. **6.** £36100000. **7.** 31·6 in.^{2}, 79 Km.^{2}, 36 yds.^{2}, 9214 yds.^{2}, 5558 dm.^{2} **8.** 28 in. **9.** 37 cm. **10.** 1032. **11.** 499 secs. **12.** 0·24, 160. **13.** 12. **14.** 13·4.

306. **15.** £5. 3*s.*, £5. 17*s.* **16.** £4. 2*s.*, £3. 19*s.*, £16. 4*s.*, £12. 15*s.*, £10. 1*s.* **17.** 42, 6, 25, 26. **18.** 8·0, 16·5. **19.** 0·34. **20.** 72·4.

307. **21.** 84 %. **22.** 308. **23.** 4·4, 5·9. **24.** 3·2. **25.** 59, 74, 11*s.* 8*d.*, 5*s.* 10*d.*, 7*s.* 7*d.* **26.** 2·8, 1·1, 2·4.

308. **27.** £3. 8*s.* 6¾*d.* **28.** 37, 79. **29.** 91. **30.** 7, 30, 25, 33. **31.** 7·9*d.*, £7. 6*s.* 0*d.*

309. **32.** 3·6, 4·0, 5·1. **33.** 1*s.* 10*d.* or 1*s.* 11*d.*, 2*s.* 2*d.* or 2*s.* 3*d.* **34.** 7*s.* 6*d.*, 7*s.* 11*d.*, 38 %. **35.** 100 oz. **36.** 1*s.* 2½*d.*, 9*s.*

310. **37.** 1·3, 2·7. **38.** 38 %, 22 %, 166 %, 138 %, 126 %. **39.** 80 million miles. **40.** 45,660,000. **41.** 243, 258, 10¾*d.*, 10*d.*

311. **42.** £1. 2*s.* 0*d.*, 15*s.* 4*d.*, 70 %. **43.** 0·917, 0·900, 0·900. **44.** 25·22 f., M. 20·43.

EXERCISE XV a

314. **1.** (i) 6·130, (ii) 28·88, (iii) 17·33, (iv) 0·001535, (v) 613·0, (vi) 2888, (vii) 4295, (viii) 0·1612, (ix) 61300, (x) 288800, (xi) 5·359, (xii) 0·1607, (xiii) 0·06130, (xiv) 0·2888, (xv) 520800, (xvi) 0·01017, (xvii) 0·0006130, (xviii) 0·002888. **2.** (i) 590·5 in.^{2}, (ii) 29960 yds.^{2}, (iii) 32·83 ft.^{2}, (iv) 862·6 in.^{2}, (v) 1826 cm.^{2}, (vi) 9966 cm.^{2} **3.** (i) 7573, (ii) 9455, (iii) 72·92.

315. **4.** (i) 87·53, (ii) 8·384, (iii) 798·6.

EXERCISE XV d and e

PAGE

320. **1.** 29. **2.** 39. **3.** 93. **4.** 246. **5.** 378.
6. 735. **7.** 817. **8.** 108. **9.** 409. **10.** 806.
11. 1001. **12.** 180. **13.** 102. **14.** 904. **15.** 7009.
16. 2009. **17.** 9089. **18.** 8734. **19.** 5·7. **20.** 1·19.
21. 3·58. **22.** 61·5. **23.** 93·002. **24.** 10·01.
25. 0·037. **26.** 0·0053. **27.** 0·131. **28.** 0·0185.
29. 0·0907. **30.** 0·7008.

321. **31.** 14·1. **32.** 4·47. **33.** 1·41. **34.** 5·48. **35.** 1·73.
36. 2·24. **37.** 3·61. **38.** 11·4. **39.** 6·47. **40.** 2·05.
41. 0·647. **42.** 0·853. **43.** 0·270. **44.** 0·0187.
45. 0·816. **46.** 2·58. **47.** 0·577. **48.** 1·83.
49. 1·22. **50.** 2·10.

EXERCISE XV f

322. **1.** 13′, 13′, 1·41′ = 1′ 5″. **2.** 69·6, 98·4, 120·5, 139, 696, 1244. **3.** 3′ 7″. **4.** 39, 24. **5.** 2′, 2′ 5″, 2′ 6″.
6. 43 in.², 7″. **7.** 10·4″ square. **8.** 59.

323. **9.** 440, 880, 1760. **10.** $\frac{49}{16}=3\frac{1}{16}$, $\frac{5}{4}=1\frac{1}{4}$.
11. $1\frac{1}{12}$, $1\frac{2}{15}$, $2\frac{1}{6}$, $3\frac{1}{8}$. **12.** 0·721. **13.** Each = 4·14.
14. Each = 0·32. **15.** $2\sqrt{6}=4{\cdot}9$. **16.** 2·49. **17.** 0.
18. 1·015.

EXERCISE XV g

1. 12·07 cm. **2.** 3·845(6)″. **3.** 27·70 cm.
4. 8·325″. **5.** 33·94 cm. **6.** 9·590 cm.

324. **7.** 7·746 cm. **8.** 7·506 cm. **9.** 23·38 cm.
10. 127·4 cm. **11.** 16·16 cm. **12.** 8·185″.
13. 58·74 cm. **14.** 22·48 cm. **15.** 5·378 cm.
16. 27·25 cm. **17.** 3·606, 5·385, 5·831, 6·164″.
18. 5·820, 8·203, 8·724, 9·415 cm. **19.** 33·60.
20. 12′ 11·6″. **21.** 13·05″. **22.** 1018′.

EXERCISE XV h

327. **1.** 0·3377. **2.** 0·1359. **3.** 0·1029. **4.** 0·1830.
5. 0·1419. **6.** 0·03162. **7.** 0·004062. **8.** 0·0001440.
9. 0·004986. **10.** 0·0009001. **11.** 1·999.
12. 2·001. **13.** 1·380. **14.** 21·10. **15.** 43·37.
16. 0·001603. **17.** 0·02534. **18.** 2·730.
19. 0·00003177. **20.** 0·07358.

EXERCISE XV i

328. **1.** 1·23. **2.** 3·76. **3.** 0·318. **4.** 0·703. **5.** 0·04025.
6. 0·00217. **7.** 0·000196. **8.** 15·2. **9.** 14·8.
10. 0·0111. **11.** 0·105. **12.** 5·63. **13.** 5·49.
14. 0·743. **15.** 2·24. **16.** 2·52. **17.** 236·4.
18. 0·3994.

329. **19.** 1·16. **20.** 17·3. **21.** 0·334. **22.** 0·0344.

EXERCISE XV j

PAGE

329. **1.** 2·54, 0·915, 119. **2.** 2·205. **3.** 0·00289. **4.** 0·0229. **5.** 15*s*. 9*d*. **6.** 1·60. **7.** 73·5. **8.** 2·78. **9.** 13·6. **10.** 56·9. **11.** 1·84 c.c. **12.** 91·7 c.c., 8·18. **13.** 8·61. **14.** 1·18. **15.** 0·702. **16.** 4·08.

EXERCISE XVI a

332. **1.** 44 m., 11 ft., 11 cm., 22 yds., 52·8 Km., 0·88 mm., 12 mi. 1320 yds. **2.** 88 cm., 44 mi., 12′ 10″, 9 mi. 946 yds. **3.** 14 m., 1·75″, 70 yds., 0·1925 cm., 574 yds., 2·975 Km. **4.** 70′, 3′ $2\frac{1}{2}$″, 0·0049 cm., 5 mi. 720 yds. **5.** 18′ 4″.

333. **6.** 54′. **7.** 28″. **8.** 180. **9.** 168. **10.** 163 cm. **11.** 34·4. **12.** 2·07″. **13.** 0·636″. **14.** 24900 m. **15.** 189″. **16.** 208. **17.** 775.

EXERCISE XVI b

335. **1.** 154 m.^{2}, 4 ft.^{2} 40 in.^{2}, 38·5 cm.^{2}, 0·1386 Km.^{2} **2.** 616 mm.^{2}, 38 ft.^{2} 72 in.^{2}, 0·7546 cm.^{2} **3.** 7 cm., 1·4 in., 1′ 2″, 154 yds., 1540 yds. **4.** 22 mm., 66″. **5.** 108 in.^{2} **6.** 77 Km.^{2}

336. **7.** 42″. **8.** 6·03 cm.^{2} **9.** 1260 in.^{2}, 1190 in.^{2} **10.** 1 acre 3740 yds.^{2} **11.** 7·28 cm.^{2}

EXERCISE XVI c

337. **1.** 1·91″. **2.** (i) 484 in.^{2}, (ii) 528 cm.^{2}, (iii) 6600 m.^{2}, (iv) 99 ft.^{2}, (v) 6·16 mm.^{2}, (vi) 140 cm.^{2}, (vii) 39·6 ft.^{2}, (viii) 1940 ft.^{2}

338. **3.** 2 cm., 2′ 9″, 2″, 0·0375 mm. **4.** 2 m., 12′, 0·003125″. **5.** 6″, 8′. **6.** (i) 44 in.^{2}, (ii) 88 m.^{2}, (iii) 154 ft.^{2}, (iv) 29·7 m.^{2}, (v) 110 ft.^{2}, (vi) 55·4 cm.^{2} **7.** 6″, $11\frac{1}{8}$″. **8.** £4. 2*s*. 6*d*. **9.** 321 ft.^{2} **10.** 35500 ft.^{2} **11.** 4·50″.

EXERCISE XVI d

340. **1.** 728 in.^{3}, 84 c.c. **2.** (i) 462 in.^{3}, (ii) 176 c.c., (iii) 2380 in.^{3}, (iv) 1510 m.^{3}, (v) 13900 c.c., (vi) 1580 in.^{3}, (vii) 481 in.^{3}

341. **3.** 2″, 2′ 11″, 2·7 mm. **4.** 1 m. **5.** 288 in.^{3} **6.** 792 in.^{3} **7.** 7700 ft.^{3} **8.** 43560 in.^{3} **9.** 1·82 oz. **10.** 552000 tons. **11.** 7·17 in.^{3} **12.** 46·1 ft.^{3} **13.** 389 lbs. **14.** 138 in.^{2}, 4·29 tons.

342. **15.** $14\frac{1}{2}$ in. **16.** No. No. **17.** 7·29 gm. Yes. **18.** 113 c.c., 31 c.c. **19.** 10·7 ft.^{3}, 4·42 ft.^{3}, 58·7 %. **20.** 75·4 c.c., 37·7 c.c. **21.** 3·77 c.c., 0·079 c.c., 2·44 gm.

343. **22.** *I*. 226 ft.^{3} *J*. 157 ft.^{3} *K*. 190 ft.^{3} *L*. 579 in.^{3} *M*. 7540 in.^{3} *N*. 141 ft.^{3} *O*. 106 in.^{3}

EXERCISE XVI e

PAGE

343. **1.** 10 mi./hr. **2.** 192·5. **3.** 3 ft.^{2} 74 in.^{2}

344. **4.** 25·1 Kgm. **5.** 1′ 9″. **6.** 880 gm. **7.** 40 ft.^{2} 40 in.^{2}
8. 2530 in.^{3} **9.** 360. **10.** 62·5 ft.^{3} **11.** 4440 ft.^{2}
12. 1′ 2″. **13.** 2·1 m. **14.** 8″. **15.** 12,700,000.
16. 1040. **17.** 2·11 Kgm.

345. **18.** 1960. **19.** 872,000. **20.** 0·059″. **21.** 6·28 in.^{2}
22. £970. **23.** 1·77 mins. **24.** Circum. doubled, area $\times 4$.
25. $48\frac{3}{4}$ secs. **26.** The square. **27.** 48 in.^{2} **28.** 754 in.^{2}
29. 18·98 in.^{2}, 5·97 in.^{3} **30.** 1·35 c.c., 371 times.

346. **31.** Pipe, 8·03 sq. m. **32.** 286. **33.** Section area is halved.
34. 1·26, 0·0785, 0·000785 c.c.

EXERCISE XVII b

348. **1.** 8·572. **2.** 8·179. **3.** 8·164. **4.** 8·684.
5. 5·198. **6.** 7·423. **7.** 37·62. **8.** 1390.
9. 1143. **10.** 149000. **11.** 9243. **12.** 87000.
13. 804·2. **14.** 2637000. **15.** 317800000. **16.** 1·408.
17. 2·752. **18.** 3·766. **19.** 3·178. **20.** 5·690.
21. 1·375. **22.** 463·1. **23.** 266·1. **24.** 1·395.
25. 155·2. **26.** 4083. **27.** 2·867. **28.** 1·006.
29. 9·779. **30.** 5·342. **31.** 20·39. **32.** 2691.
33. 84720. **34.** 12·60. **35.** 4·641. **36.** 2·763.

EXERCISE XVII c

349. Answers are given to 4 figures, but the 4th figure may be different when the order of operations is different.

1. 215·8. **2.** 11270. **3.** 234·3. **4.** 26160.
5. 755000. **6.** 203·8. **7.** 643·4.

350. **8.** 45·28. **9.** 80·37. **10.** 3·806. **11.** 9·200.
12. 9·141. **13.** 1·187. **14.** 15·88. **15.** 232·2.
16. 207·5. **17.** 53·70. **18.** 2184. **19.** 14·67.
20. 0·4240. **21.** 1·01. **22.** 2560. **23.** 3·386.
24. 1350. **25.** 1·423. **26.** 17·91. **27.** 1·490.
28. 1·082. **29.** 2·613. **30.** 1·009. **31.** 2·586.
32. 1·014.

EXERCISE XVII d

1. 15·85 in.^{3}, 835 c.c., 27000 c.c.
2. 65·4 in.^{3}, 22100 c.c., 133000 c.c. **3.** 6·20 cm., 1·06′, 1·51″.

351. **4.** 427 c.c., 201 in.^{3}, 19700 c.c. **5.** 2·07. **6.** 24·1.
7. 23·83. **8.** 116 yds. **9.** 99·3 m.^{3}, 17·2.
10. 4955 ft.^{3} **11.** 5·27 c.dm. **12.** 6·45, 16·4.
13. 24900. **14.** 4·43. **15.** 2·205. **16.** 692. **17.** 9*d.*

352. **18.** 175 yds. **19.** 76·1 mi., 4·38′. **20.** 24900 mi.
21. 103 cm.^{2} **22.** 49,700,000 mi.^{2} **23.** 78·5 yds.
24. 1·38, 8·68 in.; 15·6, 98·0 cm. **25.** £23. 18*s.*, £16. 2*s.*
26. $7{\cdot}36 \times 10^{19}$. **27.** 849 lbs. **28.** 568 gm.
29. 2·22 lbs. **30.** 458 gm. **31.** 3·74.

353. **32.** 7 tons 7 cwts. **33.** 90·3 c.c.

EXERCISE XVII f

PAGE

353. **1.** 0·7. **2.** $\bar{4}$·7. **3.** 3·8. **4.** $\bar{1}$·7. **5.** 2·2.
6. 2·3. **7.** 4·1. **8.** 2·8. **9.** $\bar{1}$·2.

354. **10.** 0·3. **11.** $\bar{2}$·3. **12.** $\bar{2}$·3. **13.** $\bar{1}$·3. **14.** $\bar{2}$·9.
15. $\bar{3}$·8. **16.** 7·2. **17.** 8·0. **18.** 2·2. **19.** 6·1.
20. 2·9. **21.** 2·8. **22.** 4·6. **23.** $\bar{3}$·2. **24.** $\bar{5}$·2.
25. $\bar{5}$·8. **26.** $\bar{6}$·8. **27.** 1·2. **28.** 1·2. **29.** 1·8.
30. 0·8. **31.** $\bar{2}$·6. **32.** $\bar{1}$·2. **33.** $\overline{10}$·5. **34.** $\bar{9}$·0.
35. $\bar{2}$·8. **36.** $\bar{5}$·8. **37.** $\bar{7}$·7. **38.** $\bar{5}$·0. **39.** $\overline{13}$·0.
40. $\bar{1}$·1. **41.** $\bar{1}$·2. **42.** $\bar{1}$·4. **43.** $\bar{2}$·9. **44.** $\bar{1}$·9.
45. $\bar{1}$·9. **46.** $\bar{1}$·4. **47.** $\bar{1}$·3. **48.** $\bar{1}$·2. **49.** $\bar{1}$·4.
50. $\bar{1}$·8. **51.** $\bar{1}$·84. **52.** $\bar{1}$·3. **53.** $\bar{3}$·7. **54.** $\bar{2}$·4.

EXERCISE XVII g

1. 21·11. **2.** 0·07568. **3.** 0·003762. **4.** 0·01490.
5. 0·01143. **6.** 0·001390. **7.** 92·43. **8.** 1·023.
9. 0·02726. **10.** 0·3014. **11.** 0·005690. **12.** 46·31.
13. 0·003758. **14.** 137·5. **15.** 0·1031. **16.** 10·28.
17. 51·02. **18.** 3·386. **19.** 7·548. **20.** 0·7881.

EXERCISE XVII h

355. **1.** 0·0003168. **2.** 0·01875. **3.** 0·2149.
4. 0·006292. **5.** 0·00005524. **6.** 0·00146.
7. 0·5400. **8.** 0·3098. **9.** 0·0003321.
10. 0·5546. **11.** 0·02802. **12.** 0·008287.
13. 0·0008287. **14.** 0·05527. **15.** 0·01613.
16. 0·4714. **17.** 0·2090. **18.** 0·08742.
19. 0·02307. **20.** 0·8980. **21.** 0·3199.
22. 0·1284. **23.** 0·09177. **24.** 0·7729.
25. 0·5044. **26.** 0·2755. **27.** 0·1492.
28. 0·09456. **29.** 0·7605. **30.** 0·3773.

The 4th figure in Answers **31–41** depends in some cases upon the order in which operations are performed.

31. 0·2497. **32.** 0·01129. **33.** 0·4158. **34.** 0·6457.
35. 0·8037. **36.** 0·0412. **37.** 616900. **38.** 7·896.
39. 0·3366. **40.** 8·208. **41.** 0·8982.

EXERCISE XVII i

356. Answers are given to 4 figures, but the 4th figure may be different when the order of operations is different.

1. 40·92. **2.** 480·6. **3.** 0·0006059. **4.** 0·4458.
5. 67·07. **6.** 2·369. **7.** 63·87. **8.** 0·2611.
9. 23·15. **10.** 0·9642. **11.** 4·649. **12.** 0·04739.
13. 152·8. **14.** 2·107. **15.** 7949. **16.** 9·559.
17. 729·7. **18.** 0·1316. **19.** 0·3776. **20.** 1673.
21. 81·32. **22.** 0·8980. **23.** 0·008277. **24.** 0·6625.
25. 16·35. **26.** 21·05. **27.** 2·652. **28.** 3·510.
29. 11130. **30.** 0·4346. **31.** 0·03215. **32.** 0·2608.
33. 20·67. **34.** 59·05.

357. **35.** 381·5, 10630. **36.** 2·613, 0·2004. **37.** 0·2571, 0·0008324. **38.** 34·66. **39.** 0·07985. **40.** 8·003.

EXERCISE XVII j

PAGE

357. **1.** 398 in.^{3} **2.** 437 lbs. **3.** 15450 yds.^{3} **4.** 152000. **5.** 1210 tons. **6.** 5·37 tons. **7.** 214 in.^{3}, 391·5 in.^{2}; 15500 in.^{3}, 4440 in.^{2}

358. **8.** 6·115 in. **9.** 121 tons. **10.** 1410, 576·3, 0·6215, 2·54, 0·9143, 119·4. **11.** 2·208. **12.** 1·76. **13.** 398·6 in.^{3}, 377 in.^{2} **14.** 6·113 in. **15.** 3·063 in. **16.** 1·33 cm. **17.** 72·6 lbs. **18.** 3·15 mm.^{2}, 2·00 mm. **19.** 13·0 in.

359. **20.** 1·86 cm.^{2} **21.** 331. **22.** 0·336 oz. **23.** 1·08 mm. **24.** 3·49. **25.** 0·790. **26.** 7·44 in. **27.** 8·58 in. **28.** 2·62 lbs. **29.** 2·15 ft. **30.** 4·84.

360. **31.** 0·00926. **32.** 7·87 in. **33.** 154. **34.** 0·025. **35.** 8·181. **36.** 8·613. **37.** 1·95 secs. **38.** 7·41 secs. **39.** 41·4 cwts.

361. **40.** 471 in.^{3} **41.** 2·01 secs. **42.** 375 ton-in. **43.** 450.

EXERCISE XVIII b

364. **1.** £50. **2.** £27. **3.** £7. 10*s.* **4.** £11. 4*s.* **5.** £18. 7*s.* 6*d.* **6.** £77. 3*s.* 9*d.* **7.** £4. 13*s.* 4*d.* **8.** £44. 19*s.* 2*d.* **9.** £15. 1*s.* 4*d.* **10.** £66. 19*s.* 8*d.* **11.** £108. 5*s.* 2*d.* **12.** £54. 14*s.* 1*d.* **13.** £82. 8*s.* 7*d.* **14.** 3*s.* **15.** 6*s.* 11*d.* **16.** £84. 14*s.* 3*d.* **17.** £43. 2*s.* 5*d.* **18.** £2862. 19*s.* 4*d.* **19.** £297. 4*s.* 3*d.* **20.** £1. 19*s.* 5*d.* **21.** £787. 1*s.* 11*d.* **22.** £25. 15*s.* 10*d.* **23.** £79. 4*s.* 8*d.* **24.** £451. 17*s.* 10*d.*

EXERCISE XVIII d

365. **1.** 2 yrs. 6 mths. **2.** 1 yr. 6 mths. **3.** 6 yrs. **4.** 8 yrs. **5.** 20 yrs. **6.** 3 yrs. 6 mths. **7.** 16 yrs. **8.** 2 yrs. 9 mths.

366. **9.** 1 yr. 9 mths. **10.** 2 yrs. 1 mth. **11.** 2 yrs. 4 mths. **12.** 5 yrs. 6 mths. **13.** 2 yrs. 5 mths. **14.** 3 yrs. 6 mths. **15.** 5 yrs. 4 mths. **16.** 2 yrs. 10 mths. **17.** 2 yrs. 2 mths. **18.** 2 yrs. 5 mths.

EXERCISE XVIII f

367. **1.** 5. **2.** 2·5. **3.** 2·6. **4.** 4·0. **5.** 4·7. **6.** 4·0. **7.** 3·8. **8.** 4·9. **9.** 4·3. **10.** 3·6. **11.** 4·7. **12.** 5·25. **13.** 3·1. **14.** 5·5. **15.** 10·7. **16.** 3·9.

EXERCISE XVIII h

368. **1.** £310. **2.** £273. **3.** £707.

369. **4.** £166. **5.** £401. **6.** £624. **7.** £532. **8.** £241. **9.** £247. **10.** £202·5.

EXERCISE XVIII i

PAGE

369. **1.** £58. **2.** £119. **3.** £490. **4.** £40.
5. £146. **6.** £302. **7.** £72. **8.** £45.

EXERCISE XVIII j

370. **1.** £59. 4*s.* 5*d.* **2.** 3·1. **3.** 2·7. **4.** £1531. 10*s.* 7*d.*
5. £565. 19*s.* 9*d.* **6.** 8 yrs. 8 mths. **7.** 3·5.
8. £351. 13*s.* **9.** 3·6. **10.** 18*s.* 7*d.* **11.** £44. 17*s.* 11*d.*
12. £580. **13.** £4000.

EXERCISE XVIII k

372. **1.** (1) £129. 15*s.* 10*d.*, £9. 15*s.* 10*d.* (2) £122. 0*s.* 9*d.*, £7. 0*s.* 9*d.* (3) £121. 15*s.* 8*d.*, £16. 11*s.* 8*d.* (4) £131. 17*s.* 8*d.*, £23. 7*s.* 8*d.* (5) £248. 15*s.* 4*d.*, £18. 15*s.* 4*d.* (6) £453. 18*s.* 11*d.*, £38. 10*s.* 5*d.* (7) £837. 1*s.* 3*d.*, £121. 10*s.* 9*d.* (8) £11070. 8*s.* 9*d.*, £533. 8*s.* 9*d.* (9) £11. 0*s.* 2*d.*, £1. 0*s.* 2*d.* (10) £145. 4*s.* 7*d.*, £26. 13*s.* 1*d.* (11) £159. 10*s.* 1*d.*, £31. 11*s.* 4*d.* (12) £34. 19*s.* 11*d.*, £3. 17*s.* 8*d.* (13) £2267. 2*s.* 9*d.*, £150. 14*s.* 10*d.*

373. **2.** £8. 2*s.* 9*d.*, £1190. 18*s.*, £2952. 1*s.* **3.** £41. 19*s.* 3*d.*, £6096. 19*s.*, £229220. **4.** £2. 11*s.* 7*d.*, £2. 13*s.* 8*d.*
5. 2756853, 2560000. **6.** £76300. **7.** £5. 7*s.*

EXERCISE XIX b

379. **1.** £2280, £555. 15*s.*, £402. 10*s.*, £766. 10*s.* **2.** £647. 16*s.*, £244. 11*s.* 3*d.* **3.** (i) £3947. 7*s.* 4*d.*, (ii) £364. 7*s.* 5*d.*, (iii) £304. 7*s.*, (iv) £695. 4*s.* 9*d.*, (v) £189. 0*s.* 3*d.*, (vi) £685. 14*s.* 3*d.*

4. (*a*) £3. 5*s.* 9*d.*, (*b*) £3. 14*s.* 6*d.*, (*c*) £4. 8*s.* 7*d.*,
(*d*) £3. 16*s.* 1*d.*, (*e*) £4. 6*s.* 11*d.*, (*f*) £4. 17*s.* 1*d.*,
(*g*) £4. 1*s.* 0*d.*, (*h*) £3. 16*s.* 11*d.*, (*i*) £3. 15*s.* 2*d.*,
(*j*) £4. 2*s.* 7*d.*, (*k*) £4. 1*s.* 11*d.*, (*l*) £4. 6*s.* 11*d.*,
(*m*) £4. 3*s.* 10*d.*, (*n*) £4. 7*s.* 11*d.*, (*o*) £4. 11*s.* 5*d.*,
(*p*) £4. 15*s.* 3*d.*

380. **5.** (i) $66\frac{2}{3}$, (ii) $87\frac{1}{2}$, (iii) $88\frac{8}{9}$, (iv) $111\frac{1}{9}$, (v) 80, (vi) $58\frac{14}{17}$, (vii) $126\frac{6}{19}$, (viii) 110, (ix) $76\frac{4}{21}$. **6.** £75, £22. 10*s.*, £17. 10*s.*, £36. 10*s.*; £28. 14*s.*, £10. 15*s.* **7.** (i) £197. 4*s.* 2*d.*, (ii) £18. 4*s.* 4*d.*, (iii) £15. 4*s.* 4*d.*, (iv) £34. 15*s.* 3*d.*, (v) £9. 9*s.*, (vi) £34. 5*s.* 9*d.* **8.** (*a*) £12160, (*b*) £10743, (*c*) £9029, (*d*) £10514, (*e*) £9200, (*f*) £8240, (*g*) £9880, (*h*) £10400, (*i*) £10640, (*j*) £9700, (*k*) £9767, (*l*) £9200, (*m*) £9545, (*n*) £9100, (*o*) £8750, (*p*) £8400.
9. £891. 13*s.* 4*d.*, £27. 10*s.* **10.** £21. 19*s.* 7*d.*, £27. 13*s.* 10*d.* **11.** £100. **12.** £3. 7*s.* 1*d.*

PAGE

EXERCISE XIX c

380. **1.** £1815. **2.** 6000. **3.** £82. 10*s*. 0*d*. **4.** £150.

381. **5.** £4. 11*s*. **6.** $66\frac{2}{3}$. **7.** £8800.
8. £382. 10*s*. **9.** £682. 13*s*. **10.** £523. 10*s*.
11. £1151. **12.** 320. **13.** 160.
14. £3. 10*s*. **15.** £22. 17*s*. **16.** 23*s*. 5*d*.
17. £3. 12*s*. **18.** £3. 2*s*. 6*d*. **19.** £1766.

382. **20.** £79. 6*s*. 8*d*.; £110. 13*s*. 11*d*.
21. £351. 17*s*.; £12. 10*s*.; £14. 1*s*. 6*d*.
22. 115; £36. 1*s*. 0*d*.; £25. 17*s*. 6*d*.
23. £4140; £170. 15*s*. 6*d*.; £165. 12*s*. 0*d*.
24. £1125; £4. 13*s*. 4*d*.; £7. 2*s*. 10*d*.
25. £96; £73; £1825. **26.** £565. **27.** £26. 5*s*. 0*d*.
28. 362; £21. 12*s*. 7*d*.; £25. 6*s*. 10*d*. **29.** £75; 3*s*. 4*d*.

383. **30.** £77. 5*s*. 5*d*.; 1581; £110. 13*s*. 5*d*. **31.** £2105. 5*s*. 3*d*.; 250; £52. 12*s*. 8*d*. + £56. 5*s*. 0*d*. **32.** £1095; £5. 9*s*. 7*d*.
33. £4. 7*s*. 9*d*.; £7. 6*s*. 1*d*. **34.** £19. 4*s*. 0*d*.
35. £55. 8*s*. 3*d*. **36.** £7. 10*s*. **37.** £20; $\frac{9}{31}$.
38. £8. **39.** £50. 0*s*. 10*d*.

384. **40.** 10 %; £120; $7\frac{1}{2}$ %.

REVISION PAPERS

Paper 55

385. **1.** 74·57, 854400, 0·000006615. **2.** 0·8451, 1·3802, $\bar{1}$·6712, $\bar{3}$·8057. **3.** 1234. **4.** $12\frac{1}{2}$ %. **5.** $\frac{3}{4}$ mi., 57 Km.
6. $\frac{p}{2}=\frac{22}{7}$. **7.** £17. 1*s*. 1*d*. **8.** 13200. **9.** 6·875.
10. 3320. **11.** 7·7.

Paper 56

386. **1.** 8·638, 1·329, 0·1964. **2.** $\bar{1}$·3701, 2·2375, 0·4972, $\bar{3}$·8457, 1·8452. **3.** 371. **4.** £57. 15*s*. 7*d*. **5.** 56·8 %.
6. 1·6 %. **7.** 50·27 sq. in. **8.** 1000. **9.** £85.
10. 6 lbs. **11.** 32·41 Kgm.

Paper 57

1. 21·87. **2.** 68·28, 525300, 0·3172; 2·875, 26·92, 0·7504.
3. 0·1716, 758·6, 10160, 0·0005768, 1000.

387. **4.** £15. 17*s*. 9*d*. **5.** 738 %. **6.** 3·5 %. **7.** 7·04 c.c., 35·2 cm.^{2} **8.** 2·96 %. **9.** £168. 14*s*. 7*d*. **10.** 35·39 yds.
11. 11 ft. 8 in.

Paper 58

1. 5·934, 0·2066, 240·6, 275600. **2.** 1701, 1·308, 0·01403.
3. 18*s*. 7*d*. **4.** 56·5 %. **5.** Copper 83 %, Zinc 8 %, Tin 9 %. **6.** 9·027, $94\frac{9}{100}$.

388. **7.** 29·3 ins. **8.** 383000 ft.^{3}, 23200 ft.^{2} **9.** 17·0 cm.
10. Cost of current the same for the same illumination, but lamp renewals greater for the latter lamp.
11. 86455000, £1081000.

Paper 59

PAGE

388. **1.** 113·7 (4 s. figs.), 0·1610, 0·3225. **2.** 0·849. **3.** 16 cwts. **4.** 643·5 cm.^{2}, $78\frac{4}{7}$ %. **5.** 3·6 %.

389. **6.** 15·3 %, 19·0 %. **7.** 48·6 %. **8.** 6 years. **9.** 12·32. **10.** 47·2, 4·3, 8·3, 40·2 %.

Paper 60

1. 19·73 (4 s. figs.), $3\frac{1}{8}$, 3·376. **2.** 2 cwts. 70 lbs. **3.** 4 %. **4.** 12 %, 12 %.

390. **5.** (*a*) +0·5, (*b*) −0·3, (*c*) −90, (*d*) −2, (*e*) +2, (*f*) +9, (*g*) −1. **6.** 1·76. **7.** £3. 6*s*. 8*d*. **8.** 155000 galls. **9.** 7*s*. $1\frac{1}{2}$*d*., 9*s*. 3*d*.

Paper 61

1. 0·008, 0·00809. **2.** 0·1907. **3.** 28·036, 28·04, 0·01 %. **4.** 55·47 lbs., 6·93 lbs.

391. **5.** 22400, 0·0739. **6.** 4·641, 3·96 %. **7.** 1676, 231, 311, 22 lbs. **8.** 17·1 %. **9.** 76. **10.** 1420 c.c.

Paper 62

1. 100000, 0·0000100, 2·85, 214, 5·10, 30·0, 0·2381, 1·59. **2.** 11·824 m. **3.** 420 yds.

392. **4.** 13·0 in. **5.** 647. **6.** 151 tons. **7.** 10, 0·83, 89 %. **8.** 25800 lbs. **9.** £2942 to within £1.

Paper 63

1. 29·9. **2.** 66·91, 17·04, 0·001875. **3.** $11\frac{4}{7}$ mi./hr. **4.** 838·5 c.c.

393. **5.** 3 h. 54 m. **6.** 105553, 85288. **7.** $2\frac{1}{2}$ %. **8.** 20 m. **9.** £2. 6*s*. 6*d*. **10.** 1*s*. 7·1*d*.

Paper 64

1. 783·7, 0·6263. **2.** Twice. **3.** £170. 18*s*. **4.** 11·3 cm.

394. **5.** 9·77 and 9·97. **6.** 1500, 3750. **7.** 89·6, 3·12, 7·28 %. **8.** 16·32 y. **9.** £6150000 to nearest £10000.

Paper 65

1. 22810, 1·48, 343·3. **2.** 6·9 cm. **3.** 22·19 c.c., 5·2 cm. **4.** 110·88 in.^{2}, 11·88″.

395. **5.** 137·3 lbs. **6.** £1068 about. **7.** 0·2 %. **8.** 62·7. **9.** 5 yrs. 8 mths.

Paper 66

1. 3·32. **2.** 21·05, 0·6463, 191·8. **3.** 200 tons. **4.** 12943. **5.** £255. 15*s*., £63. 15*s*., £2. 8*s*. 9*d*. **6.** £12. 13*s*. 9*d*. − 14*s*. 10*d*. = £11. 18*s*. 11*d*. **7.** 15.

396. **8.** 0·39 cm. **9.** 6 in./min., 15 mins.

Paper 67

PAGE

396. **1.** 18700000, 0·000002993, 546·4, 0·1402, 0·5274. **2.** 884 yds. **3.** 3·36″. **4.** 98 yds. **5.** 29,000,000, 10·9 %. **6.** 1578400. **7.** £43. 1*s*. 10½*d*., £60. 6*s*. 7½*d*., £36. 11*s*. 6*d*.

397. **8.** 3½ %. **9.** 2·0″.

Paper 68

1. 23·4. **2.** 58·51, 0·851, 0·00002567. **3.** £263. 19*s*. 3*d*. **4.** 2 tons 10 cwts. 1 qr. **5.** 0·2 %. **6.** 51·6 %. **7.** 17*s*. 6*d*. **8.** £18230000. **9.** $\frac{135}{5234}$.

Paper 69

398. **1.** 8, 3, 81, 100000, 0·04, 0·4, 1·44, 0·1. **2.** 0·1859, 0·00003224, 12·99, $7{\cdot}60 \times 10^{-9}$. **3.** 12 %. **4.** 3 cwts. 3 qrs. 17 lbs. **5.** 0·095. **6.** 243 $m.^2$ **7.** 113·1 c.c., 30·9 c.c. **8.** 146. **9.** £139771. 19*s*., 4·6 %.

Paper 70

399. **1.** 3717, 27·40. **2.** 7·75. **3.** £1000. **4.** 80 %, 20 %. **5.** 210 wks. **6.** £98. 10*s*. 7½*d*. **7.** 70, 2545. **8.** 1076. **9.** £5510, £2755. **10.** £960. 15*s*., £3. 12*s*. 10*d*., £4. 3*s*. 4*d*.

Paper 71

400. **1.** 15·61, 0·01166, 1·534. **2.** 2196. **3.** 2600 %. **4.** £2. 6*s*. 3*d*. **5.** 15 mins. **6.** 10,404,000. **7.** Nationalist by 8. **8.** 104×10^6 galls. **9.** £182. **10.** £2637 about.

Paper 72

401. **1.** 11·3. **2.** 80·71. **3.** 47·3 Kgm. **4.** 201×10^6, 7·23. **5.** £144, £108, £126, £54. **6.** 659 ft., 646 ft. **7.** £246. 4*s*. 10*d*. **8.** 275 %. **9.** 19·3 tons, 24 ft. 6 in.

402. **10.** 4 %.

Paper 73

1. 8100 $in.^3$, 3055 $in.^3$ **2.** 10·11, 3·864. **3.** £93. 10*s*. **4.** 74 ft. **5.** Nearly 22 cwts. **6.** 62½ %. **7.** 152, 171, 361. **8.** £50. 16*s*. 5*d*.

403. **9.** £18. 4*s*. **10.** £180, £270, 4½ %.

Paper 74

1. 55·00, 0·5456, 1·646. **2.** 6·4″ about. **3.** 7168 cu. ft. **4.** 7 tons 3 cwts. **5.** 4*s*. 2*d*. **6.** £960, £1207, £270. **7.** 1·26, 6·1 mm.

404. **8.** 0·735″. **9.** 88. **10.** £474, £53.

Paper 75

PAGE

404. **1.** 17·1. **2.** 3·629, 2·367, 9400. **3.** 576.
4. 465 lbs. **5.** £60. **6.** 400. **7.** 0·203 cm.
8. 2·35″, 1·84″, 1·82″.

405. **10.** 2·67 %.

Paper 76

1. 11·03. **2.** 211. **3.** 1·6 %, 12 c.c., 4·7 %.
4. 7502. **5.** 13·3 %.

406. **6.** 36. **7.** 223·5 l., 74·7 Kgm., 298 Kgm. **8.** 36.
9. 3900. **10.** £2953. **11.** 23192.

MISCELLANEOUS EXERCISES

407. **1.** 200. **2.** 112 per min. **3.** 1176. **4.** $\frac{17}{90}$=0·189 mile.
5. 40 : 36 : 25. **6.** 40 : 3.

408. **7.** 1885 c.c. **8.** 1173. **9.** 4 m. 21 s. **10.** 0·975 in.
11. 2″, $\sqrt{6}$=2·45″. **12.** 0·46 gm. **13.** £5, £5. 10*s.*
14. £1. 5*s.* 5½*d.*, £1. 9*s.* 3¼*d.*

409. **15.** 8·256. **16.** £840, £1050. **17.** 94 tons 11 cwts.
18. 23·7 k., 24·9 k. **19.** 57. **20.** 2 %. **21.** 12 %.
22. 18⅔ %. **23.** £40.

410. **24.** 17·1 %. **25.** 25 lbs. **26.** £3977. **27.** £12.
28. 137 %. **29.** 55*s.* **30.** $3\frac{9}{17}$ %. **31.** 67½ %, 66¼ %.

411. **32.** 1·02 m.^{2}, 0·0044 m.^{2} **33.** 3 %, 7 %. **34.** +2·35, −2·32; +2·48, −2·45; +4·35, −4·32; +8·18, −8·15; +20·08, −20·05. **36.** 27·1 %. **37.** 4 %. **38.** 0·2 : 1.

412. **39.** 860,000, 92,000,000 miles; 12,000 : 1, 1,300,000 : 1.
40. 21760. **41.** 41° nearly; 9. **42.** 512 in.^{3}, 437⅓ in.^{2}; 8″, 8·5″. **43.** 68 mo. **44.** 92. **45.** 0·4 %, 1·2 %.
46. £160, £200, £400.

413. **47.** £593. **48.** 1*s.* 8*d.* **49.** 6, nearly £636. **50.** 2*s.* 9*d.*
51. £4039. 13*s.* 6*d.* + £782. 16*s.* 1*d.* + £380. 16*s.* 5*d.* + £172. 4*s.* 5*d.* = £5375. 10*s.* 5*d.*; 16*s.*

414. **52.** 5*s.* 10½*d.*, 6*s.* 1*d.*, 4*s.* 7¼*d.* **53.** £5163536, 4¼ %.
54. £48. 8*s.* 4*d.*, £41. 10*s.* **55.** $206·81, £4. 5*s.* 1*d.*
56. 83¾. **57.** £5480, £5820.

415. **58.** £3080, £5920. **59.** +$4\frac{4}{9}$ %, $\frac{8}{47}$=0·17 %. **60.** £2412.
62. £580. **63.** 32.

416. **64.** £5462230. **65.** £18595. 14*s.* 7*d.* **66.** £70.
67. £12000. **68.** 95·15 ft.^{3}

417. **69.** 59 ft. **70.** 662 millions; 3 %.

TABLES

	0	1	2	3	4	5	6	7	8	9	1	2	3	4	5	6	7	8	9
10	·0000	0043	0086	0128	0170	0212	0253	0294	0334	0374	4	8	12	17	21	25	29	33	3[illegible]
11	·0414	0453	0492	0531	0569	0607	0645	0682	0719	0755	4	8	11	15	19	23	26	30	3[illegible]
12	·0792	0828	0864	0899	0934	0969	1004	1038	1072	1106	3	7	10	14	17	21	24	28	3[illegible]
13	·1139	1173	1206	1239	1271	1303	1335	1367	1399	1430	3	6	10	13	16	19	23	26	2[illegible]
14	·1461	1492	1523	1553	1584	1614	1644	1673	1703	1732	3	6	9	12	15	18	21	24	2[illegible]
15	·1761	1790	1818	1847	1875	1903	1931	1959	1987	2014	3	6	8	11	14	17	20	22	2[illegible]
16	·2041	2068	2095	2122	2148	2175	2201	2227	2253	2279	3	5	8	11	13	16	18	21	2[illegible]
17	·2304	2330	2355	2380	2405	2430	2455	2480	2504	2529	2	5	7	10	12	15	17	20	[illegible]
18	·2553	2577	2601	2625	2648	2672	2695	2718	2742	2765	2	5	7	9	12	14	16	19	[illegible]
19	·2788	2810	2833	2856	2878	2900	2923	2945	2967	2989	2	4	7	9	11	13	16	18	[illegible]
20	·3010	3032	3054	3075	3096	3118	3139	3160	3181	3201	2	4	6	8	11	13	15	17	
21	·3222	3243	3263	3284	3304	3324	3345	3365	3385	3404	2	4	6	8	10	12	14	16	
22	·3424	3444	3464	3483	3502	3522	3541	3560	3579	3598	2	4	6	8	10	12	14	15	
23	·3617	3636	3655	3674	3692	3711	3729	3747	3766	3784	2	4	6	7	9	11	13	15	
24	·3802	3820	3838	3856	3874	3892	3909	3927	3945	3962	2	4	5	7	9	11	12	14	
25	·3979	3997	4014	4031	4048	4065	4082	4099	4116	4133	2	3	5	7	9	10	12	14	
26	·4150	4166	4183	4200	4216	4232	4249	4265	4281	4298	2	3	5	7	8	10	11	13	
27	·4314	4330	4346	4362	4378	4393	4409	4425	4440	4456	2	3	5	6	8	9	11	13	
28	·4472	4487	4502	4518	4533	4548	4564	4579	4594	4609	2	3	5	6	8	9	11	12	
29	·4624	4639	4654	4669	4683	4698	4713	4728	4742	4757	1	3	4	6	7	9	10	12	
30	·4771	4786	4800	4814	4829	4843	4857	4871	4886	4900	1	3	4	6	7	9	10	11	
31	·4914	4928	4942	4955	4969	4983	4997	5011	5024	5038	1	3	4	6	7	8	10	11	
32	·5051	5065	5079	5092	5105	5119	5132	5145	5159	5172	1	3	4	5	7	8	9	11	
33	·5185	5198	5211	5224	5237	5250	5263	5276	5289	5302	1	3	4	5	6	8	9	10	
34	·5315	5328	5340	5353	5366	5378	5391	5403	5416	5428	1	3	4	5	6	8	9	1[illegible]	
35	·5441	5453	5465	5478	5490	5502	5514	5527	5539	5551	1	2	4	5	6	7	9	1[illegible]	
36	·5563	5575	5587	5599	5611	5623	5635	5647	5658	5670	1	2	4	5	6	7	8	1[illegible]	
37	·5682	5694	5705	5717	5729	5740	5752	5763	5775	5786	1	2	3	5	6	7	8	[illegible]	
38	·5798	5809	5821	5832	5843	5855	5866	5877	5888	5899	1	2	3	5	6	7	8	[illegible]	
39	·5911	5922	5933	5944	5955	5966	5977	5988	5999	6010	1	2	3	4	5	7	8	[illegible]	
40	·6021	6031	6042	6053	6064	6075	6085	6096	6107	6117	1	2	3	4	5	6	8		
41	·6128	6138	6149	6160	6170	6180	6191	6201	6212	6222	1	2	3	4	5	6	7		
42	·6232	6243	6253	6263	6274	6284	6294	6304	6314	6325	1	2	3	4	5	6	7		
43	·6335	6345	6355	6365	6375	6385	6395	6405	6415	6425	1	2	3	4	5	6	7		
44	·6435	6444	6454	6464	6474	6484	6493	6503	6513	6522	1	2	3	4	5	6	7		
45	·6532	6542	6551	6561	6571	6580	6590	6599	6609	6618	1	2	3	4	5	6	7		
46	·6628	6637	6646	6656	6665	6675	6684	6693	6702	6712	1	2	3	4	5	6	7		
47	·6721	6730	6739	6749	6758	6767	6776	6785	6794	6803	1	2	3	4	5	5	6		
48	·6812	6821	6830	6839	6848	6857	6866	6875	6884	6893	1	2	3	4	4	5	6		
49	·6902	6911	6920	6928	6937	6946	6955	6964	6972	6981	1	2	3	4	4	5	6		
50	·6990	6998	7007	7016	7024	7033	7042	7050	7059	7067	1	2	3	3	4	5	6		
51	·7076	7084	7093	7101	7110	7118	7126	7135	7143	7152	1	2	3	3	4	5	6		
52	·7160	7168	7177	7185	7193	7202	7210	7218	7226	7235	1	2	2	3	4	5	6		
53	·7243	7251	7259	7267	7275	7284	7292	7300	7308	7316	1	2	2	3	4	5	6		
54	·7324	7332	7340	7348	7356	7364	7372	7380	7388	7396	1	2	2	3	4	5	6		

	0	1	2	3	4	5	6	7	8	9	1	2	3	4	5	6	7	8	9
5	·7404	7412	7419	7427	7435	7443	7451	7459	7466	7474	1	2	2	3	4	5	5	6	7
6	·7482	7490	7497	7505	7513	7520	7528	7536	7543	7551	1	2	2	3	4	5	5	6	7
7	·7559	7566	7574	7582	7589	7597	7604	7612	7619	7627	1	2	2	3	4	5	5	6	7
8	·7634	7642	7649	7657	7664	7672	7679	7686	7694	7701	1	1	2	3	4	4	5	6	7
9	·7709	7716	7723	7731	7738	7745	7752	7760	7767	7774	1	1	2	3	4	4	5	6	7
0	·7782	7789	7796	7803	7810	7818	7825	7832	7839	7846	1	1	2	3	4	4	5	6	6
	·7853	7860	7868	7875	7882	7889	7896	7903	7910	7917	1	1	2	3	4	4	5	6	6
	·7924	7931	7938	7945	7952	7959	7966	7973	7980	7987	1	1	2	3	3	4	5	6	6
	·7993	8000	8007	8014	8021	8028	8035	8041	8048	8055	1	1	2	3	3	4	5	5	6
	·8062	8069	8075	8082	8089	8096	8102	8109	8116	8122	1	1	2	3	3	4	5	5	6
	·8129	8136	8142	8149	8156	8162	8169	8176	8182	8189	1	1	2	3	3	4	5	5	6
	·8195	8202	8209	8215	8222	8228	8235	8241	8248	8254	1	1	2	3	3	4	5	5	6
	·8261	8267	8274	8280	8287	8293	8299	8306	8312	8319	1	1	2	3	3	4	5	5	6
	·8325	8331	8338	8344	8351	8357	8363	8370	8376	8382	1	1	2	3	3	4	4	5	6
	·8388	8395	8401	8407	8414	8420	8426	8432	8439	8445	1	1	2	2	3	4	4	5	6
	·8451	8457	8463	8470	8476	8482	8488	8494	8500	8506	1	1	2	2	3	4	4	5	6
	·8513	8519	8525	8531	8537	8543	8549	8555	8561	8567	1	1	2	2	3	4	4	5	5
	·8573	8579	8585	8591	8597	8603	8609	8615	8621	8627	1	1	2	2	3	4	4	5	5
	·8633	8639	8645	8651	8657	8663	8669	8675	8681	8686	1	1	2	2	3	4	4	5	5
	·8692	8698	8704	8710	8716	8722	8727	8733	8739	8745	1	1	2	2	3	4	4	5	5
	·8751	8756	8762	8768	8774	8779	8785	8791	8797	8802	1	1	2	2	3	3	4	5	5
	·8808	8814	8820	8825	8831	8837	8842	8848	8854	8859	1	1	2	2	3	3	4	5	5
	·8865	8871	8876	8882	8887	8893	8899	8904	8910	8915	1	1	2	2	3	3	4	4	5
	·8921	8927	8932	8938	8943	8949	8954	8960	8965	8971	1	1	2	2	3	3	4	4	5
	·8976	8982	8987	8993	8998	9004	9009	9015	9020	9025	1	1	2	2	3	3	4	4	5
	·9031	9036	9042	9047	9053	9058	9063	9069	9074	9079	1	1	2	2	3	3	4	4	5
	·9085	9090	9096	9101	9106	9112	9117	9122	9128	9133	1	1	2	2	3	3	4	4	5
	·9138	9143	9149	9154	9159	9165	9170	9175	9180	9186	1	1	2	2	3	3	4	4	5
	·9191	9196	9201	9206	9212	9217	9222	9227	9232	9238	1	1	2	2	3	3	4	4	5
	·9243	9248	9253	9258	9263	9269	9274	9279	9284	9289	1	1	2	2	3	3	4	4	5
	·9294	9299	9304	9309	9315	9320	9325	9330	9335	9340	1	1	2	2	3	3	4	4	5
	·9345	9350	9355	9360	9365	9370	9375	9380	9385	9390	1	1	1	2	3	3	4	4	5
	9395	9400	9405	9410	9415	9420	9425	9430	9435	9440	0	1	1	2	2	3	3	4	4
	9445	9450	9455	9460	9465	9469	9474	9479	9484	9489	0	1	1	2	2	3	3	4	4
	9494	9499	9504	9509	9513	9518	9523	9528	9533	9538	0	1	1	2	2	3	3	4	4
	9542	9547	9552	9557	9562	9566	9571	9576	9581	9586	0	1	1	2	2	3	3	4	4
	9590	9595	9600	9605	9609	9614	9619	9624	9628	9633	0	1	1	2	2	3	3	4	4
	9638	9643	9647	9652	9657	9661	9666	9671	9675	9680	0	1	1	2	2	3	3	4	4
	9685	9689	9694	9699	9703	9708	9713	9717	9722	9727	0	1	1	2	2	3	3	4	4
	9731	9736	9741	9745	9750	9754	9759	9763	9768	9773	0	1	1	2	2	3	3	4	4
	9777	9782	9786	9791	9795	9800	9805	9809	9814	9818	0	1	1	2	2	3	3	4	4
	9823	9827	9832	9836	9841	9845	9850	9854	9859	9863	0	1	1	2	2	3	3	4	4
	9868	9872	9877	9881	9886	9890	9894	9899	9903	9908	0	1	1	2	2	3	3	4	4
	9912	9917	9921	9926	9930	9934	9939	9943	9948	9952	0	1	1	2	2	3	3	4	4
	9956	9961	9965	9969	9974	9978	9983	9987	9991	9996	0	1	1	2	2	3	3	3	4

	0	1	2	3	4	5	6	7	8	9	1	2	3	4	5	6	7	8	9
·00	1000	1002	1005	1007	1009	1012	1014	1016	1019	1021	0	0	1	1	1	1	2	2	2
·01	1023	1026	1028	1030	1033	1035	1038	1040	1042	1045	0	0	1	1	1	1	2	2	2
·02	1047	1050	1052	1054	1057	1059	1062	1064	1067	1069	0	0	1	1	1	1	2	2	
·03	1072	1074	1076	1079	1081	1084	1086	1089	1091	1094	0	0	1	1	1	1	2	2	
·04	1096	1099	1102	1104	1107	1109	1112	1114	1117	1119	0	1	1	1	1	2	2	2	
·05	1122	1125	1127	1130	1132	1135	1138	1140	1143	1146	0	1	1	1	1	2	2	2	
·06	1148	1151	1153	1156	1159	1161	1164	1167	1169	1172	0	1	1	1	1	2	2	2	
·07	1175	1178	1180	1183	1186	1189	1191	1194	1197	1199	0	1	1	1	1	2	2	2	
·08	1202	1205	1208	1211	1213	1216	1219	1222	1225	1227	0	1	1	1	1	2	2	2	
·09	1230	1233	1236	1239	1242	1245	1247	1250	1253	1256	0	1	1	1	1	2	2	2	
·10	1259	1262	1265	1268	1271	1274	1276	1279	1282	1285	0	1	1	1	1	2	2	2	
·11	1288	1291	1294	1297	1300	1303	1306	1309	1312	1315	0	1	1	1	2	2	2	2	
·12	1318	1321	1324	1327	1330	1334	1337	1340	1343	1346	0	1	1	1	2	2	2	2	
·13	1349	1352	1355	1358	1361	1365	1368	1371	1374	1377	0	1	1	1	2	2	2	3	
·14	1380	1384	1387	1390	1393	1396	1400	1403	1406	1409	0	1	1	1	2	2	2	3	
·15	1413	1416	1419	1422	1426	1429	1432	1435	1439	1442	0	1	1	1	2	2	2	3	
·16	1445	1449	1452	1455	1459	1462	1466	1469	1472	1476	0	1	1	1	2	2	2	3	
·17	1479	1483	1486	1489	1493	1496	1500	1503	1507	1510	0	1	1	1	2	2	2	3	
·18	1514	1517	1521	1524	1528	1531	1535	1538	1542	1545	0	1	1	1	2	2	2	3	
·19	1549	1552	1556	1560	1563	1567	1570	1574	1578	1581	0	1	1	1	2	2	3	3	
·20	1585	1589	1592	1596	1600	1603	1607	1611	1614	1618	0	1	1	1	2	2	3	3	
·21	1622	1626	1629	1633	1637	1641	1644	1648	1652	1656	0	1	1	2	2	2	3	3	
·22	1660	1663	1667	1671	1675	1679	1683	1687	1690	1694	0	1	1	2	2	2	3	3	
·23	1698	1702	1706	1710	1714	1718	1722	1726	1730	1734	0	1	1	2	2	2	3	3	
·24	1738	1742	1746	1750	1754	1758	1762	1766	1770	1774	0	1	1	2	2	2	3	3	
·25	1778	1782	1786	1791	1795	1799	1803	1807	1811	1816	0	1	1	2	2	2	3	3	
·26	1820	1824	1828	1832	1837	1841	1845	1849	1854	1858	0	1	1	2	2	3	3	3	
·27	1862	1866	1871	1875	1879	1884	1888	1892	1897	1901	0	1	1	2	2	3	3	3	
·28	1905	1910	1914	1919	1923	1928	1932	1936	1941	1945	0	1	1	2	2	3	3	4	
·29	1950	1954	1959	1963	1968	1972	1977	1982	1986	1991	0	1	1	2	2	3	3	4	
·30	1995	2000	2004	2009	2014	2018	2023	2028	2032	2037	0	1	1	2	2	3	3	4	
·31	2042	2046	2051	2056	2061	2065	2070	2075	2080	2084	0	1	1	2	2	3	3	4	
·32	2089	2094	2099	2104	2109	2113	2118	2123	2128	2133	0	1	1	2	2	3	3	[illegible]	
·33	2138	2143	2148	2153	2158	2163	2168	2173	2178	2183	0	1	1	2	2	3	3		
·34	2188	2193	2198	2203	2208	2213	2218	2223	2228	2234	1	1	2	2	3	3	4		
·35	2239	2244	2249	2254	2259	2265	2270	2275	2280	2286	1	1	2	2	3	3	4		
·36	2291	2296	2301	2307	2312	2317	2323	2328	2333	2339	1	1	2	2	3	3	4		
·37	2344	2350	2355	2360	2366	2371	2377	2382	2388	2393	1	1	2	2	3	3	4		
·38	2399	2404	2410	2415	2421	2427	2432	2438	2443	2449	1	1	2	2	3	3	4		
·39	2455	2460	2466	2472	2477	2483	2489	2495	2500	2506	1	1	2	2	3	3	4		
·40	2512	2518	2523	2529	2535	2541	2547	2553	2559	2564	1	1	2	2	3	4	4		
·41	2570	2576	2582	2588	2594	2600	2606	2612	2618	2624	1	1	2	2	3	4	4		
·42	2630	2636	2642	2649	2655	2661	2667	2673	2679	2685	1	1	2	2	3	4	4		
·43	2692	2698	2704	2710	2716	2723	2729	2735	2742	2748	1	1	2	3	3	4	4		
·44	2754	2761	2767	2773	2780	2786	2793	2799	2805	2812	1	1	2	3	3	4	4		
·45	2818	2825	2831	2838	2844	2851	2858	2864	2871	2877	1	1	2	3	3	4	5		
·46	2884	2891	2897	2904	2911	2917	2924	2931	2938	2944	1	1	2	3	3	4	5		
·47	2951	2958	2965	2972	2979	2985	2992	2999	3006	3013	1	1	2	3	3	4	5		
·48	3020	3027	3034	3041	3048	3055	3062	3069	3076	3083	1	1	2	3	4	4	5		
·49	3090	3097	3105	3112	3119	3126	3133	3141	3148	3155	1	1	2	3	4	4	5		

	0	1	2	3	4	5	6	7	8	9	1	2	3	4	5	6	7	8	9
50	3162	3170	3177	3184	3192	3199	3206	3214	3221	3228	1	1	2	3	4	4	5	6	7
51	3236	3243	3251	3258	3266	3273	3281	3289	3296	3304	1	2	2	3	4	5	5	6	7
52	3311	3319	3327	3334	3342	3350	3357	3365	3373	3381	1	2	2	3	4	5	5	6	7
53	3388	3396	3404	3412	3420	3428	3436	3443	3451	3459	1	2	2	3	4	5	6	6	7
54	3467	3475	3483	3491	3499	3508	3516	3524	3532	3540	1	2	2	3	4	5	6	6	7
55	3548	3556	3565	3573	3581	3589	3597	3606	3614	3622	1	2	2	3	4	5	6	7	7
56	3631	3639	3648	3656	3664	3673	3681	3690	3698	3707	1	2	3	3	4	5	6	7	8
57	3715	3724	3733	3741	3750	3758	3767	3776	3784	3793	1	2	3	3	4	5	6	7	8
58	3802	3811	3819	3828	3837	3846	3855	3864	3873	3882	1	2	3	4	4	5	6	7	8
59	3890	3899	3908	3917	3926	3936	3945	3954	3963	3972	1	2	3	4	5	5	6	7	8
60	3981	3990	3999	4009	4018	4027	4036	4046	4055	4064	1	2	3	4	5	6	6	7	8
	4074	4083	4093	4102	4111	4121	4130	4140	4150	4159	1	2	3	4	5	6	7	8	9
	4169	4178	4188	4198	4207	4217	4227	4236	4246	4256	1	2	3	4	5	6	7	8	9
	4266	4276	4285	4295	4305	4315	4325	4335	4345	4355	1	2	3	4	5	6	7	8	9
	4365	4375	4385	4395	4406	4416	4426	4436	4446	4457	1	2	3	4	5	6	7	8	9
	4467	4477	4487	4498	4508	4519	4529	4539	4550	4560	1	2	3	4	5	6	7	8	9
	4571	4581	4592	4603	4613	4624	4634	4645	4656	4667	1	2	3	4	5	6	7	9	10
	4677	4688	4699	4710	4721	4732	4742	4753	4764	4775	1	2	3	4	5	7	8	9	10
	4786	4797	4808	4819	4831	4842	4853	4864	4875	4887	1	2	3	4	6	7	8	9	10
	4898	4909	4920	4932	4943	4955	4966	4977	4989	5000	1	2	3	5	6	7	8	9	10
	5012	5023	5035	5047	5058	5070	5082	5093	5105	5117	1	2	4	5	6	7	8	9	11
	5129	5140	5152	5164	5176	5188	5200	5212	5224	5236	1	2	4	5	6	7	8	10	11
	5248	5260	5272	5284	5297	5309	5321	5333	5346	5358	1	2	4	5	6	7	9	10	11
	5370	5383	5395	5408	5420	5433	5445	5458	5470	5483	1	3	4	5	6	8	9	10	11
	5495	5508	5521	5534	5546	5559	5572	5585	5598	5610	1	3	4	5	6	8	9	10	12
	5623	5636	5649	5662	5675	5689	5702	5715	5728	5741	1	3	4	5	7	8	9	10	12
	5754	5768	5781	5794	5808	5821	5834	5848	5861	5875	1	3	4	5	7	8	9	11	12
	5888	5902	5916	5929	5943	5957	5970	5984	5998	6012	1	3	4	5	7	8	10	11	12
	6026	6039	6053	6067	6081	6095	6109	6124	6138	6152	1	3	4	6	7	8	10	11	13
	6166	6180	6194	6209	6223	6237	6252	6266	6281	6295	1	3	4	6	7	9	10	11	13
	6310	6324	6339	6353	6368	6383	6397	6412	6427	6442	1	3	4	6	7	9	10	12	13
	6457	6471	6486	6501	6516	6531	6546	6561	6577	6592	2	3	5	6	8	9	11	12	14
	6607	6622	6637	6653	6668	6683	6699	6714	6730	6745	2	3	5	6	8	9	11	12	14
	6761	6776	6792	6808	6823	6839	6855	6871	6887	6902	2	3	5	6	8	9	11	13	14
	6918	6934	6950	6966	6982	6998	7015	7031	7047	7063	2	3	5	6	8	10	11	13	15
	7079	7096	7112	7129	7145	7161	7178	7194	7211	7228	2	3	5	7	8	10	12	13	15
	7244	7261	7278	7295	7311	7328	7345	7362	7379	7396	2	3	5	7	8	10	12	13	15
	7413	7430	7447	7464	7482	7499	7516	7534	7551	7568	2	3	5	7	9	10	12	14	16
	7586	7603	7621	7638	7656	7674	7691	7709	7727	7745	2	4	5	7	9	11	12	14	16
	7762	7780	7798	7816	7834	7852	7870	7889	7907	7925	2	4	5	7	9	11	13	14	16
	7943	7962	7980	7998	8017	8035	8054	8072	8091	8110	2	4	6	7	9	11	13	15	17
	8128	8147	8166	8185	8204	8222	8241	8260	8279	8299	2	4	6	8	9	11	13	15	17
	8318	8337	8356	8375	8395	8414	8433	8453	8472	8492	2	4	6	8	10	12	14	15	17
	8511	8531	8551	8570	8590	8610	8630	8650	8670	8690	2	4	6	8	10	12	14	16	18
	8710	8730	8750	8770	8790	8810	8831	8851	8872	8892	2	4	6	8	10	12	14	16	18
	8913	8933	8954	8974	8995	9016	9036	9057	9078	9099	2	4	6	8	10	12	15	17	19
	9120	9141	9162	9183	9204	9226	9247	9268	9290	9311	2	4	6	8	11	13	15	17	19
	9333	9354	9376	9397	9419	9441	9462	9484	9506	9528	2	4	7	9	11	13	15	17	20
	9550	9572	9594	9616	9638	9661	9683	9705	9727	9750	2	4	7	9	11	13	16	18	20
	9772	9795	9817	9840	9863	9886	9908	9931	9954	9977	2	5	7	9	11	14	16	18	20

	0	1	2	3	4	5	6	7	8	9	1	2	3	4	5	6	7	8	9
10	1000	1020	1040	1061	1082	1103	1124	1145	1166	1188	2	4	6	8	10	13	15	17	[illegible]
11	1210	1232	1254	1277	1300	1323	1346	1369	1392	1416	2	5	7	9	11	14	16	18	[illegible]
12	1440	1464	1488	1513	1538	1563	1588	1613	1638	1664	2	5	7	10	12	15	17	20	[illegible]
13	1690	1716	1742	1769	1796	1823	1850	1877	1904	1932	3	5	8	11	13	16	19	22	[illegible]
14	1960	1988	2016	2045	2074	2103	2132	2161	2190	2220	3	6	9	12	14	17	20	23	[illegible]
15	2250	2280	2310	2341	2372	2403	2434	2465	2496	2528	3	6	9	12	15	19	22	25	[illegible]
16	2560	2592	2624	2657	2690	2723	2756	2789	2822	2856	3	7	10	13	16	20	23	26	[illegible]
17	2890	2924	2958	2993	3028	3063	3098	3133	3168	3204	3	7	10	14	17	21	24	28	[illegible]
18	3240	3276	3312	3349	3386	3423	3460	3497	3534	3572	4	7	11	15	18	22	26	30	[illegible]
19	3610	3648	3686	3725	3764	3803	3842	3881	3920	3960	4	8	12	16	19	23	27	31	[illegible]
20	4000	4040	4080	4121	4162	4203	4244	4285	4326	4368	4	8	12	16	20	25	29	33	[illegible]
21	4410	4452	4494	4537	4580	4623	4666	4709	4752	4796	4	9	13	17	21	26	30	34	[illegible]
22	4840	4884	4928	4973	5018	5063	5108	5153	5198	5244	4	9	13	18	22	27	31	36	[illegible]
23	5290	5336	5382	5429	5476	5523	5570	5617	5664	5712	5	9	14	19	23	28	33	38	[illegible]
24	5760	5808	5856	5905	5954	6003	6052	6101	6150	6200	5	10	15	20	24	29	34	39	[illegible]
25	6250	6300	6350	6401	6452	6503	6554	6605	6656	6708	5	10	15	20	25	31	36	41	[illegible]
26	6760	6812	6864	6917	6970	7023	7076	7129	7182	7236	5	11	16	21	26	32	37	42	[illegible]
27	7290	7344	7398	7453	7508	7563	7618	7673	7728	7784	5	11	16	22	28	33	38	44	[illegible]
28	7840	7896	7952	8009	8066	8123	8180	8237	8294	8352	6	12	17	23	29	34	40	46	[illegible]
29	8410	8468	8526	8585	8644	8703	8762	8821	8880	8940	6	12	18	24	30	35	41	47	[illegible]
30	9000	9060	9120	9181	9242	9303	9364	9425	9486	9548	6	12	18	24	31	37	43	[illegible]	[illegible]
31	9610	9672	9734	9797	9860	9923	9986				6	13	19	25	31	38	46	[illegible]	[illegible]
31								1005	1011	1018	1	1	2	3	3	4	5	[illegible]	[illegible]
32	1024	1030	1037	1043	1050	1056	1063	1069	1076	1082	1	1	2	3	3	4	5	[illegible]	[illegible]
33	1089	1096	1102	1109	1116	1122	1129	1136	1142	1149	1	1	2	3	3	4	5	[illegible]	[illegible]
34	1156	1163	1170	1176	1183	1190	1197	1204	1211	1218	1	1	2	3	3	4	5	[illegible]	[illegible]
35	1225	1232	1239	1246	1253	1260	1267	1274	1282	1289	1	1	2	3	4	4	5	[illegible]	[illegible]
36	1296	1303	1310	1318	1325	1332	1340	1347	1354	1362	1	1	2	3	4	4	5	[illegible]	[illegible]
37	1369	1376	1384	1391	1399	1406	1414	1421	1429	1436	1	2	2	3	4	5	5	[illegible]	[illegible]
38	1444	1452	1459	1467	1475	1482	1490	1498	1505	1513	1	2	2	3	4	5	5	[illegible]	[illegible]
39	1521	1529	1537	1544	1552	1560	1568	1576	1584	1592	1	2	2	3	4	5	6	[illegible]	[illegible]
40	1600	1608	1616	1624	1632	1640	1648	1656	1665	1673	1	2	2	3	4	5	6	[illegible]	[illegible]
41	1681	1689	1697	1706	1714	1722	1731	1739	1747	1756	1	2	2	3	4	5	6	[illegible]	[illegible]
42	1764	1772	1781	1789	1798	1806	1815	1823	1832	1840	1	2	3	3	4	5	6	[illegible]	[illegible]
43	1849	1858	1866	1875	1884	1892	1901	1910	1918	1927	1	2	3	3	4	5	6	[illegible]	[illegible]
44	1936	1945	1954	1962	1971	1980	1989	1998	2007	2016	1	2	3	4	5	5	6	[illegible]	[illegible]
45	2025	2034	2043	2052	2061	2070	2079	2088	2098	2107	1	2	3	4	5	5	6	[illegible]	[illegible]
46	2116	2125	2134	2144	2153	2162	2172	2181	2190	2200	1	2	3	4	5	6	7	[illegible]	[illegible]
47	2209	2218	2228	2237	2247	2256	2266	2275	2285	2294	1	2	3	4	5	6	7	[illegible]	[illegible]
48	2304	2314	2323	2333	2343	2352	2362	2372	2381	2391	1	2	3	4	5	6	7	[illegible]	[illegible]
49	2401	2411	2421	2430	2440	2450	2460	2470	2480	2490	1	2	3	4	5	6	7	[illegible]	[illegible]
50	2500	2510	2520	2530	2540	2550	2560	2570	2581	2591	1	2	3	4	5	6	7	[illegible]	[illegible]
51	2601	2611	2621	2632	2642	2652	2663	2673	2683	2694	1	2	3	4	5	6	7	[illegible]	[illegible]
52	2704	2714	2725	2735	2746	2756	2767	2777	2788	2798	1	2	3	4	5	6	7	[illegible]	[illegible]
53	2809	2820	2830	2841	2852	2862	2873	2884	2894	2905	1	2	3	4	5	6	7	[illegible]	[illegible]
54	2916	2927	2938	2948	2959	2970	2981	2992	3003	3014	1	2	3	4	6	7	8	[illegible]	[illegible]

The position of the decimal point must be determined by inspection.

	0	1	2	3	4	5	6	7	8	9	1	2	3	4	5	6	7	8	9
55	3025	3036	3047	3058	3069	3080	3091	3102	3114	3125	1	2	3	4	6	7	8	9	10
56	3136	3147	3158	3170	3181	3192	3204	3215	3226	3238	1	2	3	5	6	7	8	9	10
7	3249	3260	3272	3283	3295	3306	3318	3329	3341	3352	1	2	3	5	6	7	8	9	10
8	3364	3376	3387	3399	3411	3422	3434	3446	3457	3469	1	2	4	5	6	7	8	9	11
9	3481	3493	3505	3516	3528	3540	3552	3564	3576	3588	1	2	4	5	6	7	8	10	11
0	3600	3612	3624	3636	3648	3660	3672	3684	3697	3709	1	2	4	5	6	7	8	10	11
1	3721	3733	3745	3758	3770	3782	3795	3807	3819	3832	1	2	4	5	6	7	9	10	11
2	3844	3856	3869	3881	3894	3906	3919	3931	3944	3956	1	3	4	5	6	8	9	10	11
3	3969	3982	3994	4007	4020	4032	4045	4058	4070	4083	1	3	4	5	6	8	9	10	11
4	4096	4109	4122	4134	4147	4160	4173	4186	4199	4212	1	3	4	5	6	8	9	10	12
5	4225	4238	4251	4264	4277	4290	4303	4316	4330	4343	1	3	4	5	7	8	9	10	12
6	4356	4369	4382	4396	4409	4422	4436	4449	4462	4476	1	3	4	5	7	8	9	11	12
7	4489	4502	4516	4529	4543	4556	4570	4583	4597	4610	1	3	4	5	7	8	9	11	12
	4624	4638	4651	4665	4679	4692	4706	4720	4733	4747	1	3	4	5	7	8	10	11	12
	4761	4775	4789	4802	4816	4830	4844	4858	4872	4886	1	3	4	6	7	8	10	11	13
	4900	4914	4928	4942	4956	4970	4984	4998	5013	5027	1	3	4	6	7	8	10	11	13
	5041	5055	5069	5084	5098	5112	5127	5141	5155	5170	1	3	4	6	7	9	10	11	13
	5184	5198	5213	5227	5242	5256	5271	5285	5300	5314	1	3	4	6	7	9	10	11	13
	5329	5344	5358	5373	5388	5402	5417	5432	5446	5461	1	3	4	6	7	9	10	12	13
	5476	5491	5506	5520	5535	5550	5565	5580	5595	5610	1	3	4	6	7	9	10	12	13
	5625	5640	5655	5670	5685	5700	5715	5730	5746	5761	2	3	5	6	8	9	11	12	14
	5776	5791	5806	5822	5837	5852	5868	5883	5898	5914	2	3	5	6	8	9	11	12	14
	5929	5944	5960	5975	5991	6006	6022	6037	6053	6068	2	3	5	6	8	9	11	12	14
	6084	6100	6115	6131	6147	6162	6178	6194	6209	6225	2	3	5	6	8	9	11	13	14
	6241	6257	6273	6288	6304	6320	6336	6352	6368	6384	2	3	5	6	8	10	11	13	14
	6400	6416	6432	6448	6464	6480	6496	6512	6529	6545	2	3	5	6	8	10	11	13	14
	6561	6577	6593	6610	6626	6642	6659	6675	6691	6708	2	3	5	7	8	10	11	13	15
	6724	6740	6757	6773	6790	6806	6823	6839	6856	6872	2	3	5	7	8	10	12	13	15
	6889	6906	6922	6939	6956	6972	6989	7006	7022	7039	2	3	5	7	8	10	12	13	15
	7056	7073	7090	7106	7123	7140	7157	7174	7191	7208	2	3	5	7	8	10	12	14	15
	7225	7242	7259	7276	7293	7310	7327	7344	7362	7379	2	3	5	7	9	10	12	14	15
	7396	7413	7430	7448	7465	7482	7500	7517	7534	7552	2	3	5	7	9	10	12	14	16
	7569	7586	7604	7621	7639	7656	7674	7691	7709	7726	2	4	5	7	9	11	12	14	16
	7744	7762	7779	7797	7815	7832	7850	7868	7885	7903	2	4	5	7	9	11	12	14	16
	7921	7939	7957	7974	7992	8010	8028	8046	8064	8082	2	4	5	7	9	11	13	14	16
	8100	8118	8136	8154	8172	8190	8208	8226	8245	8263	2	4	5	7	9	11	13	14	16
	8281	8299	8317	8336	8354	8372	8391	8409	8427	8446	2	4	5	7	9	11	13	15	16
	8464	8482	8501	8519	8538	8556	8575	8593	8612	8630	2	4	6	7	9	11	13	15	17
	8649	8668	8686	8705	8724	8742	8761	8780	8798	8817	2	4	6	7	9	11	13	15	17
	8836	8855	8874	8892	8911	8930	8949	8968	8987	9006	2	4	6	8	9	11	13	15	17
	025	9044	9063	9082	9101	9120	9139	9158	9178	9197	2	4	6	8	10	11	13	15	17
	216	9235	9254	9274	9293	9312	9332	9351	9370	9390	2	4	6	8	10	12	14	15	17
	409	9428	9448	9467	9487	9506	9526	9545	9565	9584	2	4	6	8	10	12	14	16	18
	604	9624	9643	9663	9683	9702	9722	9742	9761	9781	2	4	6	8	10	12	14	16	18
	801	9821	9841	9860	9880	9900	9920	9940	9960	9980	2	4	6	8	10	12	14	16	18

The position of the decimal point must be determined by inspection.

SQUARE ROOTS

	0	1	2	3	4	5	6	7	8	9	1	2	3	4	5	6	7	8	9
10	1000	1005	1010	1015	1020	1025	1030	1034	1039	1044	0	1	1	2	2	3	3	4	4
	3162	3178	3194	3209	3225	3240	3256	3271	3286	3302	2	3	5	6	8	9	11	12	1
11	1049	1054	1058	1063	1068	1072	1077	1082	1086	1091	0	1	1	2	2	3	3	4	4
	3317	3332	3347	3362	3376	3391	3406	3421	3435	3450	1	3	4	6	7	9	10	12	1
12	1095	1100	1105	1109	1114	1118	1122	1127	1131	1136	0	1	1	2	2	3	3	4	
	3464	3479	3493	3507	3521	3536	3550	3564	3578	3592	1	3	4	6	7	8	10	11	1
13	1140	1145	1149	1153	1158	1162	1166	1170	1175	1179	0	1	1	2	2	3	3	3	
	3606	3619	3633	3647	3661	3674	3688	3701	3715	3728	1	3	4	5	7	8	10	11	1
14	1183	1187	1192	1196	1200	1204	1208	1212	1217	1221	0	1	1	2	2	3	3	3	
	3742	3755	3768	3782	3795	3808	3821	3834	3847	3860	1	3	4	5	7	8	9	11	
15	1225	1229	1233	1237	1241	1245	1249	1253	1257	1261	0	1	1	2	2	3	3	3	
	3873	3886	3899	3912	3924	3937	3950	3962	3975	3987	1	3	4	5	6	8	9	10	
16	1265	1269	1273	1277	1281	1285	1288	1292	1296	1300	0	1	1	2	2	3	3	3	
	4000	4012	4025	4037	4050	4062	4074	4087	4099	4111	1	2	4	5	6	7	9	10	
17	1304	1308	1311	1315	1319	1323	1327	1330	1334	1338	0	1	1	2	2	2	3	3	
	4123	4135	4147	4159	4171	4183	4195	4207	4219	4231	1	2	4	5	6	7	8	10	
18	1342	1345	1349	1353	1356	1360	1364	1367	1371	1375	0	1	1	1	2	2	3	3	
	4243	4254	4266	4278	4290	4301	4313	4324	4336	4347	1	2	3	5	6	7	8	9	
19	1378	1382	1386	1389	1393	1396	1400	1404	1407	1411	0	1	1	1	2	2	3	3	
	4359	4370	4382	4393	4405	4416	4427	4438	4450	4461	1	2	3	5	6	7	8	9	
20	1414	1418	1421	1425	1428	1432	1435	1439	1442	1446	0	1	1	1	2	2	2	3	
	4472	4483	4494	4506	4517	4528	4539	4550	4561	4572	1	2	3	4	5	7	8	9	
21	1449	1453	1456	1459	1463	1466	1470	1473	1476	1480	0	1	1	1	2	2	2	3	
	4583	4593	4604	4615	4626	4637	4648	4658	4669	4680	1	2	3	4	5	6	8	9	
22	1483	1487	1490	1493	1497	1500	1503	1507	1510	1513	0	1	1	1	2	2	2	3	
	4690	4701	4712	4722	4733	4743	4754	4764	4775	4785	1	2	3	4	5	6	7	8	
23	1517	1520	1523	1526	1530	1533	1536	1539	1543	1546	0	1	1	1	2	2	2	3	
	4796	4806	4817	4827	4837	4848	4858	4868	4879	4889	1	2	3	4	5	6	7	8	
24	1549	1552	1556	1559	1562	1565	1568	1572	1575	1578	0	1	1	1	2	2	2	3	
	4899	4909	4919	4930	4940	4950	4960	4970	4980	4990	1	2	3	4	5	6	7	8	
25	1581	1584	1587	1591	1594	1597	1600	1603	1606	1609	0	1	1	1	2	2	2		
	5000	5010	5020	5030	5040	5050	5060	5070	5079	5089	1	2	3	4	5	6	7		
26	1612	1616	1619	1622	1625	1628	1631	1634	1637	1640	0	1	1	1	2	2	2		
	5099	5109	5119	5128	5138	5148	5158	5167	5177	5187	1	2	3	4	5	6	7		
27	1643	1646	1649	1652	1655	1658	1661	1664	1667	1670	0	1	1	1	2	2	2		
	5196	5206	5215	5225	5235	5244	5254	5263	5273	5282	1	2	3	4	5	6	7		
28	1673	1676	1679	1682	1685	1688	1691	1694	1697	1700	0	1	1	1	1	2	2		
	5292	5301	5310	5320	5329	5339	5348	5357	5367	5376	1	2	3	4	5	6	7		
29	1703	1706	1709	1712	1715	1718	1720	1723	1726	1729	0	1	1	1	1	2	2		
	5385	5394	5404	5413	5422	5431	5441	5450	5459	5468	1	2	3	4	5	5	6		
30	1732	1735	1738	1741	1744	1746	1749	1752	1755	1758	0	1	1	1	1	2	2		
	5477	5486	5495	5505	5514	5523	5532	5541	5550	5559	1	2	3	4	4	5	6		
31	1761	1764	1766	1769	1772	1775	1778	1780	1783	1786	0	1	1	1	1	2	2		
	5568	5577	5586	5595	5604	5612	5621	5630	5639	5648	1	2	3	3	4	5	6		
32	1789	1792	1794	1797	1800	1803	1806	1808	1811	1814	0	1	1	1	1	2	2		
	5657	5666	5675	5683	5692	5701	5710	5718	5727	5736	1	2	3	3	4	5	6		

The first significant figure and the position of the decimal point must be determined by inspection.

	0	1	2	3	4	5	6	7	8	9	1	2	3	4	5	6	7	8	9
33	1817	1819	1822	1825	1828	1830	1833	1836	1838	1841	0	1	1	1	1	2	2	2	2
	5745	5753	5762	5771	5779	5788	5797	5805	5814	5822	1	2	3	3	4	5	6	7	8
34	1844	1847	1849	1852	1855	1857	1860	1863	1865	1868	0	1	1	1	1	2	2	2	2
	5831	5840	5848	5857	5865	5874	5882	5891	5899	5908	1	2	3	3	4	5	6	7	8
35	1871	1873	1876	1879	1881	1884	1887	1889	1892	1895	0	1	1	1	1	2	2	2	2
	5916	5925	5933	5941	5950	5958	5967	5975	5983	5992	1	2	2	3	4	5	6	7	8
36	1897	1900	1903	1905	1908	1910	1913	1916	1918	1921	0	1	1	1	1	2	2	2	2
	6000	6008	6017	6025	6033	6042	6050	6058	6066	6075	1	2	2	3	4	5	6	7	7
7	1924	1926	1929	1931	1934	1936	1939	1942	1944	1947	0	1	1	1	1	2	2	2	2
	6083	6091	6099	6107	6116	6124	6132	6140	6148	6156	1	2	2	3	4	5	6	7	7
8	1949	1952	1954	1957	1960	1962	1965	1967	1970	1972	0	1	1	1	1	2	2	2	2
	6164	6173	6181	6189	6197	6205	6213	6221	6229	6237	1	2	2	3	4	5	6	6	7
9	1975	1977	1980	1982	1985	1987	1990	1992	1995	1997	0	1	1	1	1	2	2	2	2
	6245	6253	6261	6269	6277	6285	6293	6301	6309	6317	1	2	2	3	4	5	6	6	7
	2000	2002	2005	2007	2010	2012	2015	2017	2020	2022	0	0	1	1	1	1	2	2	2
	6325	6332	6340	6348	6356	6364	6372	6380	6387	6395	1	2	2	3	4	5	6	6	7
	2025	2027	2030	2032	2035	2037	2040	2042	2045	2047	0	0	1	1	1	1	2	2	2
	6403	6411	6419	6427	6434	6442	6450	6458	6465	6473	1	2	2	3	4	5	5	6	7
	2049	2052	2054	2057	2059	2062	2064	2066	2069	2071	0	0	1	1	1	1	2	2	2
	6481	6488	6496	6504	6512	6519	6527	6535	6542	6550	1	2	2	3	4	5	5	6	7
	2074	2076	2078	2081	2083	2086	2088	2090	2093	2095	0	0	1	1	1	1	2	2	2
	6557	6565	6573	6580	6588	6595	6603	6611	6618	6626	1	2	2	3	4	5	5	6	7
	2098	2100	2102	2105	2107	2110	2112	2114	2117	2119	0	0	1	1	1	1	2	2	2
	6633	6641	6648	6656	6663	6671	6678	6686	6693	6701	1	2	2	3	4	4	5	6	7
	2121	2124	2126	2128	2131	2133	2135	2138	2140	2142	0	0	1	1	1	1	2	2	2
	6708	6716	6723	6731	6738	6745	6753	6760	6768	6775	1	1	2	3	4	4	5	6	7
	2145	2147	2149	2152	2154	2156	2159	2161	2163	2166	0	0	1	1	1	1	2	2	2
	6782	6790	6797	6804	6812	6819	6826	6834	6841	6848	1	1	2	3	4	4	5	6	7
	2168	2170	2173	2175	2177	2179	2182	2184	2186	2189	0	0	1	1	1	1	2	2	2
	6856	6863	6870	6877	6885	6892	6899	6907	6914	6921	1	1	2	3	4	4	5	6	7
	2191	2193	2195	2198	2200	2202	2205	2207	2209	2211	0	0	1	1	1	1	2	2	2
	6928	6935	6943	6950	6957	6964	6971	6979	6986	6993	1	1	2	3	4	4	5	6	6
	2214	2216	2218	2220	2223	2225	2227	2229	2232	2234	0	0	1	1	1	1	2	2	2
	7000	7007	7014	7021	7029	7036	7043	7050	7057	7064	1	1	2	3	4	4	5	6	6
	2236	2238	2241	2243	2245	2247	2249	2252	2254	2256	0	0	1	1	1	1	2	2	2
	7071	7078	7085	7092	7099	7106	7113	7120	7127	7134	1	1	2	3	4	4	5	6	6
	2258	2261	2263	2265	2267	2269	2272	2274	2276	2278	0	0	1	1	1	1	2	2	2
	7141	7148	7155	7162	7169	7176	7183	7190	7197	7204	1	1	2	3	4	4	5	6	6
	2280	2283	2285	2287	2289	2291	2293	2296	2298	2300	0	0	1	1	1	1	2	2	2
	7211	7218	7225	7232	7239	7246	7253	7259	7266	7273	1	1	2	3	3	4	5	6	6
	2302	2304	2307	2309	2311	2313	2315	2317	2319	2322	0	0	1	1	1	1	2	2	2
	7280	7287	7294	7301	7308	7314	7321	7328	7335	7342	1	1	2	3	3	4	5	5	6
	2324	2326	2328	2330	2332	2335	2337	2339	2341	2343	0	0	1	1	1	1	1	2	2
	7348	7355	7362	7369	7376	7382	7389	7396	7403	7409	1	1	2	3	3	4	5	5	6

The first significant figure and the position of the decimal point must be determined by inspection.

	0	1	2	3	4	5	6	7	8	9	1	2	3	4	5	6	7	8	9
55	2345	2347	2349	2352	2354	2356	2358	2360	2362	2364	0	0	1	1	1	1	1	2	[illegible]
	7416	7423	7430	7436	7443	7450	7457	7463	7470	7477	1	1	2	3	3	4	5	5	[illegible]
56	2366	2369	2371	2373	2375	2377	2379	2381	2383	2385	0	0	1	1	1	1	1	2	[illegible]
	7483	7490	7497	7503	7510	7517	7523	7530	7537	7543	1	1	2	3	3	4	5	5	[illegible]
57	2387	2390	2392	2394	2396	2398	2400	2402	2404	2406	0	0	1	1	1	1	1	2	[illegible]
	7550	7556	7563	7570	7576	7583	7589	7596	7603	7609	1	1	2	3	3	4	5	5	[illegible]
58	2408	2410	2412	2415	2417	2419	2421	2423	2425	2427	0	0	1	1	1	1	1	2	[illegible]
	7616	7622	7629	7635	7642	7649	7655	7662	7668	7675	1	1	2	3	3	4	5	5	[illegible]
59	2429	2431	2433	2435	2437	2439	2441	2443	2445	2447	0	0	1	1	1	1	1	2	[illegible]
	7681	7688	7694	7701	7707	7714	7720	7727	7733	7740	1	1	2	3	3	4	5	5	[illegible]
60	2449	2452	2454	2456	2458	2460	2462	2464	2466	2468	0	0	1	1	1	1	1	2	[illegible]
	7746	7752	7759	7765	7772	7778	7785	7791	7797	7804	1	1	2	3	3	4	4	5	[illegible]
61	2470	2472	2474	2476	2478	2480	2482	2484	2486	2488	0	0	1	1	1	1	1	2	[illegible]
	7810	7817	7823	7829	7836	7842	7849	7855	7861	7868	1	1	2	3	3	4	4	5	[illegible]
62	2490	2492	2494	2496	2498	2500	2502	2504	2506	2508	0	0	1	1	1	1	1	2	[illegible]
	7874	7880	7887	7893	7899	7906	7912	7918	7925	7931	1	1	2	3	3	4	4	5	[illegible]
63	2510	2512	2514	2516	2518	2520	2522	2524	2526	2528	0	0	1	1	1	1	1	2	[illegible]
	7937	7944	7950	7956	7962	7969	7975	7981	7987	7994	1	1	2	3	3	4	4	5	[illegible]
64	2530	2532	2534	2536	2538	2540	2542	2544	2546	2548	0	0	1	1	1	1	1	2	[illegible]
	8000	8006	8012	8019	8025	8031	8037	8044	8050	8056	1	1	2	2	3	4	4	5	[illegible]
65	2550	2551	2553	2555	2557	2559	2561	2563	2565	2567	0	0	1	1	1	1	1	2	[illegible]
	8062	8068	8075	8081	8087	8093	8099	8106	8112	8118	1	1	2	2	3	4	4	5	[illegible]
66	2569	2571	2573	2575	2577	2579	2581	2583	2585	2587	0	0	1	1	1	1	1	2	[illegible]
	8124	8130	8136	8142	8149	8155	8161	8167	8173	8179	1	1	2	2	3	4	4	5	[illegible]
67	2588	2590	2592	2594	2596	2598	2600	2602	2604	2606	0	0	1	1	1	1	1	2	[illegible]
	8185	8191	8198	8204	8210	8216	8222	8228	8234	8240	1	1	2	2	3	4	4	5	[illegible]
68	2608	2610	2612	2613	2615	2617	2619	2621	2623	2625	0	0	1	1	1	1	1	2	[illegible]
	8246	8252	8258	8264	8270	8276	8283	8289	8295	8301	1	1	2	2	3	4	4	5	[illegible]
69	2627	2629	2631	2632	2634	2636	2638	2640	2642	2644	0	0	1	1	1	1	1	2	[illegible]
	8307	8313	8319	8325	8331	8337	8343	8349	8355	8361	1	1	2	2	3	4	4	5	[illegible]
70	2646	2648	2650	2651	2653	2655	2657	2659	2661	2663	0	0	1	1	1	1	1	2	[illegible]
	8367	8373	8379	8385	8390	8396	8402	8408	8414	8420	1	1	2	2	3	4	4	5	[illegible]
71	2665	2666	2668	2670	2672	2674	2676	2678	2680	2681	0	0	1	1	1	1	1	[illegible]	[illegible]
	8426	8432	8438	8444	8450	8456	8462	8468	8473	8479	1	1	2	2	3	3	4	[illegible]	[illegible]
72	2683	2685	2687	2689	2691	2693	2694	2696	2698	2700	0	0	1	1	1	1	1	[illegible]	[illegible]
	8485	8491	8497	8503	8509	8515	8521	8526	8532	8538	1	1	2	2	3	3	4	[illegible]	[illegible]
73	2702	2704	2706	2707	2709	2711	2713	2715	2717	2718	0	0	1	1	1	1	1	[illegible]	[illegible]
	8544	8550	8556	8562	8567	8573	8579	8585	8591	8597	1	1	2	2	3	3	4	[illegible]	[illegible]
74	2720	2722	2724	2726	2728	2729	2731	2733	2735	2737	0	0	1	1	1	1	1	[illegible]	[illegible]
	8602	8608	8614	8620	8626	8631	8637	8643	8649	8654	1	1	2	2	3	3	4	[illegible]	[illegible]
75	2739	2740	2742	2744	2746	2748	2750	2751	2753	2755	0	0	1	1	1	1	1	[illegible]	[illegible]
	8660	8666	8672	8678	8683	8689	8695	8701	8706	8712	1	1	2	2	3	3	4	[illegible]	[illegible]
76	2757	2759	2760	2762	2764	2766	2768	2769	2771	2773	0	0	1	1	1	1	1	[illegible]	[illegible]
	8718	8724	8729	8735	8741	8746	8752	8758	8764	8769	1	1	2	2	3	3	4	[illegible]	[illegible]
77	2775	2777	2778	2780	2782	2784	2786	2787	2789	2791	0	0	1	1	1	1	1	[illegible]	[illegible]
	8775	8781	8786	8792	8798	8803	8809	8815	8820	8826	1	1	2	2	3	3	4	[illegible]	[illegible]

The first significant figure and the position of the decimal point must be determined by inspection.

	0	1	2	3	4	5	6	7	8	9	1	2	3	4	5	6	7	8	9
78	2793	2795	2796	2798	2800	2802	2804	2805	2807	2809	0	0	1	1	1	1	1	1	2
	8832	8837	8843	8849	8854	8860	8866	8871	8877	8883	1	1	2	2	3	3	4	4	5
79	2811	2812	2814	2816	2818	2820	2821	2823	2825	2827	0	0	1	1	1	1	1	1	2
	8888	8894	8899	8905	8911	8916	8922	8927	8933	8939	1	1	2	2	3	3	4	4	5
80	2828	2830	2832	2834	2835	2837	2839	2841	2843	2844	0	0	1	1	1	1	1	1	2
	8944	8950	8955	8961	8967	8972	8978	8983	8989	8994	1	1	2	2	3	3	4	4	5
81	2846	2848	2850	2851	2853	2855	2857	2858	2860	2862	0	0	1	1	1	1	1	1	2
	9000	9006	9011	9017	9022	9028	9033	9039	9044	9050	1	1	2	2	3	3	4	4	5
2	2864	2865	2867	2869	2871	2872	2874	2876	2877	2879	0	0	1	1	1	1	1	1	2
	9055	9061	9066	9072	9077	9083	9088	9094	9099	9105	1	1	2	2	3	3	4	4	5
3	2881	2883	2884	2886	2888	2890	2891	2893	2895	2897	0	0	1	1	1	1	1	1	2
	9110	9116	9121	9127	9132	9138	9143	9149	9154	9160	1	1	2	2	3	3	4	4	5
4	2898	2900	2902	2903	2905	2907	2909	2910	2912	2914	0	0	1	1	1	1	1	1	2
	9165	9171	9176	9182	9187	9192	9198	9203	9209	9214	1	1	2	2	3	3	4	4	5
5	2915	2917	2919	2921	2922	2924	2926	2927	2929	2931	0	0	1	1	1	1	1	1	2
	9220	9225	9230	9236	9241	9247	9252	9257	9263	9268	1	1	2	2	3	3	4	4	5
	2933	2934	2936	2938	2939	2941	2943	2944	2946	2948	0	0	1	1	1	1	1	1	2
	9274	9279	9284	9290	9295	9301	9306	9311	9317	9322	1	1	2	2	3	3	4	4	5
	2950	2951	2953	2955	2956	2958	2960	2961	2963	2965	0	0	1	1	1	1	1	1	2
	9327	9333	9338	9343	9349	9354	9359	9365	9370	9375	1	1	2	2	3	3	4	4	5
	2966	2968	2970	2972	2973	2975	2977	2978	2980	2982	0	0	1	1	1	1	1	1	2
	9381	9386	9391	9397	9402	9407	8413	9418	9423	9429	1	1	2	2	3	3	4	4	5
	2983	2985	2987	2988	2990	2992	2993	2995	2997	2998	0	0	1	1	1	1	1	1	2
	9434	9439	9445	9450	9455	9460	9466	9471	9476	9482	1	1	2	2	3	3	4	4	5
	3000	3002	3003	3005	3007	3008	3010	3012	3013	3015	0	0	0	1	1	1	1	1	1
	9487	9492	9497	9503	9508	9513	9518	9524	9529	9534	1	1	2	2	3	3	4	4	5
	3017	3018	3020	3022	3023	3025	3027	3028	3030	3032	0	0	0	1	1	1	1	1	1
	9539	9545	9550	9555	9560	9566	9571	9576	9581	9586	1	1	2	2	3	3	4	4	5
	3033	3035	3036	3038	3040	3041	3043	3045	3046	3048	0	0	0	1	1	1	1	1	1
	9592	9597	9602	9607	9612	9618	9623	9628	9633	9638	1	1	2	2	3	3	4	4	5
	3050	3051	3053	3055	3056	3058	3059	3061	3063	3064	0	0	0	1	1	1	1	1	1
	9644	9649	9654	9659	9664	9670	9675	9680	9685	9690	1	1	2	2	3	3	4	4	5
	3066	3068	3069	3071	3072	3074	3076	3077	3079	3081	0	0	0	1	1	1	1	1	1
	9695	9701	9706	9711	9716	9721	9726	9731	9737	9742	1	1	2	2	3	3	4	4	5
	3082	3084	3085	3087	3089	3090	3092	3094	3095	3097	0	0	0	1	1	1	1	1	1
	9747	9752	9757	9762	9767	9772	9778	9783	9788	9793	1	1	2	2	3	3	4	4	5
	3098	3100	3102	3103	3105	3106	3108	3110	3111	3113	0	0	0	1	1	1	1	1	1
	9798	9803	9808	9813	9818	9823	9829	9834	9839	9844	1	1	2	2	3	3	4	4	5
	3114	3116	3118	3119	3121	3122	3124	3126	3127	3129	0	0	0	1	1	1	1	1	1
	9849	9854	9859	9864	9869	9874	9879	9884	9889	9894	1	1	2	2	3	3	4	4	5
	3130	3132	3134	3135	3137	3138	3140	3142	3143	3145	0	0	0	1	1	1	1	1	1
	9899	9905	9910	9915	9920	9925	9930	9935	9940	9945	0	1	1	2	2	3	3	4	4
	3146	3148	3150	3151	3153	3154	3156	3158	3159	3161	0	0	0	1	1	1	1	1	1
	9950	9955	9960	9965	9970	9975	9980	9985	9990	9995	0	1	1	2	2	3	3	4	4

The first significant figure and the position of the decimal point must be determined by inspection.

RECIPROCALS

SUBTRACT

	0	1	2	3	4	5	6	7	8	9	1	2	3	4	5	6	7	8	9
1·0	1·0000	·9901	·9804	·9709	·9615	·9524	·9434	·9346	·9259	·9174	9	18	27	36	45	55	64	73	82
1·1	·9091	·9009	·8929	·8850	·8772	·8696	·8621	·8547	·8475	·8403	8	15	23	30	38	45	53	61	6[illegible]
1·2	·8333	·8264	·8197	·8130	·8065	·8000	·7937	·7874	·7813	·7752	6	13	19	26	32	38	45	51	5[illegible]
1·3	·7692	·7634	·7576	·7519	·7463	·7407	·7353	·7299	·7246	·7194	5	11	16	22	27	33	38	44	4[illegible]
1·4	·7143	·7092	·7042	·6993	·6944	·6897	·6849	·6803	·6757	·6711	5	10	14	19	24	29	33	38	4[illegible]
1·5	·6667	·6623	·6579	·6536	·6494	·6452	·6410	·6369	·6329	·6289	4	8	13	17	21	25	29	33	3[illegible]
1·6	·6250	·6211	·6173	·6135	·6098	·6061	·6024	·5988	·5952	·5917	4	7	11	15	18	22	26	29	3[illegible]
1·7	·5882	·5848	·5814	·5780	·5747	·5714	·5682	·5650	·5618	·5587	3	6	10	13	16	20	23	26	[illegible]
1·8	·5556	·5525	·5495	·5464	·5435	·5405	·5376	·5348	·5319	·5291	3	6	9	12	15	18	20	23	[illegible]
1·9	·5263	·5236	·5208	·5181	·5155	·5128	·5102	·5076	·5051	·5025	3	5	8	11	13	16	18	21	[illegible]
2·0	·5000	·4975	·4950	·4926	·4902	·4878	·4854	·4831	·4808	·4785	2	5	7	10	12	14	17	19	[illegible]
2·1	·4762	·4739	·4717	·4695	·4673	·4651	·4630	·4608	·4587	·4566	2	4	7	9	11	13	15	17	
2·2	·4545	·4525	·4505	·4484	·4464	·4444	·4425	·4405	·4386	·4367	2	4	6	8	10	12	14	16	
2·3	·4348	·4329	·4310	·4292	·4274	·4255	·4237	·4219	·4202	·4184	2	4	5	7	9	11	13	14	
2·4	·4167	·4149	·4132	·4115	·4098	·4082	·4065	·4049	·4032	·4016	2	3	5	7	8	10	12	13	
2·5	·4000	·3984	·3968	·3953	·3937	·3922	·3906	·3891	·3876	·3861	2	3	5	6	8	9	11	12	
2·6	·3846	·3831	·3817	·3802	·3788	·3774	·3759	·3745	·3731	·3717	1	3	4	6	7	8	10	11	
2·7	·3704	·3690	·3676	·3663	·3650	·3636	·3623	·3610	·3597	·3584	1	3	4	5	7	8	9	11	
2·8	·3571	·3559	·3546	·3534	·3521	·3509	·3497	·3484	·3472	·3460	1	2	4	5	6	7	9	10	
2·9	·3448	·3436	·3425	·3413	·3401	·3390	·3378	·3367	·3356	·3344	1	2	3	5	6	7	8	9	
3·0	·3333	·3322	·3311	·3300	·3289	·3279	·3268	·3257	·3247	·3236	1	2	3	4	5	6	7	9	
3·1	·3226	·3215	·3205	·3195	·3185	·3175	·3165	·3155	·3145	·3135	1	2	3	4	5	6	7	8	
3·2	·3125	·3115	·3106	·3096	·3086	·3077	·3067	·3058	·3049	·3040	1	2	3	4	5	6	7	8	
3·3	·3030	·3021	·3012	·3003	·2994	·2985	·2976	·2967	·2959	·2950	1	2	3	4	4	5	6	7	
3·4	·2941	·2933	·2924	·2915	·2907	·2899	·2890	·2882	·2874	·2865	1	2	3	3	4	5	6	[illegible]	
3·5	·2857	·2849	·2841	·2833	·2825	·2817	·2809	·2801	·2793	·2786	1	2	2	3	4	5	6	[illegible]	
3·6	·2778	·2770	·2762	·2755	·2747	·2740	·2732	·2725	·2717	·2710	1	2	2	3	4	5	5	[illegible]	
3·7	·2703	·2695	·2688	·2681	·2674	·2667	·2660	·2653	·2646	·2639	1	1	2	3	4	4	5	[illegible]	
3·8	·2632	·2625	·2618	·2611	·2604	·2597	·2591	·2584	·2577	·2571	1	1	2	3	3	4	5	[illegible]	
3·9	·2564	·2558	·2551	·2545	·2538	·2532	·2525	·2519	·2513	·2506	1	1	2	3	3	4	4	[illegible]	
4·0	·2500	·2494	·2488	·2481	·2475	·2469	·2463	·2457	·2451	·2445	1	1	2	2	3	4	4		
4·1	·2439	·2433	·2427	·2421	·2415	·2410	·2404	·2398	·2392	·2387	1	1	2	2	3	3	4		
4·2	·2381	·2375	·2370	·2364	·2358	·2353	·2347	·2342	·2336	·2331	1	1	2	2	3	3	4		
4·3	·2326	·2320	·2315	·2309	·2304	·2299	·2294	·2288	·2283	·2278	1	1	2	2	3	3	4		
4·4	·2273	·2268	·2262	·2257	·2252	·2247	·2242	·2237	·2232	·2227	1	1	2	2	3	3	4		
4·5	·2222	·2217	·2212	·2208	·2203	·2198	·2193	·2188	·2183	·2179	0	1	1	2	2	3	3		
4·6	·2174	·2169	·2165	·2160	·2155	·2151	·2146	·2141	·2137	·2132	0	1	1	2	2	3	3		
4·7	·2128	·2123	·2119	·2114	·2110	·2105	·2101	·2096	·2092	·2088	0	1	1	2	2	3	3		
4·8	·2083	·2079	·2075	·2070	·2066	·2062	·2058	·2053	·2049	·2045	0	1	1	2	2	3	3		
4·9	·2041	·2037	·2033	·2028	·2024	·2020	·2016	·2012	·2008	·2004	0	1	1	2	2	2	3		
5·0	·2000	·1996	·1992	·1988	·1984	·1980	·1976	·1972	·1969	·1965	0	1	1	2	2	2	3		
5·1	·1961	·1957	·1953	·1949	·1946	·1942	·1938	·1934	·1931	·1927	0	1	1	2	2	2	3		
5·2	·1923	·1919	·1916	·1912	·1908	·1905	·1901	·1898	·1894	·1890	0	1	1	1	2	2	3		
5·3	·1887	·1883	·1880	·1876	·1873	·1869	·1866	·1862	·1859	·1855	0	1	1	1	2	2	3		
5·4	·1852	·1848	·1845	·1842	·1838	·1835	·1832	·1828	·1825	·1821	0	1	1	1	2	2	2		

	0	1	2	3	4	5	6	7	8	9	1	2	3	4	5	6	7	8	9
·5	·1818	·1815	·1812	·1808	·1805	·1802	·1799	·1795	·1792	·1789	0	1	1	1	2	2	2	3	3
·6	·1786	·1783	·1779	·1776	·1773	·1770	·1767	·1764	·1761	·1757	0	1	1	1	2	2	2	3	3
·7	·1754	·1751	·1748	·1745	·1742	·1739	·1736	·1733	·1730	·1727	0	1	1	1	2	2	2	2	3
·8	·1724	·1721	·1718	·1715	·1712	·1709	·1706	·1704	·1701	·1698	0	1	1	1	1	2	2	2	3
·9	·1695	·1692	·1689	·1686	·1684	·1681	·1678	·1675	·1672	·1669	0	1	1	1	1	2	2	2	3
·0	·1667	·1664	·1661	·1658	·1656	·1653	·1650	·1647	·1645	·1642	0	1	1	1	1	2	2	2	3
·1	·1639	·1637	·1634	·1631	·1629	·1626	·1623	·1621	·1618	·1616	0	1	1	1	1	2	2	2	2
·2	·1613	·1610	·1608	·1605	·1603	·1600	·1597	·1595	·1592	·1590	0	1	1	1	1	2	2	2	2
3	·1587	·1585	·1582	·1580	·1577	·1575	·1572	·1570	·1567	·1565	0	0	1	1	1	1	2	2	2
4	·1563	·1560	·1558	·1555	·1553	·1550	·1548	·1546	·1543	·1541	0	0	1	1	1	1	2	2	2
5	·1538	·1536	·1534	·1531	·1529	·1527	·1524	·1522	·1520	·1517	0	0	1	1	1	1	2	2	2
6	·1515	·1513	·1511	·1508	·1506	·1504	·1502	·1499	·1497	·1495	0	0	1	1	1	1	2	2	2
7	·1493	·1490	·1488	·1486	·1484	·1481	·1479	·1477	·1475	·1473	0	0	1	1	1	1	2	2	2
8	·1471	·1468	·1466	·1464	·1462	·1460	·1458	·1456	·1453	·1451	0	0	1	1	1	1	2	2	2
9	·1449	·1447	·1445	·1443	·1441	·1439	·1437	·1435	·1433	·1431	0	0	1	1	1	1	1	2	2
0	·1429	·1427	·1425	·1422	·1420	·1418	·1416	·1414	·1412	·1410	0	0	1	1	1	1	1	2	2
	·1408	·1406	·1404	·1403	·1401	·1399	·1397	·1395	·1393	·1391	0	0	1	1	1	1	1	2	2
	·1389	·1387	·1385	·1383	·1381	·1379	·1377	·1376	·1374	·1372	0	0	1	1	1	1	1	2	2
	·1370	·1368	·1366	·1364	·1362	·1361	·1359	·1357	·1355	·1353	0	0	1	1	1	1	1	2	2
	·1351	·1350	·1348	·1346	·1344	·1342	·1340	·1339	·1337	·1335	0	0	1	1	1	1	1	1	2
	·1333	·1332	·1330	·1328	·1326	·1325	·1323	·1321	·1319	·1318	0	0	1	1	1	1	1	1	2
	·1316	·1314	·1312	·1311	·1309	·1307	·1305	·1304	·1302	·1300	0	0	1	1	1	1	1	1	2
	·1299	·1297	·1295	·1294	·1292	·1290	·1289	·1287	·1285	·1284	0	0	0	1	1	1	1	1	1
	·1282	·1280	·1279	·1277	·1276	·1274	·1272	·1271	·1269	·1267	0	0	0	1	1	1	1	1	1
	·1266	·1264	·1263	·1261	·1259	·1258	·1256	·1255	·1253	·1252	0	0	0	1	1	1	1	1	1
	·1250	·1248	·1247	·1245	·1244	·1242	·1241	·1239	·1238	·1236	0	0	0	1	1	1	1	1	1
	·1235	·1233	·1232	·1230	·1229	·1227	·1225	·1224	·1222	·1221	0	0	0	1	1	1	1	1	1
	·1220	·1218	·1217	·1215	·1214	·1212	·1211	·1209	·1208	·1206	0	0	0	1	1	1	1	1	1
	·1205	·1203	·1202	·1200	·1199	·1198	·1196	·1195	·1193	·1192	0	0	0	1	1	1	1	1	1
	·1190	·1189	·1188	·1186	·1185	·1183	·1182	·1181	·1179	·1178	0	0	0	1	1	1	1	1	1
	·1176	·1175	·1174	·1172	·1171	·1170	·1168	·1167	·1166	·1164	0	0	0	1	1	1	1	1	1
	·1163	·1161	·1160	·1159	·1157	·1156	·1155	·1153	·1152	·1151	0	0	0	1	1	1	1	1	1
	·1149	·1148	·1147	·1145	·1144	·1143	·1142	·1140	·1139	·1138	0	0	0	1	1	1	1	1	1
	·1136	·1135	·1134	·1133	·1131	·1130	·1129	·1127	·1126	·1125	0	0	0	1	1	1	1	1	1
	·1124	·1122	·1121	·1120	·1119	·1117	·1116	·1115	·1114	·1112	0	0	0	1	1	1	1	1	1
	·1111	·1110	·1109	·1107	·1106	·1105	·1104	·1103	·1101	·1100	0	0	0	1	1	1	1	1	1
	·1099	·1098	·1096	·1095	·1094	·1093	·1092	·1091	·1089	·1088	0	0	0	0	1	1	1	1	1
	·1087	·1086	·1085	·1083	·1082	·1081	·1080	·1079	·1078	·1076	0	0	0	0	1	1	1	1	1
	·1075	·1074	·1073	·1072	·1071	·1070	·1068	·1067	·1066	·1065	0	0	0	0	1	1	1	1	1
	·1064	·1063	·1062	·1060	·1059	·1058	·1057	·1056	·1055	·1054	0	0	0	0	1	1	1	1	1
	·1053	·1052	·1050	·1049	·1048	·1047	·1046	·1045	·1044	·1043	0	0	0	0	1	1	1	1	1
	·1042	·1041	·1040	·1038	·1037	·1036	·1035	·1034	·1033	·1032	0	0	0	0	1	1	1	1	1
	·1031	·1030	·1029	·1028	·1027	·1026	·1025	·1024	·1022	·1021	0	0	0	0	1	1	1	1	1
	·1020	·1019	·1018	·1017	·1016	·1015	·1014	·1013	·1012	·1011	0	0	0	0	1	1	1	1	1
	·1010	·1009	·1008	·1007	·1006	·1005	·1004	·1003	·1002	·1001	0	0	0	0	0	1	1	1	1

Printed in the United States
By Bookmasters